The Bridge to Dalmatia

The Album. *Carmen Violich-Goodin*

The Bridge to *Dalmatia*

A SEARCH FOR THE MEANING OF PLACE

Francis Violich

*Cartography and Drawings
in Collaboration with
Nicholas Ancel*

THE JOHNS HOPKINS UNIVERSITY PRESS
Baltimore and London

Published in cooperation with the Center for American Places, Harrisonburg, Virginia
© 1998 The Johns Hopkins University Press
All rights reserved. Published 1998
Printed in the United States of America on acid-free paper
10 9 8 7 6 5 4 3 2 1

The Johns Hopkins University Press
2715 North Charles Street
Baltimore, Maryland 21218–4363
The Johns Hopkins Press Ltd., London

Library of Congress Cataloging-in-Publication Data will be found at the end of this book.
A catalog record for this book is available from the British Library.

ISBN 0–8018–5554–3

For my Father and Mother,
my Grandparents and those
before them, for their heritage
of love of place

Contents

Preface and Acknowledgments

This work represents a personal search for meaning in urban places, an endeavor generated by my years of professional and academic involvement with the environment. I have witnessed the striving of people for identity with their own "turf" as built environments lacking human qualities have encroached over those given us by nature or by a cultural heritage. Our own well-being has been threatened by the failure to recognize the ecological interdependence between these two facets of the increasingly urban world. My experiencing of the cities, towns, and villages of the Dalmatian coast in 1937 produced within me an environmental framework for the rich cultural identity I had been born into. Later, these two components joined and provided me with a broadened basis for understanding Latin American urban places and those of my own California and America. A lifelong devotion to the interdependence of people and the environment took hold.

Following the paths this work evoked, I sought to explore through instinct and intuition the phenomenon of the human need to find symbolic meaning in physical urban places that have practical value in our fast-paced lives of today. These may be places that in the past formed a native homeland or, as in the case of my forebears, places newly adopted, whether on the same continent or a different one, which occurs so often these days. They may also be places briefly visited in the interest of stimulating travel and thereby gaining a perspective on the uniqueness of one's daily abode. A deep sense of nostalgia can persist for places remembered for their imageable environments and the mental refreshment gratefully received. This new environmental awareness can facilitate our putting down roots in places lived in over time, thus enriching our daily lives with deeply felt identity. Seen as an "existential crisis" of modern man, finding meaning in places might lead to finding meaning in life; and if that is an elusive concept, this cannot be said of the meaning of a place as experienced daily. Properties of places can be grasped and studied and their linkages to our lives made manifest by depths of concern, by loving and caring for them and fostering creative communities within them.

Toward this end, I direct my main attention to the physical, spatial character of urban places, each understood as a whole and an integral part of its natural site, landscape and urban scale combined, through an open, intuitive, and reflective frame of mind. A certain depth of observation of how they function, look, and have evolved over time through turning points of history can provide a perspective that leads to ways of more sensitively shaping them for the future in philosophical as well as physical terms. Here, the important theme of associations with places as a common source of identity comes into play. Personal and ethnic relationships that develop from the experiences one has in places become a direct and recognized source of identity. My purpose here is to show how the two sources—the physical environment, with its rich array of potential human responses, and our own personal associations of an ethnic, religious, or family nature—work together in establishing identity with place. They can enhance the quality of urban places for their dwellers through cultural interchange of a productive and humanistic nature from neighborhood to neighborhood, city to city, region to region, or nation to nation.

PURPOSES AND DIRECTIONS

This work is directed to four audiences. First, to the various types of general readers who would value building within their own perception, either through the printed page or through travel, bridges between their own urban environment and others. They could then evolve new dimensions to stereotyped tourism, especially for Dalmatia's first-time visitors. The fresh ways of experiencing urban places through increased awareness and release of the powers of intuition can prove rewarding to residents or travelers in other cultures. Among these readers, I am interested in reaching those Americans who share with me a cherished family heritage in Dalmatia and who seek to gain identity with their cultural origins. I refer especially to the large numbers of them who descend from immigrants to the West Coast, motivated by its similarities to Dalmatia. I have also in mind those who found new homes in Canada, Australia, and New Zealand, as well as in Chile and other coastal locales of Latin America. Those Americans of other foreign backgrounds might find my story and its concept of identity with place valuable in bridging their native culture with that of America. Finally, my book will enrich those visitors who will be attracted once more to the beauties and stimulus of Dalmatia's unique coastal environment when peace returns.

Second, to professionals and their collaborators in environmental design and urban planning who are increasingly seeking to broaden and refine approaches for working with the communities of America's metropolitan areas and smaller cities. In these lie opportunities for making more effective the processes of community formation through creating a collective sense of awareness and sources of identity in both their daily lives and their environmental heritage. Also, members of this group who are involved in urban development are increasingly being called upon in other parts of the world where sensitivity to other cultures is critical to the physical design of urban places.

Third, to the students and followers of Slavic cultural subject matter who may wish to better understand the increasingly critical subject of urban spatial systems and places as settings for ethnic cultural characteristics. These can be used as matrices for broadening the perception of the cultural history of the southern Slavic states, especially Croatia, of which Dalmatia is a subregion, as well as others now faced with finding compatibility of geographical and ethnic identity as independent republics.

Finally, to readers seeking to understand the issue of immigration and demographic displacement on a global scale for reasons of economic and political instability and war. The consequent process of adaptation to new environments and cultures deserves inquiries of an anthropological nature, well beyond my expertise. While the context of my material lies in the "old" immigration from Europe, at a time when newcomers could choose matching locales in the New World, the basic concepts can be drawn on in understanding the "new" immigrants from Latin America and Asia to America and especially the refugees from war-torn Croatia and Bosnia in Dalmatia.

In view of the breadth of substantive backgrounds represented by this array of readers, it should be made clear that I speak from the second group, those in environmental planning and design. Although I am not a historian, cities cannot be understood without drawing on their pasts—as I have done—to portray the shaping of urban form and its cultural symbolism. Nor am I a psychologist or a philosopher, though I deal with environmental perception and interpretation in a phenomenological sense. This awareness is gained, however, from inborn attitudes toward capturing environmental images through experiencing them rather than from a particular school of thought or analysis. Similarly I am not an anthropologist or a demographer, though I have introduced people into the work, as family and community members, indeed in a combined historical and environmental context. I do aspire to bring accuracy from these fields to enrich my contemplations of places. Ultimately this approach may bring to specialists in these fields an awareness of the connectedness of each to a wholistic way of seeing urban places. They could make critical contributions by applying their specialities to the ever-increasing complexities of urban studies in general and urban planning and design of a broadly humanistic nature in particular.

I believe that the multifaceted orientation of this book, derived from a personal and experiential context, is essential to the validity of any findings. Its breadth has grown over time from a combined academic and professional focus to the considerable autobiographical setting contained in the introduction. This personal context is recalled at various points in the substantive chapters, especially in the future-oriented messages of the closing chapter directed toward international interchange. In the legend of Odysseus's ten-year return to Ithaca, the events throughout the journey enriched his life far more than did his arrival home. For me too the process of contemplative exploration evoked such valued riches that I had no other choice but to portray broadly and in a personal way the fruitful experiences over the years of my own Dalmatian odyssey.

THEMES, SEQUENCES, AND CONTENT

The general conceptual focus of this book is the environmental setting for human life at its fullest. This setting comprises four components of the environment that are interactive and interdependent in establishing the form, scale, and quality of urban places: the *social* makeup; the *economic* process, which supports life; the *physical* structure, based on the combined natural and built environments; and the *institutional* system, which coordinates and guides the other three. Within this framework, as the work progressed, six major themes emerged. Drawing on Dalmatian urban places as case study material produced themes that I contend are applicable to all urban places and their environments.

1. *Identity with place.* At any point in time what specific determinants of the form and character of a given place provide sources of identity? How do these vary from the regional to a larger or smaller urban scale? What role does each of the four environmental components play in giving a particular quality to each place as a basis for identity?

2. *Evolution of places.* What evolutionary processes led to the present form? In what way did the turning points in history—social, economic, physical, and institutional—combine to bring a particular pattern of urban places to Dalmatia as a region and an identifiably distinctive quality to individual cities, towns, and villages?

3. *Form and human use.* What does identity mean in terms of the relationship between a given place and the people who use it? In what ways do the morphology and visual character of a city, town, or village determine the nature and quality of day-to-day living experiences and the behavior of the users of that place?

4. *Intuition and experiencing places.* To what extent do our intuitive and perceptual, as against the literal and quantitative, responses to the physical form of places as we experience them contribute to identity with them? What elements of urban places, seen as phenomena, can give our perception of them a deeper social meaning? What can we learn for

general use by exploration into "urban reading," a method developed in my study that draws on phenomenology applied to the environment?

5. *Role of associations.* As a theme of lesser inquiry for this work, how do personal and family associations with places in the form of memories of events and the establishment of relations with other individuals or groups generate environmental identity?

6. *Identity and the conflicts of our times.* Drawing on a deepening understanding of these questions, what concepts and criteria can be applied in contemporary urban planning to enmesh the urban resources of the accumulated past with the powerful technological forces of our times for more humanized environments? What lessons can be learned from cities formed through incremental participation by their residents over generations that could apply in injecting identity into the environments of our generation and those of the coming century? How can these findings become useful in the awesome task of restoration of those settlements in Dalmatia that were damaged so mindlessly by the war of 1991–95 and settlements elsewhere that have been damaged by conflicts and violence over territorial identity?

The introduction tells the story of the origins of my own identity with Dalmatia and the cultural "bridge" that has facilitated international cultural exchange since the 1890s for many Californians of "dual identity." This backdrop brings to life in chapter 1 a discussion of the nature of identity as a key to the meaning of place that grew out of my explorations up to the time of completing this book. This relationship between personal origins and place identity also provides common intellectual ground between the author's experience in the "search" for the meaning of place and the reader's experience in "reading" Dalmatian urban places with me in the following chapters. The main thrust here is to bring home to the reader how the direct experiencing of places can lead to a deepened sense of relationship between the city's form and a person's own connectedness to it. For this reason, the reader will be addressed as "you" in places that call for a sharper focus, while "we" will apply to shared experiences of a more obvious nature.

Chapter 2 establishes the historical roots of identity with place at the regional level, that is, Dalmatia as a whole, the highest in the hierarchy of environmental scales but the least subject to experiencing through daily life. The principal turning points in the evolution of urban systems pass from Illyrian times, through the Greeks and Romans, to settlement mainly by the Croatian elements of the Slavic immigration to the Balkans and to Venetian and Turkish influences. The conquest by Napoleon introduces the chapter's focus on the nineteenth century, the period most responsible for the overall system of urban settlement that we find today.

Moving downward in the hierarchy of place to the principal mainland cities—Zadar, Split, and Dubrovnik—chapter 3 describes their variations, both in historical evolution and in the ultimate city form that the reader of the book will come to know in walking a prescribed route with me, "reading" it firsthand. Our "urban reading" continues in chapters 4 and 5, at the most significant level, that of towns and villages. A comparative sampling of case studies carried out in fieldwork on island sites reveals how land and sea interact in varied ways with the urban pattern. Chapter 6 presents a similar approach in a sampling of interrelated villages on the mountainous peninsula of Pelješac. In contrast to the interdependency of seaside villages in terms of identity, the mountain-village network functions as a single identity component. At all three of these levels—regions, mainland cities, island towns and villages and mountain networks of villages—I show how the diversified and dispersed geographical character of the immediate landscape setting allows for the formation of widely differing forms and qualities of urban places. From this individuality stems the traditionally strong sense of identity residents of Dalmatia feel for their own particular settlements.

With the outbreak of the war in 1991, as this book was taking shape, the final chapter took on a life of its own. Suddenly the subject of identity with place became a precious symbol of far-reaching significance. Chapter 7 reviews the original aims set forth and the paths followed to reveal what we have learned. This concluding step in my analytic method takes the form of a set of properties of identity with place generated directly from each of the places covered in the "urban readings." These are then used to evaluate the depth of the impact of the 1991 assaults on Dalmatian identity by again walking through the very places the reader has come to know and "reading" these assaulted places.

More importantly, by reinforcing environmental identity with place as a central theme, these precepts are used to develop general approaches to the reconstruction of Dalmatia and its long-range planning after the war This message is presented by me to the younger generation in the personal spirit of the introduction. Even as the writing grew more coherent over the time called for by its intriguing nature, the world seemed to be moving toward increasingly smaller national territories based on historical identities and self-determination. In this movement, identity with place evolved its own voice, clearly yearning for a future that would connect land and people for more effective human fulfillment. Under the spatial revolution transforming the globe into an integrated mosaic of would-be autonomous places, the building of "bridges" for professional and educational interchanges between nations, states, and regions becomes a critical international theme.

A focus on environmental planning, city building, and regional management anchored to the human appeal of identity with place would be a fitting way to deal with environmental issues as we move into the next millennium. In this light, the metaphor of The Bridge introduced in the introduction became a broadened vehicle for interchange beyond my original intent, based on family ties. Identity with place would serve as a unifying and mutually beneficial theme for professional and related cultural exchanges between Dalmatia and California in the context of Croatia and the United States. The program could be broadly interdisciplinary, reflecting the distinctive content of individual urban places, and deal with contemporary problems and methods in the fields of the natural and the built environments.

THE ILLUSTRATIONS

The range of types of illustrations, carefully selected and organized, reflects the structure and content of the book as a whole. All drawings not otherwise credited were done by Nicholas Ancel, who holds master's degrees in landscape architecture and city planning from the University of California at Berkeley, in close collaboration with me. The regional maps and urban plans, together with their sections, were developed from sources in Croatia and in the Map Room of the University of California at Berkeley. Considerable research and inventiveness were required to adapt the scarce information to the needs of the text. Ancel's line drawings were based on color slides taken by me as part of my fieldwork. Some of the slides were also made into prints in order to offer detailed images to supplement the line drawings. The historical prints emphasize the urban evolution described in the text. For the most part these were secured from archival sources in Zagreb by Nenad Lipovac, of the University of Zagreb, who also helped in their final selection and placement. Other contributors are indicated by credit lines.

Full-size copies of the original illustrations were exhibited along with excerpts from the text in the College of Environmental Design, University of California at Berkeley, in December 1996.

SOURCES AND PARTICIPANTS

This book draws on a broad range of sources generated over time by firsthand experiences with people and places in Dalmatia and in California. These include library, archival, and professional materials, as well as the results of field research. My own personal experiences and family records have served as an integrating element and context of authenticity and reality. In that sense, the book as a whole takes on some of the phenomenological quality of chapters 3–6, that is, direct experiential and intuitively based "urban readings" of cities, towns, and villages themselves. It is essential for the reader to understand that the sources used to supply factual information and interpretations did not grow out of a research project of predetermined design in the usual academic sense; rather, the research came into being as an inherent part of a larger context of life experiences.

The nature of this work has led me to involvement with, and support from, numerous organizations and persons whose response to my aims and whose similar attachment to Dalmatia have been a major force toward its completion. I am deeply indebted to them for their guidance, consultation, and collaboration in major and minor ways.

The institutions that facilitated travel and research provided not only funds or services but also recognition of work in an area and subject given little attention in the past. Arranged roughly in chronological order, these include the Center for Slavic and Eastern European Studies and the Farrand Fund, both of the University of California at Berkeley, 1969; the International Research Exchange Institute, 1970; the American Philosophical Society, 1970; the American Academy in Rome, 1970; Urbanistički Zavod Dalmacije (Urban Planning Institute of Dalmatia), Split, 1970; the U.S.-Yugoslavia Fulbright Commission, Zagreb and Belgrade, 1979; the Art History Faculty, University of Zagreb, 1979; the Architecture and Urban Planning Faculty, University of Zagreb, 1979–90; the Center for Architectural Restoration, Split, 1979, 1981, 1990; the Croatian Regional Archives: Zadar 1979, Split 1990, and Dubrovnik 1970; the Institute of Urban and Regional Development, 1991–93; and the International Association for the Study of Traditional Environments, University of California at Berkeley, 1988, 1990.

I am grateful for guidance in dealing with my broad subject matter from a far larger number of people than can be acknowledged here. The assistance ranges from sheer encouragement and common-sense advice to solid scholarly criticism in areas where my enthusiasm for the subject matter exceeds my environmental expertise.

Advisers from the University of Zagreb include Eugen Pusić, public administration; Bruno Milić, urban planning; Sena Sekulić, architecture; Milan Prelog and Igor Fisković, art and cultural history; and Vedrana Spajić Vrkas, a Fulbright visiting scholar at the University of California at Berkeley, in 1993–94. Nenad Lipovac, in architecture and urban planning as well as a visiting scholar at Berkeley in the spring of 1996, lent valuable creative ideas and skilled judgment to the final stages, especially in securing the historical images. Through this project he joined me in initiating exchanges between our respective departments in Zagreb and Berkeley.

Those from Split were Tomislav Marasović, architectural history and conservation; Cvito Fisković, architectural and cultural history; and Ivana Šverko, architecture and urban design, who served as my critical and faithful liaison with my colleagues in Croatia. From Dubrovnik,

Paula Brailo secured needed material. From the University of Sarajevo, a most devoted collaborator was Duško Bogunović, urban and regional planning, a Fulbright visiting scholar at Berkeley from 1991 to 1993, who for obvious reasons is presently settled in New Zealand.

From the outset, Paola Coppola Pignatelli, urban design, of the University of Rome, shared a common interest with me in exploring identity with place.

In the United States I am indebted to David Seamon, architecture and geography, of Kansas State University; Andrej Škarica, urban planning and design, formerly of Split, now in California, where his insights have guided me from the outset; Borut Prah, computer technology, a native of Slovenia, now of Berkeley; Ivan Zaknić, architecture, of Lehigh University; Ivo Banac, Slavic and Croatian history, of Yale University; Bariša Krekić, Croatian history, University of California at Los Angeles; and especially George F. Thompson, president of the Center for American Places, who from the first reading captured the sense of my manuscript and guided me with sensitivity through the substantive editing and publication processes at the Johns Hopkins University Press. In the final stages it was due to the editorial skills of Joanne Allen, staff member of the Press, in responding to my "experiential" writing that the work achieved my original aims.

I am grateful to the following individuals from the University of California at Berkeley: Donald Appleyard, urban planning and history (1928–82); Leonard Duhl, urban planning and public health; Kenneth Craik, environmental psychology; Gene Hammel, demography and anthropology, with focus on the former Yugoslavia; Michael Southworth, urban design; and T. J. Kent and Corwin Mocine, emeritus faculty members in urban planning (1911–97).

Three others from the San Francisco Bay area, with whom my relationship reaches back to my earliest identity with Dalmatia, are Jozo Tomasevich (1908–94), Rudy Pahlihnich, and Alex Vucinich. Those many devoted helpers over the years among colleagues and friends, students and staff, in both California and Croatia are to be found unnamed throughout the manuscript in the many small ways that hold the work together. Among these, Irena Rašin assured cultural accuracy in translating Croatian sources and in developing the glossary; Kaye Bock contributed in a large way through her exceptional devotion, patience, and skill in the mechanics of putting my flow of words onto disks and paper as I sought a firm footing for clarity and coherence of the work as a whole; John Banks made a generous contribution by finalizing the text and the list of illustrations; finally, Nicholas Ancel—with fine tuning from Ray Isaacs—brought some of the free spirit of the text to the drawings that give the book its own identity. To all, my heartfelt thanks.

The Bridge to Dalmatia

INTRODUCTION

Crossing the Bridge to California

Each of us . . . has his own America . . . the aggregate of people, places, traditions, ideals, institutions and diverse other forces which in one way or another have influenced one's life and contributed to one's education—as an American and as a person. —Louis Adamic, 1938

LINKING TWO CONTINENTS
AND FIVE GENERATIONS

My father's village of Kuna stood for centuries high above the Adriatic Sea on a mountainous plateau on the Dalmatian coast. Located midway up the Pelješac peninsula, this bold landscape became essential to the independent city-state of Dubrovnik. Though his life in San Francisco represented quite another world, I first became aware of the profound sense of identity my father held for this home of his youth when I was only five years old. An image stands forth clearly of my father tracing his index finger lovingly over a hand-tinted photograph of the red tile rooftops of this rugged settlement, as though he wished to engrave it in my mind and my heart. Shoulder to shoulder, the stone houses formed a strong linear pattern of urban character below the rolling flanks of an imposing ridge of aged limestone, all capped by an intensely blue sky.

This image, bright and sharp throughout my lifetime, was contained in a picture album that my father's older brother had sent to him in 1914, signed in Croatian "In memory of Kuna and Pelješac." This link between continents and generations remained always in evidence in our household then, as it does in mine now. Its publication was initiated by an enterprising native of Kuna as a vehicle for strengthening the bonds between family and place for those who had emigrated to the New World. My father seemed moved by the opportunity to connect me to the place where he had lived, even twenty years after he had left, at the age of nineteen. Was his leaving a conscious move, considering that in his heart he knew he might never return, never see his family?

This unforgettable childhood incident revealed a less tangible consciousness, a sense of tenderness and contemplation—reverence, in a way—on my father's part that in my later and more perceptive years became a permanent part of my visual image of Kuna. Today, in retrospect,

The village of Kuna: A 1909 postcard from my father's family. *Violich family archive*

Kuna.

I see this sense of personal relationship to place as an integral part of our visual consciousness. This fusing of environmental sources of identity was nurtured by Kuna's small scale, exposure to its rugged landscape, and the strong sense of family bonds inherent in Dalmatian village life.

Unlike the strong and direct introduction I had to my father's mountain village, the first images of my maternal grandmother's home on the Dalmatian coast came from my mother's romanticized descriptions of her seaside village on the island of Brač. The vivid quality of that environmental setting contrasted with the starkness of my grandfather's mountain village and became a theme of my exploration over the years. Since my maternal grandmother was the only grandmother I knew in person, her commanding presence and heroic life story reinforced and colored these images during my first fifteen years. Even to this day, after having studied her village, I find that it is her personal image, rather than Pučišća's intrinsic physical properties, that dominates my identity with the place. Thus, I have learned that people can supersede the properties of the physical place, yet when the place portrays qualities that are as memorable and striking to the eye and heart as those of Pučišća, together the impact becomes a powerful resource that heightens one's innermost identity with place.

Just as the topographic character of the peninsula of Pelješac influenced the nature of its towns and villages, so the island of Brač shaped Pučišća's qualities and the direction of my grandmother's life. In the first place, it lay just across the water from Split, home of that treasured remnant of the ancient urban world, the Palace of Diocletian, which grew into a city as the centuries passed after the fall of the Roman Empire. This prime heritage of Dalmatia records Split's cultural history in a rich assembly of architectural works by the Venetians and Croatians. Indeed, the marble for the palace was cut from the cream-colored quarries of Pučišća. It was also used to build my grandmother's house and most of the other houses of this urban village, as well as the walks, steps, and docks—an environmental foothold for identity with place not to be readily matched.

Thus, more than a century ago, through the album sent from Kuna and the stone of Pučišća, my father and maternal grandparents laid a firm foundation for a cultural Bridge from their ancestral living environments on the Dalmatian coast to new homes and new lives in the similar maritime setting of California. They brought with them a strong sense of environmental identity nurtured by intimately experiencing the inherent qualities of their native villages in striking settings of land and sea. These resources facilitated the challenging cultural transition to their adopted homes in the foothills of the Sierra Nevada and ultimately in San Francisco. Family and friends always talked of "the old country," the *stari kraj,* of my father's village of Kuna and of Pučišća. Postcards with images of these revered places arrived with stamps depicting Franz Joseph of Austria. Letters came from cousins who lived just outside the walls of the Roman emperor Diocletian's palace. Others arrived to settle in California and in time made more tangible my own sense of identity with Dalmatia.

My mother's stories of Sutter Creek and Amador County, with its gold mines, where my grandparents raised a family of eight, were filled with descriptions of places, the wonders of nature, and the events that people pursued in those places with affection and understanding. Her strong sense of connectedness and sensitivity to place replicated what my grandmother had known in the *stari kraj.* This combination of family unity, cultural identity, and spatial rootedness—house, neighborhood, city, and regional landscape setting—became my own personal heritage. I saw how these memories of places, kept alive in San Francisco, where steep streets, views to ships in the Bay, and the pedestrian world of the 1890s, recalled the Dalmatian homeland. As my generation came along, these images built strong bonds between us, the members of this pioneer family, and the larger Bay Area community of fellow Dalmatians, today's Croatian-Americans, and those elsewhere in California, in Sacramento, Watsonville, and San Pedro, for example.[1] They generated my fascination with environments and their role in enriching people's lives, including my own and those of my children. Over time The Bridge became a two-way vehicle for travel, letters, exchange of photographs and family news, and continues today via telephone and FAX. When not in actual use it served as a symbol of fruitful connections that led me toward understanding urban environments and their laudable settings as a source of human meaning, belonging, and cultural identity.

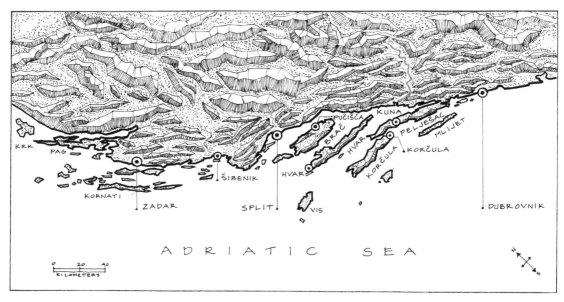

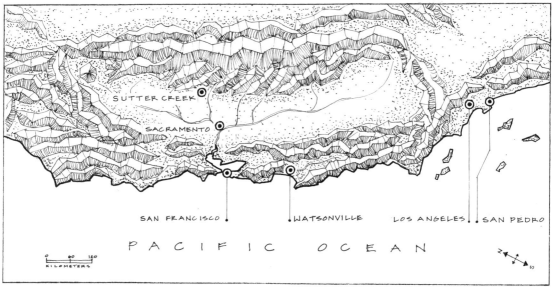

Dalmatia (top). The interaction between the sea and the highly irregular geography built identity firmly into people and place

California (bottom). Dalmatian settlers adapted well to a west-facing seashore backed by mountains and fertile coastal lands

Throughout these years in California enormous changes took place as urbanization spread over the untouched landscape, for two thousand years the bountiful source of life for many native Indian tribes, each with its own place of identity. New social patterns emerged, along with environmental changes precipitating the adaptation of their personal lives to a new culture. They could little have imagined that their Bridge would span time as well as space and maintain a continuous flow of family communication throughout the twentieth century. Reinforced through use, The Bridge has created a sense of identity with two environments of vastly differing cultural origins that has enriched my outlook as an American.

Impelled by these issues, in 1937, in my mid-twenties and having just finished graduate studies in New England, I went to Dalmatia. Deeply moved by the experience, I instinctively adopted the concept of The Bridge, a decision that shaped my personal and professional development. Two men who pioneered in clarifying the role that cultures play in urban life made important contributions to my thinking about my identity: Louis Adamic and Lewis Mumford. My first exposure to European cities in general served as a frame for the locale of my Dalmatian roots, and on my return to California these two social analysts broadened my sense of inquiry beyond my field of studies. They became my mentors, each in his own way, via their writings and the personal attention they gave me.

Louis Adamic, in *My America* and *The Native's Return*,[2] reinforced my own discovery of the richness of my parental origins and cultural heritage. These works, with their hearty Slavic personal flavor and broad vision of America, encouraged my inclination to maintain the values of my pioneer forebears and immediate family and to resist the popular melting-pot concept of the 1920s. Through other of his works on the bridging of cultures in general, I saw the opportunity to blend the humanistic resources of both "Americanization" and the *stari kraj*.[3] Thus, by choosing continuity with my own heritage rather than conformity to values brought to America with the English language, I could create my own identity. This would be a more rooted and indigenous basis for contributing to the promising environmental future of our generation in California.

Adamic, a Slovenian by birth, had come to America in 1913, first to southern California, later moving to New York City. In the early 1930s, having immersed himself in the social issues of American life through direct human involvement, perceptive observation, and enlightened writing, he returned to prewar Yugoslavia by means of a Guggenheim fellowship. Having all but lost his old culture in America, he discovered new qualities in the dual context of the two cultures he bridged. We met in San Francisco on several occasions, and his gift of a signed copy of *The Native's Return* made quite final my determination to make my own return—in the place of my father—to *my* Dalmatia.

In *My America,* published just after my experiences in the old country, Adamic strengthened my interest in the grassroots movements of the less well-off to improve their own lives. In his own close-up portrayal of the Great Depression, which in itself gave me direct professional experience in planning communities for housing the homeless and public works projects for the employed, he focused on the labor movement. This gave us common ground in addition to our South Slavic origins. Adamic's strong advocacy of Franklin Delano Roosevelt, whose environmental policies I supported and whose programs I had participated in, became the source of my permanent orientation toward public concerns. I, along with others of my generation, saw this as a way of maximizing the potentials for human development through shaping urban places in inventive ways.

While Adamic opened my eyes to see Dalmatia more clearly in a personal way, Lewis Mumford, in *The Culture of Cities* and other works, broadened my vision of cities as cultural expressions of their builders.[4] His ability to describe the cities of Europe in historical terms as end products of evolving social, economic, and political systems compelled me to take a more comprehensive and evolutionary view of urban places. Through Mumford's themes inspired by the work of Patrick Geddes, I learned how social patterns evolved over centuries and determined the patterns of cities, instilling a sense of individual and collective identity in the minds of residents. This lesson could be applied to guide modernization, as Geddes did in India.[5] His and Mumford's strong sense of social equity and the quality of life in cities motivated the generation of the 1930s in the San Francisco Bay Area. We were eager to adapt the new technologies to making our cities more human throughout the century of peace that we had anticipated after World War I.

Since the earliest times, as Mumford made clear, urban settlements have played a major role in building bridges between diverse cultures as people have moved about the earth. Today, our increased mobility and the displacement of ethnic groups by war have underscored the issue of retaining environmental and cultural identity. Although the unprecedented growth of cities and communications in heterogeneous America has fostered valuable fusion, it has been accompanied by social instability and cultural costs. Now as we approach the end of the twentieth century, which has been fraught with violence, it is crucial that we prevent homogenization from obliterating the fragile yet fertile ties that offer meaning and anchors to us as individuals. The clean slate of the twenty-first century and the increased immigration to cities provides impetus for enriching our culture by bringing together differing societal groups in urban environments characterized by a high level of social accomplishment.

Today, as history appears to be rewritten from week to week, it is difficult to envision what it must have been like in the nineteenth century to uproot oneself from the old country and make an entirely new beginning in the New World, given the barriers of distance, language, and culture. A sense of the reality of that experience can be brought out by portraying the environments of my forebears as they were in the nineteenth century as a sample of Dalmatia as a whole. Crossing The Bridge with them to new homes in California, we may visualize the gap in identity they had to make up for. In 1889 my father left his home for Seattle, and by 1893 he had settled in San Francisco. My maternal grandparents preceded him by about eighteen years and made their home in the gold-mining town of Sutter Creek, moving to San Francisco about the time my father arrived there.

My perception of these forebears' identity with their home environments engrained bold images in my consciousness. The intensity of their attachment, both to Dalmatia and, on my mother's side, to Sutter Creek took root early in my life. I realized in my later years that absorbing a mix of places and people fostered in my subconscious an awareness of the power of environmental imagery and the role it can play in giving people a focal point to guide their personal development. By sharing openly with the reader this chronological sequence of my own process of Dalmatian identity formation, perhaps I will stimulate the reader to seek a meaningful sense of relationship to his or her own places of origin and identity.

DALMATIA'S NINETEENTH-CENTURY LIVING ENVIRONMENT

Pelješac, the Peninsula; Kuna, the Općina

My father's album contained hand-colored photographs of most of the twenty villages and hamlets comprising the *općina* (county) of Kuna. In time I came to know these as an interconnected, identifiable system, one that was almost self-sustainable in a marginal way in the latter part of the nineteenth century. This experience expanded my vision of the region well before my first visit in 1937. Most significant were the photographs of Zagujine, located a mile or more to the south of Kuna, a *zaselak* (hamlet). My father's grandfather had moved there from nearby Potomje to start his own family in the 1830s. My father was born in 1870, and by the 1890s the old family home had become the farmhouse, the oldest son having built for his large family a three-story "city house" in Kuna proper. The "city house" was located up the street from the headquarters of the county seat, where my grandfather was the *načelnik,* or mayor. Potomje had been built in the sixteenth century by Dubrovnik as part of its settlement program on Pelješac. The census of 1673, following the devastating earthquake of 6 April 1667, which is still available in the archive of that city-state, shows five Violić families, with a total of forty-three members, living there.[6]

Also included in the album is Podobuče, a seaside village firmly integrated into a tiny protected cove at the foot of slopes that drop precipitously to the Adriatic. Down the coast stretch the vineyards that produce the fine grapes of Dingač, which in my father's time were carried by *mazge* (mules unique to Dalmatia) up and over the ridge to the winery at Potomje. In chapter 6 we shall walk this network of villages to experience them as samples of Pelješac as a whole.

Zagujine.   KUNA.

PODOBUČE.   PODOBUČE.

Images from the
Album. *Violich family
archive.* Zagujine (top),
Podobuče (middle),
The Franciscan
Monastery (bottom)

Crkva Gospa Loreta i Samostan franjevaca.   KUNA.   Eglise notre-dame Loreto et monastère Francescani.

This elongated mountain peninsula thrusts itself northwestward along the coast from a narrow neck to the mainland at Ston, some thirty miles up the coast from Dubrovnik. Two knifelike *karst* (limestone) ridges frame the relatively fertile valley of Kuna, one facing the mainland across the Gulf of Neretva and the other facing the Adriatic. Either side has the advantage of contact with the outside world via the sea, though the steep flanks of the peninsula prohibit the frequent access to the sea enjoyed by Brač, with its rounded slopes and indented shoreline. The maritime life of Dubrovnik to the south and Korčula and its adjacent island only a few miles offshore compensated for the lack of direct access to the sea.

The Village Network

To create a maximum sense of reality for a period a century ago, let us hear from two natives of the region, and California residents, whom I interviewed in Palo Alto and San Francisco in the spring of 1987. They speak in their own words about the social, economic, and local governmental systems within the framework of the physical environment. Jozo Tomasevich comes from Košarni Do, one of the smaller villages, and Rudy Palihnich is from Kuna. Born in 1908, Tomasevich received his Ph.D. in Switzerland, followed by a Guggenheim at Harvard, and he held faculty positions at Stanford and California State University in San Francisco. He became well known among Slavic scholars for his definitive writings on economic and political history in the former Yugoslavia,[7] and his close friends revered him for his sharp wit and retentive memory of his homeland. Tomasevich died in 1994. Rudy Palihnich left Kuna to study maritime engineering in Dubrovnik, then came to America with the help of relatives, ultimately serving many years as superintendent for the San Francisco Opera House and its Performing Arts Center. He has made regular visits to Kuna via his own bridge and has kept alive a strong identity with the homeland.

Tomasevich speaks on the Kuna area's marked sense of community among villages:

You must remember that on Pelješac a relatively great number of people were in seafaring. That was the Dubrovnik tradition, and Orebić on Pelješac plus Korčula across the channel were very strong in that regard. When these seagoing people retired, they would continue as what you might call gentlemen-farmers and they would spend part of their savings to build comfortable homes. Whenever you came to a village and saw bigger houses, called *kapetanske kuće,* you could identify who were the sea captains. As a result, in the second half of the nineteenth century many people from Pelješac settled in North and South America and in Australia and New Zealand. They would send money for the family; some would actually return, as my father did, from California. These Dalmatians would seek other coastal areas—California, the state of Washington, Chile, for example.

Let me point out two other ways these villages acted as a single community. My village of Košarni Do shared the cemetery of Prizdrina with four others. There were no roads for vehicles, so your dead had to be carried to the grave in a wooden box, called *odar,* on the shoulders of the people, whoever could help. Those experiences deepened the sense of relationship. Each family had its own plot.

As to transporting wine, in earlier times every family had its own cellar, the presses for the grapes, barrels to keep wine, and so on. But in order to deliver to the harbor way down at Trpanj, we had a communal arrangement. Each family had one or two pairs of goatskins and a *mazga* or two. They would share them. It would then take five or six men to guide them safely to the harbor. After you came home, the people who helped you transport your wine would all come to your house for a very, very good lunch. Loaning the skins, the mules, and helping was an obligation—you couldn't get out of this share in community's economic life.

On agricultural practices, Tomasevich pointed out that by the time of World War I only a vestige of the nineteenth-century, semi-feudal property rights remained. He described this as *kolonat,* a very old form of sharecropping:

One person would own the land, but another would work it, giving the owner a fraction of the gross production. When the French abolished the Dubrovnik Republic in 1808, the landlords no longer held the supreme power over the people that they had held since they had settled new people on Pelješac in the sixteenth and seventeenth centuries. After Austria took over Dalmatia in 1815, the Dubrovnik nobility kept some land under the sharecropping system, but it was gradually sold to the commoners.

As a matter of fact, I have a special document here in which the French administration in Dubrovnik in 1809 allows my great-grandfather and one of his friends to buy land on Pelješac from a Dubrovnik landlord. The document permitted this in spite of the fact that the law of the Old Republic still prohibits such sale. With the establishment of the new Yugoslav state in 1918, all feudal institutions were abolished.

With so little arable land, it was intensively cultivated. Each family reserved land for certain crops: some wheat, some barley, often roasted and added to coffee. There was always a place to raise potatoes and of course a few cherry, walnut, and almond trees, but many fig trees. Practically each family had a few sheep, a mule or donkey or both—my family had six or seven—and occasionally a horse. The only cash crop was red wine; the rest was straight subsistence, no produce markets.

I asked Tomasevich about local government, since in my youth I had heard much of the pros and cons of Austria's highly centralized rule. However, as I had learned from my 1937 visit, that rigid system, plus Dalmatia's spectacular coastline, serrated by mountain ranges, peninsulas, and islands, generated a strong desire for local control.

"Greetings from
Kuna's Općina":
Photograph sent to
California in the 1890s.
Violich family archive

Well, local resistance to Franz Joseph's one-way policies brought about a degree of autonomy during the 1860s. The lowest official level was the *općina*, or
komuna [municipality]. That is where your grandfather served as *načelnik*, or
mayor, after these reforms. This was an honorary and minimally paid position
and had to be filled by a man respected by the community. Daily affairs were
taken care of on a paid basis by the *tajnik*, or secretary. I believe your cousin
Josip, who had studied law in Zagreb, served in this role for a time after World
War II.

    As I understood from my family records, from about 1852 to 1878, the county
seat was in Košarni Do, and a cousin of my grandfather was the *sindik*, as the
mayor was called at this time. The *komuna* was divided into maybe six different *odlomak*s [sections]. Each one contained several villages, each represented
by a "headman."

    Tomasevich described his family as "pretty well off," pointing out hu-
morously that the Violić family was among those who had borrowed
money from his grandfather in the last century.[8] Furthermore, as evidence of the connectedness between villages, his mother was from Po-

tomje, the home base of the Violić family. When Tomasevich talks about his house and the social and economic life of the village, the image of his childhood home is so clear to him that he might have been there yesterday, a sure sign of a strong identity with place:

In the first place Košarni Do means "stable valley." *Košara* is where horses are kept. And *Do* means "valley." This place of my youth is a compact Dalmatian village. All the houses are of stone, a material that abounds on Pelješac. From the type of stonework, Košarni Do must have been built and inhabited three or four hundred years ago, about the time of Dubrovnik's settlement in Pelješac with Bosnians wishing to be free of the Turks. The village is laid out in three, not really streets, but three rows of houses lined up, about twelve in all. One of these is the Korta Tomašević. *Korta* means "court." In my time three Tomašević families lived there in separate houses forming a *zadruga,* or family enclave, entered through a gate.

Our family's house is on the downhill side, with the entry on the second floor. You then go down to the lower space, where wine pressing went on. This main building had the sleeping rooms, and across the small Korta was our kitchen, in a separate building. The fireplace had cooking utensils hanging from the chimney, and there was a separate bread-baking oven. When you heated it, you were heating the room as well. You would put the loaves in for six or eight hours and then close the door. The women knew exactly how long it took. You could also bake the bread under a large iron bell on an open fireplace. That was called the *catura.*

I'd say the house was built in 1750 or so. In two of our bedrooms upstairs we had commodes with pitcher and wash basin. Water came down from the cistern on the roof. As your cousin Vijeko has told you, pointing to the sky: "When it rained, the water comes down from right up there, just for us, *nema problema!*" To say *nema problema* in Croatian—and they say it a lot—is their way of coping with life: "No problem!"

Like many Pelješac families trained in Dubrovnik's well-established merchant marine academy, the Tomašević family had a tradition of going to sea. This fostered a drive toward education beyond the six primary grades close to home, especially for Jozo because he had "a predilection for learning."

Normally children attended school from six years to twelve, but I started when I was a little more than five because I was such a pest at home. They had to buy me a slate. We were children from six villages in one school building, which had one teacher in one room for all six grades. To have any more than six years, families had to have money to send them to either Dubrovnik or Mostar. But fortunately, most of the children in those days did receive six years of schooling, enough to equip them for learning from life experiences, either in the village or abroad, should they leave, as so many did. Your father among them, no?

After high school in Mostar, Tomasevich attended a commercial academy in Sarajevo for four years. Several years later, after working in

Belgrade and Zagreb, he had saved enough money to travel to Switzerland, where he attended the University of Basel in Switzerland, from which he received his doctoral degree in 1937. During his years in Basel he supported himself by a variety of jobs, from work in a dry-cleaning shop to farm work, making the most of the initiative and self-reliance he had learned from village life.

After making something of a name for himself in the public sector on the former Yugoslavia's economic policies and problems, he obtained a Rockefeller grant and arrived in the United States to begin studies at Harvard in 1938, just one year after my first visit to Dalmatia. It was there that I first heard of this scholarly native son of Pelješac.

While Tomasevich's case is far from typical, it is evidence that the simple facilities for education in the village and the stimulus of parents and grandparents who had had broader experiences with the world at large made these people far from rural peasants. They were urban people who sought education and personal achievement. Yet they maintained pride and a lifelong identity with the villages of their upbringing, as Tomasevich documents:

If you think of villages and peasants from the interior of the country and compare them with Dalmatians, you absolutely cannot. What made these families was the fact that they were open to the seas and, especially for centuries within the territory of the old Dubrovnik Republic, engaged so much in shipping and trade with other places.

Even as far as dancing and dress are concerned, we were city people. In my young days we were always dancing polkas and waltzes, but never the *kolo*. We would have one *kolo* in an evening, and that was all. Secondly, we were clad "a la Franka," the name for city clothes. No peasant garb. And shoes of this kind, and not *opanci* [moccasins]. One of my father's three brothers went to sea and followed the men's fashions so much that they called him *Paridjin,* or "the Parisian."

Jozo's stories were in line with my own experiences. The family photographs sent to my father in the 1890s had been taken by professionals, and members of the family, though village residents, looked like city people. In one photograph of my father taken just after he arrived in Seattle with the friend who accompanied him from Kuna, both looked like established men of the city, though my father was only nineteen.

Let us now hear from Rudy Palihnich about Kuna. Although Palihnich was born in Kuna during World War I, he and his wife, also of Kuna parentage, keep their family history of the nineteenth century very much alive. To begin our conversation, I showed him my father's album.

I can tell you this: Marko Škurla, who prepared and published your album, was my uncle. He was a fairly prosperous man and had the general store in Kuna

New Violić house next door to *kavana*, 1890s. *Violich family archive*

on its main street, next door to the new Violić home, built after your father left. In addition, he had a business buying and selling wine wholesale.

My wife's mother just died three months ago at ninety-one, and I am sad about it since, having been born in 1896, she had vivid memories, and I listened well to her stories. Now there are no old-timers left with whom to discuss the old country ways and manners. You know, the school I went to was in the house your father's brother Baldo built in the 1890s shortly before your grandfather died. It was on the upstairs floor—one classroom for all kids—and on the ground floor was the *čitaonica,* or public reading room.

As we looked through the album, Palihnich called my attention to the way Kuna stretches along the slopes overlooking the level vineyard land, with several parallel routes dividing the houses into an upper and a lower neighborhood. The home of the renowned nineteenth-century artist Celestin Medović stands in the older, western end like a small castle dilapidated from a neglected past. The monastery of the Franciscans rises out of the low-lying vineyards and, like the parish church,

Gospa Delorita, stands free of the village proper. Palihnich pointed to his father's house on the higher road of the upper section:

Typically, earlier in the nineteenth century there were few large houses. It was not until the steamship and the voyages to the New World that the economy permitted more space and new ideas were gained abroad. My father's house was one of these, built in 1907 after he came back to Kuna from some years in New Jersey.

To cover my family house, I have to go back a bit into history. In those days, when my mother was a young girl—say, from the 1870s to 1905—her family consisted of twenty-one members, a true *zadruga,* three houses in a cluster. It was like most Dalmatian families—closely knit, but particularly directed by one person, and that was my grandfather. At dinner time he would assign each member of the family their chores for the next day. One to go into the field and plow, another to collect firewood, and a third, to bake bread, and so on.

Palihnich then described how his father returned from America and was financially able to build his own family home. His mother was given—as a dowry—some of the vineyards and a family *stranj,* a large building for farm implements and fertilizer, together with the *konoba,* for storing wines in barrels. His father rebuilt this structure into the present spacious house overlooking Kuna. On the first floor were the entry, a large dining room, and a kitchen. Upstairs there were three bed-rooms and the *sala,* comparable to a living room in America, a huge room for a village house but very much in scale with the panoramic view over the valley. Palihnich pointed to a corner of his living room.

You see that painting on my wall there? That's the view from the upstairs bed-room. You can see the cemetery, church, and monastery. When I look at this scene, I see part of my past. You see that antique there to your left? That was the lantern we used in Kuna. You lift the top and pour the olive oil right in there. A wick goes into each of those three spouts and you light up. Kerosene came by the 1930s, and there was no electricity until the 1960s.

My mother would bake the bread in a side house. We had an open-hearth fireplace, primarily for smoking meat. In the wintertime we would all gather around it to create the smoke needed to dry the meat. The winters were harsh, and the houses had no heat. You had to stay around the fireplace, and you faced in front of it until your front warmed up; then your back would cool off, and you'd turn it around again to the fire. That's how it was when I was a kid, and for my father as well when he was a boy. I had to cover myself with the warm blanket to get from the kitchen to the bedroom.

One boost for Kuna came during the nineteenth century, when Austria built vehicular roads to Kuna. One, for example, the road down to Crkvice, where a quay was built as a port for Kuna. They also built the road to Trpanj, especially the beautiful little road from the town to the cemetery. Because a family friend was very close to the court of Austria and knew Franz Josef personally, the emperor actually paid a visit to Trpanj, his hometown. The road to Crkvice had seven serpentine curves, all too sharp for today's cars. Only small cars can make it down to the many *vikendice,* what they call weekend houses, that Kuna people have there.

Palihnich then spoke of social customs held over from the nineteenth century that encouraged him to find a new life in California and others that endeared him to a lifetime of identity with Kuna and its people.

My family would grow vegetables in the summertime—potatoes, cabbage, green beans, carrots, and the like. But when winter came there was only *kupus,* a rugged form of cabbage, or *blitva,* something like our Swiss chard. You yourself commented that family friends in San Francisco joked a lot about *blitva* as the mainstream of life in the old country. Your father would say that it was *blitva* and obligatory service in Franz Josef's army that drove us to leave for America. As for meat, the two butcher shops would get half an animal from Trpanj, and, with no refrigeration, they would cut it up and sell it within two days. That's all there'd be for the week. Then the next week they'd go down and get another half a carcass for the whole village.

In those days the leisure time was only for older men, not women, and they would spend it in the *čitaonica* reading newspapers and exchanging comments. For the younger people there was a dance almost every Saturday night in the *dom,* a social club, called Omladinski Klub, or Youth Club. There was a large cement plaque on that hall erected in 1925 commemorating the thousandth year of the coronation of Tomislav, the first king of independent Croatia, in 925. They called the people to the dance by having an accordian player go up and down the street for an hour before the dance started. The native *kolo* dance disappeared after World War I.

On religious training, we had catechism instructions twice a week in the public school that was located in your family's house, where your uncle Baldo's wife, Nina, was teacher. On the Day of the Ascension, in May, we would have a blessing of the vineyards. The kids of the lower and upper villages each would have a contest for who will make a more beautiful wreath of ivy leaves and flowers for the priest to use in leading the procession through the paths in the vineyards, bless them, and make the choice of which was the best.

For all their remoteness, the people of Kuna were quite aware of art. One pleasant memory from my earliest childhood was seeing the nineteenth-century painter Celestin Medović. He became an integral part of Kuna, his birthplace, after his active years living elsewhere in Croatia and in Rome. Many houses still have family portraits and his paintings of Pelješac's lovely landscapes. Going by his old stone house, he would be sitting in the portal on a chair in a black robe quietly absorbing the morning sun. Now, whether that was a priest's robe or just an ordinary robe, I don't know. He died January 26, 1920.

For myself, through these studies, the life and work of Medović became an integral part of my own family home in Berkeley. A large and profusely illustrated book on his life given to me by its Zagreb author included the portrait of my grandfather that hangs in the house in Zagujine. The wide range of subjects, from historical events to the landscapes

Pučišća: Nestled in its hidden harbor, 1890s. *Violich family archive*

of Pelješac's highlands and the seashore, demonstrates his devotion to, and strong identity with, his homeland.[9]

The Island of Brač and My Three Grandmothers

The largest of the Dalmatian islands, Brač is also identified by its distinctive loaf shape. Unlike Pelješac, with its steep walls, Brač's shore is broken with numerous natural coves, especially on the mainland side, and thus offers convenient accessibility for settlement. These were landing places for settlers seeking refuge from the Roman towns when they were driven out in the seventh and eighth centuries by the Slavs and Avars. Until the Middle Ages they served only as ports for the settlements on the relatively flat and scenic plateau land, high above the sea. My grandfather's village of Pražnica was one of about fifteen highland villages, the largest being centrally located Nerežišće. Pučišća and a dozen other harbor settlements, such as Sutivan and Bol, did not come to life until the sixteenth century, under the protection of Venice.

These geographical factors—access, terrain, and size of site—led to a relatively even distribution of settlements, each with its own identity and relative self-sufficiency, except that each seaside port settlement was destined to be linked permanently to its nearest older hill village. With the coming of the Turks to the interior, a higher social and economic class, mainly from Bosnia, settled in the more favored harbor locations, such as Pučišća. This class distinction between my two grandparents on Brač contrasted with my father's homogeneous family background in Kuna, and the role of geography in revealing social relationships has been a source of fascination for me.

By the 1920s, a half-century after my grandmother left her island

home, her visions of that environment had been displaced by the hardships she had known during twenty-five years of raising a family in the
gold-mining foothills of the Sierras. I came to know her only in her last
years. My grandmother intimidated yet drew me toward her. All those
visions of place I identified her with impelled me to understand her personality, her life, and her times by studying the other environments in
which she had lived. After school I would drop by her flat in San Francisco's Haight-Ashbury district and find her sitting at the window, erect
yet rather squat, wearing a dress in the style of her earlier years. Her
broad Slavic face with quizzical eyes set off by a wrinkled brow revealed
that she was quite sure who she was and where she came from yet expressed wonderment at the miracle that she had experienced so many
changes of place and culture. When she stood up she didn't seem much
taller, yet with head held high she had a commanding presence. As an
adult I saw her strong will tempered with a deep concern for family continuity and well-being for the three generations she had launched.

My grandmother passed on to my mother her zest for Brač and her
seaside village of Pučišća and its visual images through the stories she
told her when she was a girl in Sutter Creek. My mother had a particular ability to absorb and articulate to us these memories of my "three
grandmothers."[10] First there was the child, young girl, and bride in the
stone house occupied by her family. Then by the 1890s she had become the young pioneer mother in a small frame house in Sutter Creek,
California. Finally she became the aging matriarch whom I knew in person, contemplating her past in the bay window of her flat near Golden
Gate Park.

Urged for years by her progeny, in 1952 my mother wrote an account
of these memories. Her own words carry a sense that even though she
had a remote personal link to Pučišća, she had absorbed an identity,
in her own romantic way, by experiencing firsthand her mother telling
these stories.

My mother's life began in a 16th century castle on the shores of the blue Adriatic in far away Dalmatia. A great deal of her childhood, as she related to me,
and her greatest pleasure, was boating on the sea and trips on horseback over
the winding trails of the nearby mountains of this little seaport town of Pučišća. Her ancestors were of nobility through her grandmother on her mother's
side, hence mother was raised in luxury and had servants which was far removed from the life she was destined to encounter in America.

Her father was an attorney and carried the title of "Don" to his name and at
the age of forty-five lost his eyesight and lived fifty years blind and died at the
age of ninety-five. Her mother lived until she was ninety-three. Under those
conditions the oldest son, her brother George, took over the management of
the household. He was a sea captain and owned a vessel which required him to
travel to Vienna via Trieste often. As mother grew older and was very high spir-

Pučišća: Sailing ships, church, and the Lukinović home, 1870s. *Violich family archive*

ited, loved life and music, she persuaded her brother to take her on those trips to attend the opera. And mother had a lovely contralto voice. She soon learned to sing and became familiar with the operas and used to sing them to us.

Her brother was so very strict and kept close watch on Mother, but she used to plan parties and masquerades. On one occasion, he returned sooner than was expected and the vessel was sighted just in time for all her friends to depart before his arrival. It was very hard to dampen mother's spirits and quelch her "spunk," as you will see as the story unfolds.

Contrasting Villages: Pučišća and Pražnica

Rooted to the island of Brač, my grandmother's identity focused intimately on Pučišća. Though her isolation and the limitations of village life were marked constraints, the prominent position of her home at the water's edge and her family's prominence in the community gave her many advantages: contact with the entire Adriatic and other cultures and languages, the unique urban setting of the protective harbor the town wrapped around, and the shipping activities of her family. Split, a major cultural center, stood only a few hours away, Dubrovnik a little further to the south, and both were directly linked to Venice and the Italian side of the Adriatic. Her visits to Vienna in the 1860s took place when Franz Josef was beginning the famed Ringstrassen, the boulevards and cultural monuments that replaced the sixteenth-century city walls.

On the other hand, my grandfather came not from a leading maritime family but from Pražnica, a true farming village located on the highlands, where virtually all environmental qualities were constraints. It was a harsh land of limestone rock pushing up through the meager earth, which was sunken in places where underground caves had fallen in, and studded with *gomile,* solid domes of stones assembled over the centuries as patches of soil were being cleared for crops. Several nearby sites hold records of prehistoric settlements, and a mural in the chapel at the cemetery carries a date of 1467.

Just as each of Brač's interior villages or hamlets was connected to its more recent port town, Pražnica was tied to Pučišća, though they were isolated from one another by the primitive paths built for the *mazga,* Dalmatia's traditional beast of burden. Schooling, medical services, and economic activities were extremely limited, yet over the centuries the church lent security through faith. It was no wonder that by the end of the nineteenth century many of the young men were drawn to California, first by the lure of gold and later by the prospects for a more balanced, settled life. The sharing of the new and bountiful environment of California with wives brought from other towns on Brač generated in the newcomers to California a more collective identity with the island as a whole than existed traditionally in either Pučišća or Pražžnica. Thus, emigration to the New World not only broadened the territorial basis for identity with the homeland from village to region but also created dual environmental identities linked by the cultural bridge of my family experience.

Again, my mother tells of the circumstances that led to the joining of these two individuals from so very different urban places and their new identities at the far end of another continent. Her words suggest a sense of the persistence and strength of will in their personalities that played an important role in the adaptations they were called upon to make.

My mother was very petite, vivacious and attractive and hence as she grew to young girlhood, she had many admirers. Among them was a young student of the better families who started courting her. As the family restrictions were so great, none could visit longer than 9 o'clock when, if they did not leave, he was politely told to do so.

This particular one, whom Mother loved deeply, objected and so there was a lover's quarrel. She resorted to dropping notes from her windows on a string. He would answer by tying his reply to the string for her to draw up. This continued for some time until he resorted to courting her best girl friend, to see if the situation would change. So Mother's "spunk" arose and she said, "I will show him. I will marry the first man who asks me."

This turned out to be the adventuresome forty-two-year-old bachelor from Pražnica, who had just returned from the gold fields of California seeking a bride. With this advantage, he sought someone of higher social standing, and in the spring of 1871 he was introduced to Clementina Lukinović. His quiet, kindly insistence won her over. Within three weeks they were married in the main church, with its ornate bell tower, on the waterfront and adjacent to her home, where she had been baptized and as a girl had assisted the priests. Some of her ancestors were buried there. As she walked down the aisle in her taffeta wedding dress from Vienna, she kept asking, "Should I say yes? Should I say no?"

This story tells much about the independent spirit of this young lady. My mother concludes this story of my "first" grandmother's break with her native environmental identity as follows: "So she left her loved ones. Mr. Tadich went with them, at the age of fifteen. In England they bought their silverware. We still had some pieces up to the time of the San Francisco earthquake and they were lost in the ruins of our home. Then, from England, they set sail for the United States and were many, many days crossing the ocean and we still have a photo of the ship in which they crossed the Atlantic."

And so ended the youthful period of identity with the contrasting native environments of my grandfather and my grandmother, put aside in favor of a common bond at the opposite end of The Bridge. They now stood together on the threshold of a new environment with the prospect of bringing forth progeny who would have limited knowledge of their own origins. Out of a growing awareness of this concept, I came to recognize the intergenerational continuity of environments as a basic ingredient to identity with place.

CALIFORNIA: THE OTHER END OF THE BRIDGE

The House in Sutter Creek

The pictures of my grandmother taken in Trieste are dominated by a sense of her self-assurance; her hair is curled into ringlets atop her head, and she is standing in order to accommodate her generously bustled silk dress. How proud of her my grandfather appears, seated, in his gentlemen's coat with velvet collar, her hand resting on his shoulder. How hopeful he must have felt about a new life in the New World with this stylish, spirited young woman!

The trip was rugged and harrowing, and later my grandmother often told us that she had "shed enough tears on the trip from England to New York to float the ship back again." Yet she drew on her love of music and her compassion to comfort young Polish immigrants homesick for their families by singing the operas she had learned at home.

The couple made the trip with a party of seven from Brač, and as the only woman in the group, my grandmother took charge of a fifteen-year-old from Starigrad on nearby Hvar. This was John Tadich, who became a leading restaurant owner in San Francisco and a lively figure in social and political life. His first restaurant, located on an abandoned ship, later became a landmark in a building on Clay Street and still operates in the 1990s. Its name, The Original Cold Day Restaurant, was the result of a bet he had made on the outcome of a local election; but that is another story. Mrs. Tadich, an elegant lady whom he later brought over from Starigrad, was named my *kuma* (godmother), and she too impressed me with a strong sense of identity with her place of origin.

Mr. Tadich's own stories held me spellbound and filled my mind with images of Dalmatia to round out those of my grandmother and father. The story of his journey, written in 1932 at the age of seventy-six, is a reminder of environmental qualities that these forebears experienced in making a major shift in their identity with place.

On the twenty second day of May, in the year 1871, I left my native town of Starigrad, in Dalmatia, on the beautiful eastern coast of the Adriatic Sea. It was a day not to be forgotten, the birds were singing and the flowers were in full bloom, and I, about to depart for the promised land, "Zlatna Kalifornija," the golden California, was happy beyond all description. . . .

Bidding farewell to my mother, I started on my long journey, my father and mother accompanying me as far as Split. We had to go first to Mirca, Otok Brač [Island of Brač]. . . .

The following morning, Monday, we started for Supetar where we met a party of four men . . . [including] the pioneer mining man of Sutter Creek, Amador County . . . Mr. John Kusanovich, and his beautiful bride, Mrs. Clementina Kusanovich. We embarked on the ferry boat to Split and spent a day or two visiting friends. Finally we got on an Austrian Lloyd boat for Trieste where we stopped for some time seeing the sights. From Trieste we took the train for Vienna. That ride to Vienna was my first experience on a train and it was very fascinating to me. I have not ceased to be thrilled by the whistle of a steam engine.

On arrival in Vienna, we found ourselves in a magnificent railroad station. We spent about two days in Vienna; we wanted particularly to see the capital city of Franz Joseph, Emperor of Austria. I remember still the beautiful cathedral of St. Stephen, built many centuries ago, and the old royal palace with a beautiful square in front, paved with white stones. All this grandeur made a deep impression on me. It was Sunday and we saw there the real life of the people of Vienna. The park was filled with men, women and children. There was music, dancing, and singing everywhere.

From Vienna we took the train to Hamburg, by way of Berlin. From there we took the steamer across the channel to the port of Leeds, England, then to Glasgow, Scotland by train at sixty miles an hour—some speed! We stayed at Glasgow for two days until our ship, the 'Sidonia,' was ready to leave for New York. She was an old type steamship, but a good strong boat, and, of course, built of iron. After we were out on the ocean about the fifth day her shaft broke and she was disabled. For nearly two days the ship remained stationary, until another ship sighted us and towed us back to Glasgow. We were forced to stay on shore for about seven days; the Steamship company provided us with a boarding house

While we were in Glasgow, Sunday was a very gloomy day. Everything was closed, no place for strangers to go. The City looked as if it were in mourning.

It was a decided contrast to the customs of people in cities on the European continent, where there was much gaiety on Sundays. After the boat was repaired we were ready to proceed to New York. The prospect of the long ocean trip ahead was not glamorous as we were tired and weary. Thirty-five days had passed since we left Supetar and here we were only leaving Glasgow when we should have been at our destination.

For myself, I had been homesick ever since I left my mother and I had lost interest in California and everything else. But when I was feeling most downcast, that lovely lady, Mrs. Kusanovich, came to me and extended her sympathy saying, 'Courage my boy. Let us hope that everything will be all right when we get to our destination.' By this time I had come to know her very well. My memory of Mrs. Kusanovich is so vivid and lasting that I shall always have a very pleasant thought of her. She was the sunshine of the party, gifted with a beautiful singing voice, and she was always ready to sing and thus make everybody happy around her.

But at last the happy moment came when we sighted the American continent. Then there was much joy. When our ship approached New York, the American flag went up to the masthead. I stood gazing at the flag intently for some time. It thrilled me with its beauty. A cheerful flag with its combination of lively colors, red, white, and blue, its stars in the upper corner, a symbolic flag of the great republic of the United States of America. Then the band began to play various American national airs; and the sight before my eyes of the wonderful city of New York filled me with pride and ambition and gave me the first feeling of courage and hope which I had felt since I left home and my people.

We spent two days in New York looking over the City, resting and refreshing ourselves and visiting friends. It was now forty-eight days since we had left home. From New York we travelled on the railroads by way of Chicago, Burlington and Council Bluffs. We crossed the Missouri River on a shaky wooden bridge. From Omaha to Sacramento, the train traveled so slowly that the men of the party were becoming impatient. They were anxious to reach California as quickly as possible and start to work; they were all married men with families left in the old country. I was cheered particularly when Mrs. Kusanovich would sing the songs with which I was familiar. We would join with her in the songs although more in discord than in harmony. She always sang more beautifully when she didn't have our assistance.

To my young mind, it was a wonderful trip, filled with much pleasure. I shall never forget the beautiful scenery and I am happy that I had the opportunity to see the country as it was then. It was a great privilege to travel those immense plains, the majestic Rocky Mountains, the picturesque Sierra Nevada mountains and the romantic valleys of the beautiful golden state of California. All are always before me.

I recall now that whenever our train would stop on a side track, hundreds of Indians and their squaws, with papooses on their backs, would gather around the train. They were just as curious about us as we were about them. Another interesting thing to me was my first sight of a group of Chinese. They had on large sun hats and were repairing the railroad bed. The sight of Indians and the Chinese made a lasting impression and I enjoy the recollection to this day.

On our arrival at Sacramento, the men in the party and Mrs. Kusanovich left me and proceeded to Sutter Creek, Amador County. I traveled alone to San Francisco to my uncle, Nicholas Buja, at whose home I arrived on the 22nd of July, 1871.

John Kusanovich died many years ago in Sutter Creek, and after his death, Mrs. Clementina Kusanovich moved to San Francisco with her large family of four boys and four girls. With her characteristic determination and her motto 'Courage and Hope' she was able to raise a lovely family. The dear old lady passed away a few years ago after having lived to a good old age. I shall always revere her memory, God bless her.[11]

Several events of those times will provide a visual context for this crossing of the continent. The Statue of Liberty was in the process of being constructed under the direction of the French, both in Paris and in New York City. Only some four months after their changing to the recently opened Union Pacific Line in Chicago, that frontier city was virtually wiped out overnight by the fire of October 1871. Robert Louis Stevenson made the trip in 1879 in the company of European immigrants and wrote with pathos of these travelers who had left their varied and familiar homelands for the new environment of the Far West: "It was a troubled uncomfortable evening in the cars. Thunder in the air helped to keep us restless. A man played many aires upon the cornet, and none of them were much attended to, until he came to 'Home Sweet Home.' It was truly strange to note how the talk ceased at that and the faces began to lengthen. . . . This aire belongs to that class of art best described as a brutal assault upon the feelings." [12]

Arriving by the Central Pacific Line in Sacramento, then a comparatively large city in their experience—it had a population of some 16,300—my grandparents went by a four-horse stagecoach, passing through Ione in the foothills twelve miles below Sutter Creek. A new settler coming for the first time in 1876, the Reverend H. B. Williams, remembered that sitting beside the driver was an express messenger with a sawed-off shotgun, a precaution against banditry fostered by the presence of gold. He pointed out that the first white men in the area had been Sutter's party in 1844, when some five thousand Indians still inhabited the Sierra foothill area. This was only twenty years before the arrival of my grandparents. However, Williams found that many of the previous settlers, already experienced in meeting small-scale community needs in New England, were well along in establishing schools, medical care, churches, and law and order.[13]

Sutter Creek and its region indeed presented a new world for these pioneers of the nineteenth century. Gold and fertile soil attracted American settlers prepared to assume the positions of community leadership in business and government that my grandmother's family had held in Pučišća. Thus, not only her environmental identity but also her social

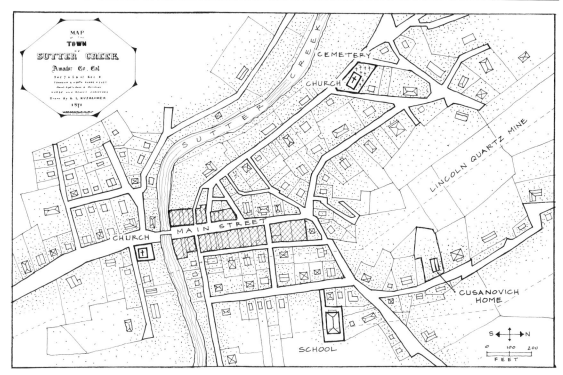

Sutter Creek in 1871: Well established only two decades after its founding. *Amador County Archive*

position was challenged. Yet, in following decades, as she raised her eight children, she established in them a sense of self-esteem and an awareness of cultural values. They found fulfillment in the richly visual world of nature around Sutter Creek and its nearby villages. The creek itself, the pines, wildflowers, and fishing and hunting provided common ground for lifelong friendships generated by neighborliness and schooling. These became the main themes of their lives and took the place of my grandmother's European-style social life in Pučišća. They were also exposed to a strong sense of community within the immediate region, generated by the network of other settlements, similar to that of my father's experience in Kuna. The names of a dozen such places became nostalgic household terms, as my mother, aunts, and uncles identified their friends from, for example, Jackson, the county seat, Volcano, Dry Town, and Mokeluomme Hill. My mother touches on the new life in Sutter Creek:

Imagine her disappointment at first with just a little two-room house to dwell in and not being able to speak English. My father working in the mines at night for all the rich mine owners and she would be alone and have a watch dog. Her life was so far removed from the sea and all the comforts she was accustomed to.

After a year her first child was born, a girl. Thereafter, it seemed every two years there was an addition to the family until she had nine of us, five boys and four girls. She was blessed with good health always, although we did not have the luxury of the homes of today. No central heating, but a wood stove in the kitchen only. But she kept up her courage and with good health and in good spirit managed to carry on.

My father had obtained some cuttings of fig trees from his homeland and they flourished and bore fruit. Many friends obtained cuttings from him and today there are many fig trees in the town. He was very fond of his neighbors and would save the best figs and grapes for them. I remember well when we all wished for the choicest fruit, he would say: "My neighbors first." And I would have to take the dish with a large fig leaf on it and the fruit nicely arranged to that neighbor.

As the children grew into adults, the Sutter Creek environment became their main source of identity, as Pučišća and Pražnica had been for my grandparents. Unlike their stone houses built wall to wall, with their eternal character, those of Sutter Creek appeared instantly from trees felled and milled nearby. In the interests of space and privacy, home builders left space between the houses and surrounded them with flower and vegetable gardens, grapevines, and fruit trees. This pattern symbolized the individualism of the West and contrasted with the pattern of clustered houses and the collective mentality of Pučišća. My grandmother's "old home," as the family called it, was simplicity itself. It had no grand salons or third-story bedrooms looking out on community life and ships from distant lands; rather, it was a straightforward, wood-frame rectangle with a shingled gable roof and a commodious front porch continuing around one side. While the house was small, its spirit was generous; and what was lacking in space was made up for by the human resources of my grandmother. She readily made a house a home, as a focal point for family identity.

The grapevines on the porch and the towering Lombardy poplars that framed the house gave the whole homestead a friendly, rustic appearance that became very dear to my mother, aunts, and uncles, who had who gained a solid start in life there. This concept of love for and identity with a place laden with meaning gathered by experience took root in my mind as I listened to their stories and myths in my early years.

Whereas Pučišća had virtually no commercial district, the principal image of Sutter Creek was its Main Street, where the consumer economy got a boost from gold. The prim white houses of wood, fronted with picket fences and gardens, recalled New England villages, representing a relatively even class pattern, with nothing like the distinction between the seagoing people and the rural, hill people of Pučišća. The public schoolhouse, a dignified brick structure which still stands in its place on a hill, spoke for education. There and in the Catholic church and its tiny cemetery, as one can see by looking at the headstones, the Irish, Italians, and Slavs mixed freely.

My grandfather, in his devotion to wife and children, turned with patience and good will to the hard life of a miner. By the 1890s the mines had been monopolized by mining companies. The first major strike, in 1891, led to labor reforms by the turn of the century. The rigors of my grandfather's work led to his death before he and my grandmother had been married twenty-five years, and my grandmother's life pattern and environment changed once more. In 1896, led by the oldest sons, the family moved to San Francisco to enter the stream of urban life of that already cosmopolitan city. Again my grandmother, now a seasoned matriarch, was alone.

My creative bachelor uncle captured the experiential source of the Cusanovich family's lifetime identity with Sutter Creek in a poem of fifty-six verses that he wrote in 1920, twenty-five years after leaving for San Francisco. Laden with nostalgic images of the environment and the way people related to it, his poem "Memories of a Sutter Creek Boy" had several printings and was popular among townspeople and visitors for years to come, as the following selected verses imply.

> Sutter Creek, the town where I was born,
> Our home still stands, 'tho muchly worn.
> It's many years since we moved, I must say
> So memories I'll quote in this zig-zag way.
>
> The creek we call "Sutter" runs through the town,
> Its waters all gougy, smudgy, murky and brown,
> Flowing through Sutter's roaring mills,
> Tho a foaming, snowy white from the upper hills.
>
> Down below the graveyard where the big creek turns,
> Gathering Johnny-jumpers, maiden-hair, and ferns;
> Spearing wood from the Sutter Bridge,
> And gathering mushrooms on Butcher Ridge.
>
> Birds that we tried our hardest to kill,
> Woodpecker, blue-jay and butcher-bill.
> The gold-finch, of little boys very shy,
> Always built their nests up on high.
>
> The big town fire we thought never would stop;
> Frightened folks scurried to the hill top.
> How our house trembled in the still of the night;
> Miners were blasting with dynamite.
>
> I can see every street, corner and landmark,
> And I can hear every song of the meadowlark.
> I feel related to every hill, tree and valley
> Even our neighbor, "String-Bean-Alley." [14]

The Move to San Francisco

Led by her two older sons, then barely twenty, my grandmother found a new home in San Francisco, a city with many of the Mediterranean

SAN FRANCISCO IN 1900

San Francisco in 1900: An international community with full economic and cultural identity. *San Francisco Public Library History Room*

qualities of Dalmatia, where other pioneers had recently come from. The fresh images of the Bay Area with its similarities to the old country nurtured our own permanent sources of identity. Yet San Francisco was an entirely new city, unlike history-laden Split, Zadar, or Dubrovnik, and with a population of some 320,000 people, far larger. The temporal fire-prone quality of wooden buildings took the place of the lasting quality of stone my forebears knew.

For a decade the family lived in the area called "South of Market"; as her children married they found their own homes. Then, in a single day the earthquake and fire of 1906 left my grandmother without a home and separated from her sons and daughters. Within a week, however, they received a postcard from her, improvised on shoebox cardboard and sent post-free, saying she was fine, having "escaped with the clothes on my back and having a wonderful time" in a temporary camp in Daly City. Even in this crisis she was quick to make a home and establish an identity.

In retreat from the burned district, like many families, my grandmother and her two bachelor sons moved westward, to a flat in Haight-Ashbury within walking distance of my mother and aunt, who had moved to the Sunset District, adjacent to Golden Gate Park. Because it was the only house in which I actually visited my grandmother, its image is strong and real. Her second-story flat was located in a block of dwellings with identical floor plans and no side yards. Their varying Classic Revivalist facades were characterized by a sedateness and dignity quite befitting the stalwart lady. San Francisco's housing had become the end product of speeded-up commercialization of land, in contrast to the centuries-long process of evolution of Pučišća or even the "instant village" of Sutter Creek.

I found excitement in the steepness of the exterior entrance steps framed by distinctive Corinthian columns, in ringing the doorbell and being greeted by my grandmother from her second-story flat. I was especially attracted by the room of one of my bachelor uncles, with its piles of sheet music from the operas he sang, the smell and colors of oil paintings under way, and pencil and charcoal drawings of nudes from the art school he attended. As manager of the Southern Pacific terminal that served the produce district close to the Ferry Building, he echoed the colorful waterfront life of the Embarcadero. A portrait he painted of my grandmother now hanging in my own home serves as a reminder of my grandmother's foreign, self-sufficient ways and the warm visits of aunts, uncles, cousins, and Dalmatian friends, who treated this pioneer matriarch as if she were a precious remnant of life in the old country.

When I made my frequent visits on my way home from Lowell High School, the aging lady would always be sitting in precisely the same position in front of the bay window, whose placement maximized the Western sun and offered a glimpse of passers-by below and Golden Gate Park a half-block away. Outside the window a billowing mass of vibrant, reddish-magenta bougainvillea all but hid the sidewalk.

Her manner commanded obedience, for example, when she would say to me, "You get me glass a *voda!*" mixing her native Croatian with the English she had learned through her family. When I was a small boy she would pat me on the head and say firmly in the Dalmatian dialect, "*Lipi mali* [nice little boy], you go be priest!" I did not become a priest; rather, I became a reformer of physical environments for human betterment. Her forceful, solid character, built in Pučišća and Sutter Creek on a foundation of values evolved over lifetimes, embedded itself in my subconscious. The image of her flat evokes a welling up of beautiful feelings that enliven my spirit and support the multiple identity that I gained in the ensuing years from the two other houses and their surrounding environments where she had spent major parts of her life.

Three houses, three environments, three cultures add up to the three grandmothers who helped shape my ideas about the intricate nature of urban places. For her there was no joining of cults and movements, no self-awareness or self-assertion programs, no straining to find a new identity. From these experiences linking generations I learned two lessons: how eloquently houses and their community settings can speak for the people who used them if we open our minds and hearts and how, in spite of the patterns of places and people that come and go, one can maintain basic values by drawing on the cultural heritage that physical environments symbolize.

Seattle in 1889. The pioneer boom town sprung to life in ten years. *U.C. Bancroft Library*

From Seattle to San Francisco

During my Brač grandparents' first eighteen years of American frontier life, my father was growing into young adulthood in Kuna. On 16 June 1889 he left for Seattle, Washington, in the company of a trusted friend to join others from his region. When I look at a studio photograph the two young men had taken to send back home, I try to find in the facial expression of this nineteen-year-old who was to become my father what he may have felt on arrival. What was it like for him to travel from his home to such a distant place, even one whose beautiful environment had many of the water, land, and mountainous characteristics of Dalmatia? In his later years he appeared to be a quiet, sensitive person, hardly one who would have been prone to adventure.

The two young men traveled by train from New York City to Chicago and through the Northwest on the new Northern Pacific Line. This journey surely gave him a firsthand view of the enormous unurbanized landscape, so different from the diminutive Kuna. At that time Seattle's abundant waterways fostered shipping and fishing activities, which, together with the lumber industry, attracted many Dalmatians from Pelješac and nearby Korčula and Orebić. In 1880, when my father was ten years old and living in Kuna, Seattle's population numbered only 3,530. On his arrival in 1889, there were some 35,000 people, who had already cleared the forests and set up sawmills and enterprises on demand to create a mushroom city. However, there is little chance that he knew that just one week before he left his homeland in mid-June of that year a fire wiped out the city's core of twenty-five city blocks of wood frame buildings within twelve hours.[15] That may explain why he worked outside of Seattle, at the Puget Mill Company.

After four years my father moved to San Francisco, shortly before my mother's family moved there from Sutter Creek. He may have been influenced by San Francisco's being a far more established city, with a population of some 35,000. Perhaps he yearned for opportunities to have the kind of social and family life he had known in Kuna, perhaps a bride, in contrast to the bachelor world of Seattle. He may also have been attracted by reports from fellow natives of Pelješac of San Francisco's milder, drier climate, its mountain landscape, the Bay, and the sea—which made it more like Dalmatia—and even the presence of people from Kuna itself with whom to identify. In the course of the social life of San Francisco's growing Dalmatian colony, he met my mother. She tells how he politely asked my grandmother for permission to dance with one of her four girls, and he chose my grandmother's namesake, Clementine Cusanovich. She says he won her with his unmatched dancing ability.

Thus, as the new century arrived, the progeny of Brač and Sutter Creek joined forces with the progeny of Pelješac and together forged their own environmental and cultural identity with that of San Francisco. Throughout my youth I observed their enthusiasm for experiencing fully the Bay Area's variety of natural environments. They would take Sunday ferry trips to Marin County and hike to Muir Woods or go to the Berkeley Hills, returning laden with wildflowers. Family photographs document how with their own countrymen in San Francisco they enjoyed the intimacy and small scale of places such as the city's busy downtown, Fisherman's Wharf, the Sutro Baths, or the Cliff House.

When their flat south of Market Street was ruined by the traumatic earthquake and fire of 1906, they camped in tent colonies set up in Golden Gate Park, where wooded hillocks stretched west to the ocean. Only a decade before the city had witnessed the Mid-Winter Fair of 1894, inspired by the 1893 Columbia Exposition in Chicago. How stimulating it must have been for natives of Pučišća, Kuna, and Sutter Creek to see the permanent remains of this fair and its Classic Revival architecture in the setting of an urban park in the style of Vienna, already familiar to my grandmother! Cultural attractions abounded: the Japanese Tea Garden, the De Young Museum, the granite bandstand at one end of the sunken esplanade with the pleached European plane trees. These became a fertile source of my own identity with San Francisco, strengthened by the 1915 Panama-Pacific World's Fair, which my family visited innumerable times and discussed for years after. All these experiences laid out before me the enormous possibilities for creating out of the imagination built environments to enrich people's lives.

Thus, in 1908 my visually oriented mother chose a home for us just a half-block from Golden Gate Park, whose meadows, wind-blown live oaks, buffalo paddock, and horticultural nurseries became a part of our daily life. On Sunday afternoons we could hear the rousing music of San Francisco's city band from our backyard garden, the pride of my mother. My father could walk to Ocean Beach, with its wild surf, Seal Rocks, and the Sutro Baths. These inviting landscape features, contrasted with the dark foreboding forest of eucalyptus planted by Mayor Sutro on the barren hill above us that was named for him. There we came to know Ishi, the Yahi Indian rescued from his devastated tribal homeland in the Sierra foothills by Alfred Kroeber and housed in the University of California's Anthropology Museum.[16] These steep slopes above us, the salty ocean breezes, and the summer fog provided a diverse array of experiences and environments to build my own composite of identity with place.

Depictions of the third generation's transitions in identifying to place are contained in a tiny leatherbound diary my father kept for five years, starting in 1918, as World War I concluded. Even thirty years after his arrival, his old-fashioned handwriting and phonetic spelling revealed his European origins. Yet his words reveal his striving to put down roots for his family and to gain a cultural and material level of life not possible in his Pelješac home.

One entry records a 1918 achievement of which he was very proud: paying in full the balance of the mortgage on the house bought in 1908. In celebration of this milestone and prompted by Dalmatian *zajednica* (togetherness), he and his family traveled "by automobile" to Los Angeles to visit my mother's older brother, who had made a name for himself as a creative display director for Bullock's department store, returning by train. My father mentioned joining the local chapter of the Hrvatsko Bratinstvo (Croatian Brotherhood)[17] and taking Sunday ferry trips to Sausalito and Mill Valley "with family and Grandma." He noted the Spanish Influenza epidemic starting in the East and moved West; thus he wrote that "Business is poor" and that on 20 October both he and my mother "come down with the flu when it was bad in San Francisco." Some entries display his political awareness, in part the result of his father's being mayor of Kuna. He bought a three-hundred-dollar Liberty bond; and he wrote, "Germany asks for a Truce and Armistice." Austria, which was where his passport was from, and Germany finalized the Armistice, and "the Allies sent Germany terms which were very stern." Finally, "Kaiser Emperor Wilhelm II of Germany abdicates" and "The Great War is over." He noted that on 4 December President Woodrow Wilson sailed for Europe for the peace conference, during which the future "Kingdom of Serbs, Croats and Slovenians" was discussed. In 1919 he witnessed President Wilson's arrival in San Francisco.

While we children took for granted our schooling, he noted with pride our entering the first grade in the nearby grammar school and our later attending Lowell High School, founded in the 1880s as the first public high school in the West. He wrote about our indoctrination at St. Anne's Church, administered by friendly Irish priests from Dublin; the funerals of relatives in San Jose; the novelty of trips in "automobiles" belonging to generous "Slav" friends.[18] He speaks reverently of a trip to Sutter Creek in the new car of my favorite uncle and of a trip to Stockton, where I spent hot summers with my sister, far from San Francisco's fog. This social life represented a close relationship with members of his own culture, but he also recognized his Irish-Italian neighbors on one side of our house and the Scots on the other.

The house itself, which had been hard-earned, was a joyful setting for get-togethers with our extended family. My father recorded payments to "Mr. Trobock," the Dalmatian contractor, for major additions, the result of my mother's highly creative mind. She replaced a common back porch with a glass-enclosed breakfast room with sun room above, both overlooking the garden she later allowed me to redesign. We acquired a "Sonora phonograph and a Farrand Cecilia player piano," on which we all were given lessons by a teacher whose parents were from Dubrovnik. My own abilities led me to accompany uncles and aunts in songfests after family dinners and to perform publicly, an experience that clarified my creative directions.

On a first trip to Monterey the many family members crowded into two cars were struck by spring in the unique environment of the California coastal region. We had picnics with Croatian friends in Watsonville, the town known in Dalmatia as Little Dalmatia because the majority of its residents were from Dubrovnik, Konavle, and Brač. There the apple orchards were in bloom and the fields were filled with yellow mustard and blue lupin. On one fishing spree, my uncle pulled out of San Francisco Bay a striped bass weighing fifty-seven and one-half pounds and had his picture in the paper. Such was the bounty of the environment, a far cry from its devastation in the 1990s.

The entries conclude in 1923, as the horizons of the family broaden beyond San Francisco. My favorite uncle and role model, the painter, made his second trip to New York City on a railroad pass from the Southern Pacific and returned with images of "big city" places. A handsome uncle in Los Angeles became an important designer for Bullock's department store, which later sent him to Europe. My grandmother grieved that he could not go to Dalmatia. I began to visualize myself as the first of the progeny to visit our family's place of origin. The rich life around me led me to identify with the ways of the old country as well as with the city of San Francisco and the Bay Area.

This regional identity flourished when I set my sights on fog-free Berkeley and immersed myself in landscape architecture at the University of California in 1929. The natural qualities of the campus became a source of knowledge and spirit for me: its creeks, glades, and rolling terrain, the oaks, and the views back to San Francisco and out to the Golden Gate. Day by day, as I experienced the reflective mood of crossing the Bay by ferry, the environmental base for my identity broadened.

The View Back to Dalmatia

Fulfilling as my parents' settled life in the neighborhood by Golden Gate Park was, they kept in close touch with Dalmatia. Letters from Kuna extended my mother's sense of the individual qualities of my father's village, and she came to know and love the members of his family. They sent pictures taken in the village by a photographer from Dubrovnik. After World War I, sending used clothes to relatives in Kuna became a monthly ritual. My mother's empathy with Dalmatia reinforced The Bridge as the years went by, and we talked of traveling there some day. But this hope was terminated when my father became an invalid in 1927 and passed away in 1934. These experiences and the realization that my father could never return made me increasingly conscious of our good fortune in contrast with the hardships the family at the other end of The Bridge. It was as if I had acquired a second pair of eyes.

During my university years, in the early 1930s, I realized that the pressures to conform to Anglo-Saxon values left little room for maintaining the cultural qualities my forebears had brought with them. I reasoned that the severe limitations of Kuna were offset by the human qualities the harsh environment produced. The Dalmatian coastal environment, with its outstanding natural grandeur, its rich history, maritime life, and the rewarding cultural resources of community life, made up for the hard life of those who lived there. Those who were able to come to California, with its environmental advantages, could readily enter into a beneficent life here. Among them many, including my family, could resist the pull of Americanization by means of The Bridge and thrive on continued contact with the environment and culture that had shaped them.

EXPERIENCING FIRSTHAND
A FARAWAY IMAGE

In the fall of 1936, traveling by car to carry out graduate studies at Harvard and MIT, I experienced the broader American environment firsthand. The visible elements of New England and its urban places brought life to the history I had learned and made more precise my identity with such contributors to our democratic ideals as Thoreau and Emerson. This reality encouraged me to pursue fully my own identity by crossing The Bridge to Dalmatia. This I did in the summer and fall of 1937. I first traveled by bicycle through France and Italy, experiencing urban history through Mumford's vision, then sailed across the Adriatic to my father's home, all with my mother's enthusiastic support.

Since there was no highway system, in my roughly six weeks on the Dalmatian coast I shifted to the universal highway of the sea, the basis of that region's history: first to medieval Dubrovnik, then to my father's Kuna and Pelješac, on to Pučišća and Pražžnica on Brač, and finally to Split and Venice. The Bridge came to life, and I gained a realistic sense of who I was, where I was from, and what America had contributed to my individuality. By the time I returned to San Francisco at year's end, the experience had illuminated and stimulated my interest in the relationship between place and culture and the need for incorporating new and meaningful social and cultural ingredients into physical urban planning. I sensed that a genuine form of this identity arises, not out of design in the contemporary sense, but out of the indigenous resources of land, mountains, and sea and the deep involvement of the people of each place and its region.

Pelješac and Kuna

My anticipation before seeing Dalmatia for the first time was heightened by the trip from Italy by boat at night. As I entered the deeply indented fjord—the Boka of Kotor—the sun rose from behind the mountains of Montenegro that separate the coast from the interior, and formed an indelible image in my mind. Further up the coast, Dubrovnik, jutting out into the sea, made the visions of medieval cities I had formed in libraries real and to be walked in. Day by day, the city's magnificent setting, the integrity of the grid street system, and its enclosing sea walls absorbed me into their presence. The historic past of the surrounding regions became the present when I saw Trebinje, with its Turkish settlement intact; Ćilipi, where costumed folks of the Konavle subregion danced the Kolo in rhythmic rings; and Cetinje, the capital of Montenegro, high above the Kotor fjord. My ethnic origins found a broader context in the human effort and the richness of nature's landscapes. At the same time I was repelled by the poverty and backwardness that had been brought to the beautiful urban place and driven people to emigrate to California in search of a better life. For two weeks I had the wondrously illuminating experience of sharing in the daily life of a small villa at the water's edge. In Gruž, the old port of Dubrovnik, the home of my uncle who had married into my mother's Sutter Creek family, I felt as if I had returned to Dalmatia in place of my father forty-eight years after he had left.

Later, I boarded a small white steamer that traced the base of the precipitous slopes of the Pelješac peninsula to leave me at Korčula, my connecting point for a smaller boat to the port village of Trstenik. Dropping into the sea with but few footholds, the mountain range looked forbidding and insurmountable. Hard-won terraces, however, planted efficiently with vineyards in the more accessible places, were a manifestation of the resourceful Dalmatian spirit. Korčula itself, a tiny Dubrovnik, stood as an example of urban history come to life.

After a sleepless night in a torrential summer storm, I was greeted by a clear morning sky, and a small motorboat took me across the channel to the tiny coastal village of Trstenik. When I asked about a bus to Kuna, gales of laughter answered me from a friendly group who pointed to a *mazga,* the local "autobus." Indeed, the owner had already been informed of my arrival; calling me the *rodjak iz Amerike,* the cousin from America, he offered me his own *mazga.*[19] Seeing the zigzag trail I had to follow up the canyon and the blue water far below, I spurned the ride in favor of attempting the trip on foot

When the loose sharp stones and the careening course of the trail among pine, laurel, and heather were beginning to get the best of me, my friend and guide excitedly pointed far above to two figures riding down the trail leading a third, saddled *mazga.* My father's niece and nephew—Dobrila and Vijeko—greeted me as if I were one of the family who had been away a few years, and the strangeness vanished as the link between my father and his family was reestablished. Together, we three cousins continued the climb up the ever-narrowing canyon to some fifteen hundred feet above the sea before entering the beautiful mountain valley of Kuna. Rocky ridges of *karst* dotted here and there with sparse clusters of pine broke the irregular, fertile floor. Vineyards, cabbage patches, olive groves, and stone walls running this way and that set off relatively small plots of farmland. The blue sky, grey-green hills, red-brown soils, and deep green vineyards set off occasional clusters of stone houses with red tile roofs, recalling the colored photographs of Kuna in my father's album.

First came the hamlet Zagujine, my father's birthplace, on the way to Kuna, where I was to stay in the three-story house my father's brother had built in the 1890s. Its location on the main street did not fit the typical image of a rural village. I came to know my father's three living sisters, the six or seven nieces and nephews, my first cousins, and their children, some of whom were on vacation from school in Split. To the older members of the family, it appeared as if my father had returned to them. Because my knowledge of Croatian was limited, family friends who had lived in California translated for me, though more often I turned to the French I had excelled at in high school, since that was the language used by the younger people. This generosity and kindness were

Kuna's main street, 1937. *Photo by author*

expressions of a genuine connectedness that had not been weakened by my father's departure to a place so far away a half-century earlier.

During my weeks in Kuna family members walked me over the network of paths where no motor-driven set of wheels had ever passed so that I would know them as my father did. One sunny day we followed the route he took from school in Kuna to Zagujine. I could picture my father as a ten-year-old boy in 1880, trudging up the path that wound as if it were in a park at the foot of the rocky hillside. Clumps of windblown pines spiced the air, and there were fragrant rosemary and cascades of heather in bloom among the rocks. Facing the setting sun, located well up the slopes of the hill, and backed by protecting mountains, the long, shallow two-story house looks out through a row of great pines. From a broad terrace that linked the home with two others, we could see beyond the little valley the mountains and the village of Potomje, my grandfather's birthplace, on the other side.

Nina, a daughter-in-law who had actually known him in person, told me that when this leading citizen decided to build a new house on the site of the outdated building, the men of Kuna showed their respect for his position by contributing fine white stone from Korčula for the doorways and window frames. When my own father was a boy, he and his mother drove the mules that hauled the local stone and timbers. When all was finished, my father helped my grandfather plant the row of pines along the edge of the terrace. Visiting merchants, clergymen, and sea captains were entertained on the terrace under the then young pines, and there my father heard of America. I could imagine his departure and how strong an image he carried with him of this home.

We walked the several miles to the villages of Piavično and Potomje to visit my father's three elderly sisters. In Piavično I visited Teta Fila. I saw in her the high forehead and quiet manners of my father, and she saw in me her own brother as he had been when she was a young girl. This village was made up of scattered houses that were like small farms, very different from the urban character of compact Kuna. Teta Kate, in Potomje, had even more of my father's softness of manners but was less withdrawn. She recounted how at age fifteen she had accompanied him down the mountain canyon to the port of Trpanj when he departed. She remembered only that she cried very much because she was sad to see him go, and now she rejoiced in my returning in his place. The eldest sister, Frane, for whom I was named, was known as my grandfather's favorite; in her I could see more of my father's impressive bearing, dignity, and stoicism. All three women seemed to have reached a state of internal peace after the arduous years following World War I.

In the evenings, after dinner, we would take turns singing of Dalmatia and California, but on my last evening there were only songs of departure; one, "Zbogom Kuna Grade" (Farewell to Kuna), I had heard from my father when I was a boy. Now I was reliving his departure. At 2:00 A.M. we arose from our beds, and the entire household accompanied me to the waiting *mazge*. Their hoofs clattered on the cobbled street as Vijeko led me toward the road to the port at Crkvice. Only the twinkling of the vast array of stars broke the stillness of the night air as we moved down the steep path. Dawn was breaking as the little steamer docked. Vijeko tossed my bag on board, and as he embraced me I felt that my father's ties to Kuna and its people had become my own. As the boat moved away, the dock became smaller and the stars disappeared. Dawn came and Kuna was gone.

Brač: Pučišća and Pražnica

In Split, I stayed several weeks with my father's niece and young family. One of eight children, she ultimately had eight of her own. Today they are an important source of the intergenerational continuity that has contributed to my search for meaning of place. This thriving port city had begun to sprawl outward from the nearby walls of Diocletian's Palace, the vibrant heart of the city, where some two thousand people lived among contrasting remnants of Rome's grand structures and Venetian architectural character. Though housed in antiquity and still free of cars and planes, Split struck me as being very much alive with twentieth-century vitality and social life.

From the windy, pine-crested top of Mount Marjan all of Split spread out before the gargantuan naked mountains that sweep upward to the clouds. From the palace walls and the *obala* the blue sea reached out across the channel to the long, rounded island of Brač. Down its flank,

The ancient pine tree: Symbol of Pučišća's fortitude. *Photo by author*

Pučišća lay hidden in its deeply indented harbor, and high on its stony slopes stood Pražnica. Fired by the gradually unfolding picture of my people's emigration, I put off seeing the phenomenal remains of classic Rome and made the three-hour trip to Pučišća across the channel the following day.

Our crowded little white ferry traced its zigzag route through the fjordlike inlet from the sea to the inner harbor around which Pučišća wraps to form an amphitheater, furnished with dwellings of its own cream-colored stone. As we docked, an emigrant recently returned from the New World pointed out a well-dressed gentleman leaning on a gold-headed cane and engaged in a merry conversation with a group of rural women. This was Tonko, my grandmother's nephew, who was awaiting my arrival. We disembarked on the creamy-white stone *obala* that surrounds the hidden harbor of intensely blue, transparent water, hidden so cleverly from the open sea. On the piazza stood a single ancient pine tree so old and gnarled that a stone crutch had been placed to hold its leaning trunk and broadly spreading branches. I saw it as a living monument to the generations that had built Pučišća.

Tonko made his way to me. But a boy when my grandmother left for California, he was an old man when her grandson returned. Dignified, he welcomed me. There it was: all the pride, formality, and urban manner that my grandmother had displayed during more than a half-century of courageous family building in America.

Just beyond stood my grandmother's great stone house, next to the parish church. It was large, but it was not the "castle" my mother had claimed. We toured the three stories from cellar to attic and inspected ancient timbers and walls, which had been dated to the eighteenth century by the stone cutters. Tonko showed me the room where he stored the musical instruments he imported for the two bands the town maintained and the "ballroom" where my grandmother had held her parties while her older brother was away at sea.

I was to sleep, not in the small wing kept for living as the family had dwindled during the years he had spent in Chile, but in the third-story bedroom overlooking the harbor. In the darkness I leaned on the stone sill and allowed myself to experience the stone *obala,* the miniature harbor of still, clear water my grandmother had grown up with. Poised on the surface of transparency, a boatman in striped shirt attracted silver fish to his net with a lantern that illuminated the depths. Other boats rocked above their own patch of clear water drenched with light. The absolute quiet, the entire village asleep, and the fisherman's oneness with the sea and nature as a primal drive to live: these images emerged as I reflected on the beauty of my grandmother's early years here in Pučišća and the strength she gained from them.

As the boats drifted out of sight, I left the stone sill and crossed the room. The fluttering candle flame exaggerated the baroque form of the marble-topped washstand. Under my bare feet the solid plank floor felt smooth and grainy from its use by many generations. Between clean rough sheets I slipped down into a deep sleep and into the centuries surrounding the huge old bed.

During my days in Pučišća I learned that those who who were familiar with New York City's Metropolitan Opera House called this waterfront location the town's "Diamond Horseshoe" because of the "front row" of large houses around the theaterlike harbor. Tonko and the Kraljević branch of the family explained how an environment of sea, ships, and travel to other cultures extended their cultural awareness. They spoke of a family ancestor, a seventeenth-century architect named Trifun Bokanić, who had designed the upper stories of the renowned bell tower in Trogir, near Split. I met members of the Dešković family next door, who had been sculptors for generations, an art that had grown out of their proximity to the giant quarry. Tonko told me of the number of pianos, musical instruments, bands, and groups Pučišća had supported in the days when no music could be heard unless it was played on an instrument or sung by a human voice. In this way I came to know my grandmother as a young girl. I could understand why she held her head high even in old age and strived to instill in her progeny a sense of connectedness to place and the life within it.

My images of Pučišća were in keeping with my grandmother's character, but I still needed an environmental context for my grandfather's identity. Tonko arranged for my visit to Pražnica. In the morning a pitcher of fresh water stood on the washstand, and at seven Tonko appeared at the door to announce that the *mazge* were ready. As if to welcome this inexperienced grandson of his uncle, my host in Pražnica had outfitted the mules with padded cushions and laundered covers. This was Martin Kusanović, who had gone to California as a young man at the turn of the century. He had done well and returned to live in luxury, but during World War I, when the banks had failed, his money had evaporated. He had become a pitiable figure almost entombed in the lifeless home of his birth, devoid of elements with which to establish any sense of identity.

My first view of the patterns of Brač's landscape impressed me because of the human resourcefulness it revealed. Over the centuries stone walls have run rampant on the island, arranged helter-skelter as the population tried to contain precious soil for vineyards, olive groves, or patches of grass for sheep or goats. As we wound our way upward, the blue inlet of Pučišća fell away. The riotous walls began to break down and become great polka dots of stone piles scattered across the rounded flank of the mountainside. These *gomile,* tossed into heaps since the time of the earliest settlers, looked like bizarre geological formations. I envisioned my grandfather descending the same path in 1871 to marry in Pučišća, never to return.

Pražnica rests at the top of these slopes, where they flatten out to become a weathered plateau broken by protruding ridges of limestone and sunken potholes. The sharper form of Mount Vidova Gora (elevation 2,400 ft.) served as a magnetic backdrop for this still untamed landscape and seemed to pull together the threatening clouds. Pražnica's wild environment was beyond my expectations. Its remoteness from the family warmth of Kuna and the comings and goings of Pučišća's marine life prompted me to depend on my inner self during the days ahead. At his California-style cottage, with a front porch like those that are common in the Sierra foothills, Martin greeted me as one whose life had ceased decades ago.

Without the commanding configuration of land and sea that created Pučišća's clear image, Pražnica's one hundred or so dwellings formed no coherent urban pattern. Only the modest bell tower of the church where family members had been baptized for centuries rose to provide a focus of village life. The stark reality of survival was evident in the stone dwellings, which seemed to be constantly in a state of repair. Grapevines or even tiny pines clung to pockets of soil in the grey stone walls or their tilting rooftops, which huddled against the flank of the mountain slopes as though seeking the comfort and security of the earth. Pražnica's streets were vacant, its youth and vigor gone. Like the ghost cities of the California gold rush days, the village seemed a sad part of the past. Except for Martin, none of my grandfather's family remained; even the original home was gone. Both family and community identity had drained away. Here indeed was a lesson in the dependence of identity of place on the quality of the built environment.

Tonko went back down the mountain, and I remained for the night. My planned return to Pučišća the next day was violently obstructed by a demonstration of the power of nature, against which humans had for centuries struggled for survival. At 4:00 A.M. I awoke. The very stones of the house seemed to rumble with thunder. Window shutters banged in a sodden wind as an Adriatic summer storm poured forth its fury on the sturdy stone of Pražnica. From the window, I watched as flashes dramatized the bleakness of the mountain village and lighted the water rushing down its cobbled streets. It seemed that all would wash away.

For three days the storm marooned me in this tired old village. During short spells of quiet we explored the stony pathways, and all the while I deplored its poverty and deathlike quiet. Each morning Martin would make the rounds seeking an egg for my breakfast, and each day I was given the only one the few hens produced. In time my host lost interest in reviving memories of his youth in San Francisco. He dropped back into his bitter lethargy of twenty years, broken only by his constant solicitude for my well-being while I waited out nature's rage. On the fourth day the sun finally greeted us. Saying goodbye to my patient though unhappy host, I observed how the characteristics of places can shape the qualities of the people who live in them. In that sense the somber environment of Pražnica reflected in my grandfather's quiet nature contrasted for me with the lively and enterprising qualities of my grandmother, which had been generated by the visual and human stimulus of Pučišća.

Those last few days allowed me time to talk with younger family members about the differences between Pučišća and Pražnica and between Dalmatia and California. In endless conversations they contrasted to their love of life and their enjoyment of the quiet and beauty of their precious home with the images they had of America's stressful way of life and its materialism. In their young Dalmatian minds, tomorrow seemed to be another matter. To sail their boats, to sing, to live simply among a community of friends and family, they said, gave them a direct personal identity with their natural environment. At that time, none of us could have envisioned how important environmental values would become for America, nor how violently their situation in Dalmatia would change with the war and revolution of the 1940s and its appalling repercussions a half-century later, in the 1990s.

The words of a Dalmatian-born New Zealander who visited the homeland years later describe well the emotions that family and environment combined can evoke: "No words can adequately capture the experiences, the feelings aroused, the impressions gained in the birthplaces of relations and parents. Places and names suddenly gained substance as earth, rock and flesh. . . . the reasons for emigration and the aspirations and courage of those who departed were brought into focus by first-hand observation of landscape and occupants, . . . the vexed questions of identity and purpose of research were fused and resolved.[20]

The Power of Environment and Culture Combined

Because of my experiences in Dalmatia, the power of identity and con-nectedness with one's home place in the Old World became a reality for me and had a fundamental impact on my professional and personal values. So much so that on my return to California I began making plans to return to Dalmatia in 1939 to spend two years living in my father's village. There I would directly experience in depth the urban places in both their regional environment and in terms of local family and com-munity qualities. I had as a model the sociological study of Middletown, Ohio, by Robert Lynd and Helen Merrell, which offered a real-world look at how people live in cities; this work broadened the outlook of city planners of that period, from the physical to the community structure of urban places.[21]

Louis Adamic's multicultural concept would protect me from the melting-pot concept of the 1920s and enrich my California identity. The power of Dalmatia's physical environment with its balance between the unique natural and built environments within a broad context of history became a force within me. I wanted to experience firsthand how these towns and villages responded to the forms of coastlines, valleys, and mountains, how they evolved in history to their present shape and distribution. I saw an opportunity to portray them as a phenomenon of history, geography, and culture on the threshold of twentieth-century change. As I outlined my study and sought support, Hitler marched into Yugoslavia and my vision was demolished.

Not until 1968 could I return to Dalmatia. In the meantime, inspired by President Franklin Delano Roosevelt's Good Neighbor Policy, I spent one year, from 1941 to 1942, doing field research on the history, prob-lems, and planning approaches of Latin American cities.[22] Many Medi-terranean cultural qualities I found there and cousins in Chile from both Kuna and Pučišća served to lessen my disappointment at not being able to visit Dalmatia. In 1961 a Fulbright award made it possible for me to study the evolution of the Spanish city as a basis for understanding more fully the cities of Latin American.[23] Happily, this experience provided a model for what I might do eventually with Dalmatian cities. Destiny favored me when I attended a conference of the International Federa-tion for Historic Preservation in Santiago de Compostela in the summer of 1961. By chance, two colleagues from the then Yugoslavia took charge of me, one of whom, Tomislav Marasović, was mainly responsible for the restoration of Diocletian's Palace in Split. He exuded a love of all things Dalmatian, especially to a direct descendant born in California, and committed himself to making sure I got there.

This seemingly preordained link gave impetus to return trips to Dalmatia, which occurred in the summers of 1968 and 1969. On a second sabbatical in 1970, with Rome as a base for my family, I commuted to Split. All of us spent the summer in Dalmatia. During these trips I explored my subject all along the coast and in Zagreb, where I found a wide array of people who were knowledgeable in my areas of concern. In 1972 I traveled to Sarajevo and Belgrade, which helped to round out my sense of the former Yugoslavia as a whole.

On my 1968 visit I found that enormous changes in the physical environment had taken place. Most devastating for me was finding the gaping shells of the two principal family houses I had slept in on my 1937 visit, my father's in Kuna and my grandmother's in Pučišća, which had been sacked and burned by the Italians. The first, rising three stories from the street and four from the slopes in back, consisted only of a roofless hull. The blue sky filled the empty window frames. Birds, vines, and even fig trees clung with their roots to the crevices where earth had collected as if to soften the violence of the attack by Nazis and Italians some twenty-eight years before. By targeting the larger houses, they could wipe out resistance by community leaders.

Now I slept in Zagujine, honored to share a room with Celestin Medović's portrait of my grandfather. My view was his, out to the vineyards and valley, framed and scented by the pines my father had helped him plant in the 1880s. But this joyful opportunity to reinforce my earlier identity with this strong and appealing home place was severely fractured the next bright morning, when Vijeko told me of the cruel and very personal violence of war committed just outside. The family showed me the wall against which the enemy had lined up the men. To stifle resistance, they had shot them in the presence of their wives and had burned farm tools and equipment to further decrease the chances for survival.

In Pučišća, my grandmother's home was a stark ruin open to the sky, a monument to war! The gnarled pine where the old men of the village would assemble in the shade at midday had been cut down, against local wishes. The creamy-white stone paving of the *obala,* worn smooth since early times, had been covered with asphalt, a sacrifice to the coming of the car and truck. Old Tonko was gone too, but somehow the family image of my grandmother remained. Not until 1979, while on my second Fulbright fellowship, was I able to sleep once more in Pučišća, this time in a similar ancient house belonging to the son of grandmother's Lukinović family who had remembered my 1937 visit. The grand villa

Two family homes in 1968 that were burned and gutted during World War II. *Photos by author.* Kuna (top), Pučišća (bottom)

they had received me in then also stood gutted; the socialist authorities later had converted it into a vacation home for youths. Martin was gone from Pražnica, but a new reservoir built by the Allies was a symbol of hope for the future.

Meanwhile, the "socialist" state of Yugoslavia had established itself independent of the highly centralized Soviet Union's political mandate and embarked on a long overdue program of modernization. The modernization program was based on decentralized local initiative and self-management, though ultimately centralized authority dominated. In

1968 I found that on the mainland road systems had been built, electrical power and telephone lines extended to virtually all villages, schools, and clinics, and tourist facilities were widely in evidence. Yet Brač and Pelješac remained quite the same as in 1937. The Adriatic Highway was well advanced, linking all urban places along the mainland. Fortunately for my study, the United Nations was working with the Urban Planning Institute of Dalmatia between 1968 and 1971 at the regional level and at the level of the *općina,* or municipality, of the larger cities and towns. I had access to their pioneering development studies. However well-intentioned in terms of improving the economy and thus improving local living conditions in the towns and villages, the overemphasis on large-scale tourism and limitations on small-scale private enterprise threatened the fragile link between their physical and cultural sources of identity as many smaller places lost population to the larger cities.

Later I learned that during the 1970s, after a bad start with over-scaled hotel and resort projects in places that were not suited to traditional urban development, conservationists were able to scale back new projects to make them more compatible with the environment. Compared with what happened to the coastlines of other European countries after World War II, Dalmatia was protected against unplanned growth by the constraints of "socialism." In Italy and Spain, for example, tourist development was given over to private enterprise; the degree of commercialization that took place was disastrous to both the natural environment and traditional settlements. Throughout the 1970s, on the other hand, Dalmatia, with its socialized economy, showed more control, and its architects demonstrated greater sensitivity. Competitive advertising was nonexistent. Dalmatian conservationists, imbued with pride in traditional cultural values, became leaders among the Europeans, working actively with UNESCO in the preservation of the traditional character of towns and villages, often financed by adapting existing structures to new uses. The natural environment suffered, however, especially along the new Adriatic Highway between Split and Dubrovnik, as the air and waters close to major ports were polluted by industries. Likewise, the ecology of the coastal landscape was damaged by forest fires and large-scale earth moving.

During the 1980s overdevelopment of tourism swept to one side the local guidelines previously set down. The imbalance between the natural and built environments continued to increase throughout that decade and up until the aggressive attacks on Dalmatia that occurred early in the war of 1991–95. This will unfold for us as we look closely at the region and its hierarchy of cities, towns, and villages strung out along a uniquely varied stretch of maritime geography.

The Impetus to Explore the Meaning of Identity with Place

In May 1990 I returned to refine my case studies of Zadar, Split, and Dubrovnik, contained in chapter 3. At that moment the newly formed democratic parties won an overwhelming majority in the Croatian Parliament. This became the most meaningful turning point in the history of these urban places; now they could fully express community values and local identity. However, the breakdown of the other republics appeared inevitable. By the fall of 1991 the optimism I had found vanished with the violent attacks on these three cities and their environs, symbols of the country's cultural identity. This stimulated me to pursue in greater depth the meaning of identity and its implications imposed by the war. The outbreak of democracy there had generated a fresh sense of enterprise, which was expressed in the look and feel of Zadar, Split, and Dubrovnik. Portions of my paper describing the new vitality I had witnessed were published and commented on in *Nedjeljna Dalmacija,* Dalmatia's leading newspaper: "We find in the words of Violich an affirmation of faith in this vitality. . . . As it has always been, is and will be, we have the right to believe in the continuity of the individual qualities of urban places, even when brutality has interrupted—as the recent history of Zadar testifies—these cities still found strength to renew their lives. . . . And so, let it be like that once again."[24]

CHAPTER ONE

Identity
Key to the Meaning of Place

Man dwells where he can orient himself to and identify with
an environment, or . . . when he experiences an environment as
meaningful. Dwelling . . . implies something more than "shelter."
It implies that the spaces where life occurs are places.
—Christian Norberg-Schulz, 1980

OUR LOSS OF ROOTEDNESS TO PLACE

Throughout the world, the global increase in urban population has come upon us at a rate unknown in history, and simultaneously with our boundless technological creativity, which almost has a life of its own. Yet, while mankind has benefited from this creativity, the energy released has all but erased that basic sense of connection between people and their environments, their identity with place, that has propelled human advancement. The intimate union between land given by nature and its development by man gave birth to the city as a highly effective sociological phenomenon. Stability came as the insecurity of roving tribal life gave way to identity and commitment to a particular piece of land within a geographically defined region. Over generations and millennia, whole cultural, economic, and governmental systems evolved out of this union.

This century of ours broke with this tradition and brought vast new spatial scales to cities and the rapid growth that caused the malfunctioning of the machine that the city has become. Alienation, social distress, economic loss, and pollution have become trademarks of urban places. Through this life-threatening experience we are learning that the elements of urban environments—land, buildings, and people—exist as one whole. Together these three elements are essentially ecological; therefore, they require a deeper understanding and awareness than their daily use allows. Only on this concept can we base a genuine exploration of identity and meaning of place.

The present century has produced technologically dominated urban environments of unprecedented force. These have been generated by a powerful economy often out of touch with the basic human need for a sense of belonging to a given place. Contrary to the corporate symbol seekers, who have claimed identity of our cities for themselves, they are places comprised "not of material objects, but of living relationships. . . .

so now is ecology basic to the stewardship of a new meta-industrial society."[1] The creators of these new, ever more massive metropolitan environments have given free reign to the collaboration of private and public entrepreneurs—builders, developers, investors, urban planners, designers, engineers, and architects—all too eager to embark on adventures in creative self-expression made possible by surplus capital and limitless technology. Enveloped in sprawl, small towns near urban centers have lost their native individuality, and the humanistic beginnings of urban planners in the enlightened post-Depression 1930s have been lost. Their true client, the public at large, has been diverted from its rightful participating role. The human and monetary cost of this dominance of modern technology demands a high degree of collaboration among these individuals and groups.

Furthermore, new communication forces have made possible the shrinking of time and space and ready access to a global system of cities. It would seem that at last, worldwide, we have the opportunity to set standards for Healthy Cities, large and small, and to employ them to re-establish our sense of connectedness to place by, as Leonard Duhl has established, "optimizing all the functions of human living."[2] Toward this goal, the overriding criterion for standards becomes the individuality of each urban place as a full reflection of its people and the site given by nature, the essence of what the city is all about.

From their beginnings, cities, towns, and villages have offered a sense of continuity between generations and changing cultures; today they serve as daily teachers and revered shrines of the history and heroism of events that formed them. As reflections of regional character, inherent qualities of specific sites, and local building materials, the urban places of our forefathers have become "treasures of the past" compared with twentieth-century urban places. We look at them with awe, sensing how their builders prided themselves on distinctive urban forms and identities that enriched the cultural and commercial interchanges between urban centers. Today these creations provide a principal link to local distinctiveness throughout the world as technology and economics spawn built environments of homogeneous character without regard to local cultures. Man's identity with place and his own capacity to build grew as he grappled with the environment on which he was dependent.

What are some paths that might lead us to this deeper meaning of place? According to Tony Hiss, experiencing places calls for inducing intimacy and connectedness with places, seeking responses based on maximizing and giving free reign to intuitive responses.[3] This approach can produce insights or "feelings" about places, which the use of abstract quantitative and analytic information gained through indirect surveys cannot do. The potential is great for deliberately recasting our perceptions of the city into images that are socially creative rather than lifeless for individuals and communities. Viewed from this new vantage point of sharpened visual awareness, increased participation between city shapers and city users can help restore connectedness with place. Expanding on these issues of scale, identity, and disconnectedness has produced a sequence of steps toward identity with place as an ecological framework for bettering the quality of urban life.

TODAY'S NEED FOR MEANING IN CITIES

The New Urban Scale

As the twentieth-century urban growth, of unprecedented scale, all but engulfs us, the smaller urban neighborhoods, country towns, and even villages and hamlets have taken on a new appeal because of the experiences of intimacy they offer. The Main Streets of some U.S. cities have been restored to compete with the anonymity of suburban malls. Within metropolitan cores, attracting the return of population to the central city has been attempted through conservation of environmental character. And the still largely untouched towns, from New England to the intermountain West, including the Mendocino coast and the gold country of California, are becoming magnets as they offer a more pleasant daily life than do central cities or suburbia.[4] Even new urban developments—the Neo-Traditional style, or the New Urbanism—are being designed to emulate the intimacy of early settlements and potentials for a sense of connectedness to place.

Yet the phenomenon of rebirth of older urban places has its darker side. Rising real-estate prices and property taxes caused by preservation and zoning restrictions drive out long-term residents. Santa Fe, New Mexico, is one example. In larger cities, lower-income and minority groups find themselves deprived of opportunities for positive and socially productive identity, either with a physical place or with a community of people. Whole districts of overgrown metropolitan areas take on visual images of abandonment, sterility, and hopelessness. These set self-destructive directions for the young people of those areas, who are in quest of a future identity of their own.

Moreover, people of the Third World within U.S. cities are far outnumbered by their counterparts in the sprawling, metropolitan cities in the regions of the world that have yet to optimize all the functions of human living. In Latin America's primate cities I have seen communities comprising hundreds of thousands of dwellings within metropolitan areas of many millions. These communities have been built entirely by the economically and socially deprived immigrants from rural areas and with their own hands.[5] Unacceptable standards also prevail in the impoverished neighborhoods of our own cities. In all cases, the energy and initiative to be found among the inhabitants speak for an intense desire for stabilized identity with a place and a role in society.

Diversity and Cultural Expression

The heterogeneous quality of our cities reflects diverse identities that were unknown to my forebears and other immigrants in their homogeneous homelands. Even in Sutter Creek and San Francisco this diversity served to build tolerance of cultural differences, a central goal of democracy that seems threatened in America today. This produced the contrast in visual quality between the Chinatown and the Italian North Beach of my youth. The dominance of the materialistic elements of the Anglo-Saxon culture, however, and the implications of economic power in city building and racial prejudice worked against a truly shared environmental identity. In most American cities that received other cultures, these factors induced bland, commercialized city patterns that washed away potentially enriching cultural sources of identity. This point is critical now since immigration is no longer from Europe but from Latin America and Asia, bringing new and different cultural and social patterns to enhance the physical environment. If we are to tap their cultural resources, a more equitable and socially fulfilling pattern of environmental design is needed in our cities and neighborhoods. Vast psychological distances, however, can separate the users of a given urban place and those who make the decisions about urban form, represented by the economic, technical, and political elites of these times. Environments are shaped by official policymakers and legislation, yet people, when they are fully recognized as participants, enrich with human meanings the policies that determine the quality of environments.

Urban Anaesthesia, Disorder, and Stress

The urban spaces in America have become hodge-podges of synthetic eclecticism and tiresome look-alikes, grown too large to allow the common citizen to claim a sense of identity with them or pride of belonging. Because insufficient opportunities are provided for participation in one's community, individuality is reinforced at the cost of a collective sense of environmental well-being. In spite of innovative architectural forms—the overrationalized qualities of the International Style and the more recent frivolous romanticism of the Postmodernism—urban design concepts still reflect the sanctity of building form per se rather than enhancing the essentially collective function of the community area or the city as a total environment. They fail to incorporate the physiological and psychological needs of users, who live and work under conditions of stress generated by our increasingly mechanized environments. Buildings express the identity of their designers and developers rather than the communities that inhabit them.

It is time to pick up the lost threads of meaning in the enormously discredited term *aesthetics* as it relates to urban quality and to adapt it for use in the new vision of cities as comprehensive wholes striving for social and ecological betterment. This implies a shift from individual exploitation in city building to an attitude of collective caring for communities as a whole. To thus reorient ourselves, we must realize that the word *experience* is inherent in the term *aesthetics,* from the Greek word meaning "to perceive." We are talking about a particular personal and firsthand experience of which all of us as human beings are capable. Looked at either individually or collectively, this capacity is a sociopsychological phenomenon; no one can perceive for someone else, nor can one feel with another's senses. This latent ability, conditioned by one's particular geography and sociocultural makeup, has been numbed by predigested education, by urban stress, and by the relentless salesmanship of mass-communication media, all based on pervading materialistic values. In medical terms, we have become "anaesthetized," and we need to rekindle the traditional qualities and capacities of the brain to give social and cultural meaning to our overly quantified environments. Deprived of the ability to feel and respond, to make use of the senses of sight, sound, and motion, the body and mind become inert and subject to whatever manipulation is desired. Our submission in the urban environment becomes as complete as if we were under the ether of the hospital operating table.

In overcoming this deadening of the senses and reducing the stressful conditions generated by it, individual intuition, fed by broadened cultural sensitivity, is the key to releasing the qualitative, poetic, and connective responses that lie unused on one side of the brain, crowded out by the descriptive, literal, and analytic responses on the other side. Our intuitive capacities somehow went adrift—particularly in the twentieth century under the influence of positivism—as Western materialism evolved away from the Eastern philosophies, through which an internalized use of the mind and contemplative attitudes establish an ecological relationship of the individual to the environment.

A Framework for Environmental Awareness

A full understanding of the potentials for dealing with the problems reviewed above requires a more specific framework than environment alone. My field experience with Dalmatian towns and villages, combined with an extended period of community activism in Berkeley and San Francisco, established three main determinants of environmental quality to use as "gathering elements," to borrow a phrase from Martin Heidegger, as we follow the "paths" leading us to a clear vision of identity sources.[6] These interdependent elements of the environment—place, buildings, and people—are inherently ecologically relationship. By looking at each element in terms of its dynamic qualities we can focus on the nature of their relationship rather than on each element individually.

The particular characteristics of the *natural environment,* its specific and regional landforms, soils and subsoils, vegetation, and climate and the ecological relationships between these, constitute the first determinant of identity in the individuality of a place.[7] The second determinant, the quality of the *built environment,* depends on the extent to which buildings, streets, and the urban fabric are adapted to the site through the sensitivity of the builders working in close relationship to the user. Christian Norberg-Schulz calls "creative participation"; that is, the formulation of aesthetic quality in vernacular places was forced to "participate" with the available building materials and methods, geographical constraints, and a rigid social structure. People had to depend on the natural environment at hand in generating urban structure and architectural character, which, as a result, expressed local cultural values. These places—for most of us in our times—are seen as objects in a museum, not readily to be "experienced" or to be sensitively adapted to modern use. Today, if we seek a similar reflection of local values in new building, we must build a new kind of wholeness and extended interpretation for the concept of "creative participation."

Beyond these two principal components, I have identified a third environmental determinant, *the web of environmental decision making.*[8] This "web" can be visualized as a lively though invisible nonstop process of negotiating over environmental-design issues that takes place between public and private entities and determines the form and identity generator of our cities. By its very nature as process, the web fills the ecological role of protecting the interrelatedness of the natural and built components of the environment against undue exploitative forces, whether private or public. The energy generated by this decision-making process, which is generally unperceived by the public, though it is well known to community activists, drives decisions on the location, scale, and design of private buildings and public places. This energy can be harnessed through "communicative action and interactive practice," an approach advanced by the younger generation of planning theorists of the 1980s and 1990s, to assure social relevance to environmental decision making and increased community awareness.[9] A new relationship would then emerge based on identity and participation in the daily use of urban places; in short, there would be an awakening from environmental "anaesthesia."

Together, these three determinants of urban quality and individuality of place form a useful and comprehensive triad from which a firm sense of collective identity with place can develop. In a recent study Paola Pignatelli called these three determinants "the generative triangle of identity . . . constituting a continuous process of cooperative action and mutual influence: The site conditions its spatial organization and the quality of life of people . . . , people reproduce in spatial terms their desired way of living and values . . . , built space defines human uses and creates its visual images. Identity of a place results from this reciprocal interaction."[10]

Through deliberate use of the web or the triangle we can evolve a macroecology of the city that would reveal the communicative interrelationships of its many human components and provide a rationale for their synthesis in our times of accelerated growth. This would replace the incremental evolution of visually rich urban centers, which in previous centuries achieved the urban integrity we admire in Europe and elsewhere. Our environmental-quality legislation has advanced us toward a common rationale to guide urban planning and management to be more responsive to the public at large. Once a city is understood not as a collection of "material objects" but in the comprehensive light of ecological human relationships, working toward the goal of "optimizing all of [its] functions" can provide the needed experience and commitment to evolve a deeper sense of identity for its citizens. Each can play a role by following a sequence of steps along "paths" leading from fully experiencing a place of residence to a commitment to participate in the web.

PATHS LEADING TO MEANING OF PLACE

Experience: The Starting Point

How do we create the attitudinal conditions necessary for individuals and groups to offset their isolation from the environment caused by forces of mechanization and homogenization? The facing up to a single critical problem of the natural or built environment by an individual can lead to a collective initiative toward community involvement within a given area. The realization that a treasured source of identity always taken for granted has been threatened by private developers or public action can become the first step toward bringing the community together. A crisis over a shared environment—street, neighborhood, school, park, greenway, city, or region and its history—can generate involvement on urban policy issues at many levels. By bringing into full play the realm of feelings, intuition, and the senses we can arrive at a personal interpretation of the environment and genuinely motivated involvement. Quantitative considerations can follow and lend support, but alone they fail to arouse a deep sense of community conviction or to lead to collective commitment.

Experience provides the starting point toward identity with a place, our first teacher of reality. As William James said, "Everything real must be experienceable somewhere and every kind of thing experienced must somewhere be real."[11] We begin during our infancy and childhood by directly experiencing, and responding solely with our senses and feelings to, our first environment, our body. Catherine Howett has shown how conditioning in childhood can influence in positive or negative ways our attitudes toward the environment as adults.[12] This experience determines our capacity to see sensual qualities in the form and color of flowers, limbs and leaves of trees, and the undulating landscapes of nature, as well as in the built environment.

Bernd Jager has pointed out that this early experience with the body provides much of the groundwork for attitudes and identity toward our rooms, houses, neighborhoods, cities, and landscapes. From it we build into our subconscious visual images from which we learn and develop sets of values. As we mature, this process unfolds into the room and then the house; in our choices of style and furnishings we can first externalize these images and values, even though we are unaware of the process. As we extend our range of experience to the street, the neighborhood, and the city, we extend our identity with, or our alienation from, urban places.[13]

The further we move from house to neighborhood and to city, the further we broaden the scale of our experiences; or, the less direct the experience becomes, the more the personal conviction and potential for leadership loses its grip. Through the centuries, direct experience in using and shaping environments—whether agricultural, village, town, city, or metropolis—has been the means of learning how to build and manage the highly complex places we now live in. How limited the chances for direct experience with the environments beyond the body and the house have become in our times.

In today's world of material objects, we need to be reminded that it is the depth of perception and feeling generated by the physical properties of an urban place, rather than the place alone, that represents the core of evocative experience. This experiential phenomenon can reveal the essence of that place. Direct exposure to all of the senses—sight, sound, touch, smell, taste—will induce a series of environmental messages that, together, support each other. By fully seeing and reflecting on the nature of the images we register in our minds—that is, by deliberately "reading" urban places and holding dialogues with them—we can take note of these responses. We have the opportunity to do this with our eyes in our everyday lives. Yet our minds are too occupied with our immediate relations with people and the role we play daily in the pursuit of living to nurture other than the most shallow sense of identity with a place used habitually.

When we take full advantage of that opportunity, however, the rewards are many and lasting. We can intensify the experience by turning inward to ourselves and dwelling, not on the place or building being viewed, but on the responses it evokes. To do this requires deliberately focusing on the mind in a state of reflection. It is experiencing the awareness of that introspective phenomenon that counts, not the environment itself. One needs to experience the experiencing and to monitor consciously the intuitive responses that in due time emerge on their own. A total immersion into this phenomenon, within a fixed period of time—hours or days—and walking a prescribed series of blocks or a single neighborhood becomes equivalent to "reading" the words, lines, sentences, and paragraphs of a "text." From this experience we "learn" accumulatively the contents and qualities of a place, the relationship of its parts to its unifying structural system. Myriad details of design—open spaces and how people use them, signs, poles, and wires—will unfold building by building, street by street, and speak to you. Together these pieces will assemble as a statement of the particular quality—in Heidegger's terms, "the thingness of a thing"—of a place and the source of its identity.[14] A wealth of understanding of the past builders and present users of the place can be gained through observing their historical styles, the quality of maintenance of buildings and landscape elements, and the way the environment accommodates interaction among people.

Intuition: A Trustworthy Guide

I have found that intuitive responses flow freely and abundantly when one's mind is open in the spirit of phenomenological inquiry. In 1979, when I came to study the nature of individual urban places in the Dalmatian system as a whole, I chose to explore on my own an analytic approach derived from systematic experiencing of each place. Out of this open frame of mind grew "urban reading." I had already learned of such "readings" of Rome from Paola Pignatelli and exchanged ideas with her on this phenomenological method.[15] This became a welcome way to conduct one's more personal search for meaning to complement and lend depth to the accepted qualitative analytic methods developed by pioneers in the field of environmental-behavior studies. Kevin Lynch, in his late 1950s research among users of urban places, established the wide variety of perceptions of urban—or visual—form that places can evoke according to the frequency and depth of experience of particular users.[16] Others, including Donald Appleyard, developed survey methods by which the urban design analyst gains valuable responses from samples of users to obtain "scientific" evidence of expressed collective attitudes to an environment. This factual information can then offset proposals—such as for demolishing a historical landmark or cutting a freeway through a park or neighborhood—that would be disruptive to residents' sense of identity.

My instinctive feeling, however, was that the urban design analyst should also probe his or her own mind, with its accumulative experience, for insights that would lend a needed synthesizing context to interpreting such information in order to make ultimate design decisions. Indeed, because of this instinctive feeling, I found it difficult at first to break with my own conditioning to accountable analysis in urban design processes. Throughout history intuitive judgment has been a major source of creativity in the great works of city building and urban design, and it should not be stifled today in favor of an overreliance on factual, descriptive, and quantifiable methods of inquiry.

Toward this end—and prior to knowing Heidegger's works—I began my experiment. Appleyard and I had been engaged by the Urban Planning Institute of Dalmatia to organize such a study in 1979. As a first step, I undertook, along with extensive fieldwork for my own research, a review of a variety of sites on the offshore islands to determine their suitability for surveys on receptivity to introducing tourist facilities. After I returned to California, I presented my findings at a gathering of Fulbright alumni, where I learned from a commentator on my presentation—Anne Buttimer, a human geographer of great breadth—that my instinctive approach matched essentially the area of phenomenology.[17] This led to my giving papers at environmental conferences and publishing, facilitated by the work of David Seamon and Robert Mugerauer in relating phenomenology to environmental design.

Appleyard was also stimulated to pursue the subject of identity further, having become aware of the contrast between the integrity of Dalmatian towns and the mismatch of people and place in today's urban agglomerations. He had developed a comprehensive study of urban sources of identity, their causes, and their outcomes, entitled "Identity, Power, and Place," before his untimely death in 1982. The extensive manuscript covers a broad range of identities other than the environmental focus of this work. This unfinished manuscript is a compendium of rich resources to be explored by others, as it was for me, in seeking one's own directions.[18]

By the sequence of steps I had followed in intuitively experiencing place—that is, by viewing an object, a room, a dwelling, a street, or an urban place as a phenomenon free of previous concepts and by seeking to grasp firsthand the intrinsic nature of the subject—phenomenology became simplified. Seamon summarizes the approach in finite terms:

Phenomenology is a descriptive science, the heart of which is concern, openness, and seeing clearly. It is part of a philosophical tradition which has variably been labeled "humanist," . . . "reflexive," or "philosophies of meaning." . . . Arising largely in continental Europe at the turn of the century, this tradition includes, besides phenomenology, the related perspectives of hermeneutics and existentialism. Key figures associated with the tradition include . . . Husserl, Heidegger, Merleau-Ponty. . . . All of this work is interpretive and grounded in careful looking, seeing and understanding. Its primary substantive focus is on description of human experience and meaning *as they are lived.* . . . Architects and other designers have become interested in phenomenology by a *practical* crisis: the frequent failure of architectural formalism and functionalism to create vital, humane environments.[19]

The steps I had followed spontaneously released intuitive wellsprings of understanding. This led me to delve more comprehensively into the relationships of the parts that make up the whole and the processes— native landscapes and history, for example—that give the subject its form. Such a "phenomenology of place" clearly stands in an ecological context. David Seamon expresses this approach in pragmatic terms closely related to my own experience and to the role of urban places in everyday life: "Phenomenological ecology supposes that beneath the seeming disorder and chaos of our world and daily life are a series of underlying patterns, relationships and processes that can be described qualitatively through heartfelt concern, sustained effort, and moments of inspired seeing and interpretation." Especially related to my "urban reading" method, this approach, in Seaman's view, "not only widens and deepens our knowledge of the world *outside* ourselves, but also facilitates our *own* growth as individuals whose abilities to see and understand can become keener and more refined . . . . and [who can] see things in a more perceptive, multi-dimensional way."[20]

Essentially, I found myself seeking to describe these urban places in their bold and distinctive natural settings in human-oriented and qualitative, rather than factually quantitative, terms. In the 1980s a number of geographers pursued a range of scholarly approaches to description and interpretation of the environment as given to us by nature. One work, published well after my field studies, is especially valuable for understanding the multifaceted landscape of the Dalmatian Coast and, indeed, the complexities of geographical, political, and ethnic identity in postwar Croatia and neighboring countries is *Qualitative Methods in Human Geography.* This book presents a wide variety of case studies of selected localities, though not in the usual regional-cultural context. They pose the disintegrating effect of positivist thinking against the comprehensive and holistic value of ecological analysis, especially in human terms. As John Eyles and David Smith state in the preface, John Pickles, in concluding the volume "puts forward phenomenological method as the key to the approaches that we have termed 'qualitative,' bringing together questions of philosophy and technique,"[21] which was what I had found in Dalmatia. Thus, I did not fully embrace phenomenology in its purist philosophical sense since that would have limited the comprehensive scope of experiences these places of urban scale, from small villages and towns to their regional and historical settings, require. My attraction to phenomenology became a major step in broadening the context of my own predilection for identity with place and the horizons that are potential within all of us.

Kenneth Craik has interpreted my approach as "a guided phenomenology of places" as against a "spontaneous phenomenology of places."[22] I drew on the concept of phenomenology instinctively as my field studies developed rather than applied it as a formula for determining methodological effectiveness. I then turned from those responses to their interpretation—or hermeneutics, to use the philosophical term—to see fully the places lighted by my rounded-out knowledge and exposure to them. Less-informed persons might use "pure" phenomenology to spontaneously build personal meanings for themselves. In contrast, I was able to synthesize my observations of physical forms with my knowledge of the sociological context to reveal the coherence and order underlying the uniqueness of each city, town, and village. A less-informed viewer could not readily do this but ultimately might be assisted by applying my "urban reading" method. Indeed, even the lifelong residents of the various places could not have acquired a similar depth and breadth of knowledge of the sources of identity. In following these "paths," the reader should bear in mind that identity with place is gained deliberately; it does not take root full-blown.

Identity: Self and Group

In our competitive and heterogeneous society, we tend to seek environments that are compatible with our desired self-image. In the more homogeneous urban cultures of the past, patterns of values and behavior provided established ways of living. Standardized site planning of dwellings and their settlements contributed to holding social values in place. Along with this approach came a sense of security and belonging; there was no need to develop a permanent self-image in the way one dressed, in the choice or design of one's house, one's street, one's neighborhood, or one's city. Today, in contrast, with great individual opportunities for regional mobility, one can choose among diverse physical and cultural environments in various parts of a large city or metropolitan area as a way of individualizing one's cultural expression. In the search for housing, neighborhood identity has a high priority.[23]

Identity must also be regarded as a manifestation of community; neighborhoods tend to be made up of persons who hold a common image of identity. A threat to this sense of identity held by "insiders" of a given area in the form of damaging proposals by "outsiders" for alterations of the environment can generate community formation and an activist attitude of collective identity. Appleyard's extensive experience in places having a varied culture broadens the dimensions of the basic concept of insiders versus outsiders: "While familiar environments like home and neighborhood are pregnant with social meaning, and central to our everyday lives, evoking powerful emotions, psychologically distant environments which are rarely encountered are symbolically and emotively light. People are less participants, more observers—Outsiders rather than Insiders."[24]

Connectedness: Our Goal

From identity with place derived from experience we move to a sense of connectedness, proportional to the depth of our experience and intuitive responses. This step becomes manifest when the close linkages people establish with the physical features of a place lead to the tightening of bonds between those people. Through this awareness of the meaning of place we reaffirm the essentially ecological nature of our environment. We rediscover that connection between environment and one's inner self and, more intimately, the way we feel about a given place.

In this respect Martin Heidegger's thinking illuminated my Dalmatian field studies with a new awareness not just of our connectedness to the environment but of our dependence on it. In his essay "Dwelling, Building, and Thinking," Heidegger speaks of how connected we are to earth and sky in the vertical position of everyday life.[25] Gravity holds our feet to the earth and our heads to the dome of the sky. The asphalt of the streets and the floors of our homes and skyscrapers replace the earth of early man, and the smog of our cities and the ceilings of our homes, offices, and stores replace the sky.

Our eyes provide a horizontal connection to interior walls of rooms and to exterior facades of buildings and provide us with the ability to seek familiarity and security. As our feet move us across the earth, our eyes can lead us toward those places and people that fulfill the need for belonging. An expanded sense of environmental awareness can accumulate experience and help us to make choices about directions to follow and places and people with whom to make social contact. We can become conscious of the arrangement of our streets and circulation routes in relation to patterns of land uses and thus "optimiz[e] all the functions" of our social and economic daily lives.

The phenomenon of connectedness is explored in *Place Attachment— A Conceptual Inquiry*, a collection of case studies that represent the forefront of thinking on the subject in the early 1990s. The work supports my interdisciplinary approach by drawing on a range of interacting fields that together forge a sense of connectedness to place. The editors—Setha Low, an anthropologist, and Irwin Altman, an environmental psychologist—aim to illustrate the multidisciplinary foundations of place attachment. Using a phenomenological method of inquiry, the contributors examine attachment to differing types of places, such as homes, neighborhoods, plazas, and landscapes. These case studies deal with interacting forces of territoriality and ethnicity as in the former Yugoslavia, a point I shall return to in chapter 7.[26]

In Low's own essay, "Symbolic Ties That Bind," the definition of *place attachment* complements the key points expressed above:

Place attachment is the symbolic relationship formed by people giving culturally shared meanings to a particular . . . piece of land that provides the basis for . . . understanding of and relation to the environment. . . . It is argued that while there are often strong individualistic feelings that may be unique to specific people, these feelings are embedded in a cultural milieu. Thus, place attachment is more than an emotional and cognitive experience, and includes cultural beliefs and practices that link people to place.[27]

Involvement: Commitment and Community

Involvement in a given place grows out of *commitment* to it, a sense of responsibility or stewardship that goes well beyond the physical properties of the place. Individuals working in unison can take initiatives to correct, refine, and improve their adopted place, and in the process commitment becomes a continuing collective force through time. Maximizing this sense of responsibility, the users are no longer passive bystanders but active participants in the formation and management of the urban environment. Through my "web" of environmental decision making, people become the ecological agent of the place. Viewed in this light, the users of the cities can give vernacular meaning to a given place and, at the same time, receive from it a sense of fulfillment and collective stability.

This awareness inevitably brings those most attached to a particular environment opportunities for direct involvement in the planning and design of urban places. Whether from a neighborhood, a park, a city, or a region, their leadership will generate energy and initiatives among those who ordinarily would not have become involved nor been introduced to the first step of experience. In time, they too will take the steps we have laid out and become members of a community committed to protecting the unique qualities of a shared identity and offset the increasing environmental anonymity and alienation of our times.

"URBAN READING" UNFOLDS

Learning to "Read" Dalmatian Villages

The diverse environmental situations of the Dalmatian coast—the innumerable bays, harbors, coves, peninsulas, promontories, peaks, and islands—held strong meaning for their dwellers. For that reason, these places served as productive subjects for "urban reading" and understanding how the form of urban places can create distinctive identities and patterns of human use. The step-by-step process that evolved as I

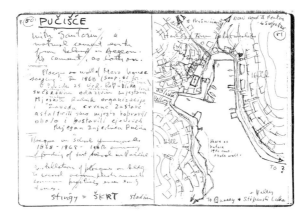

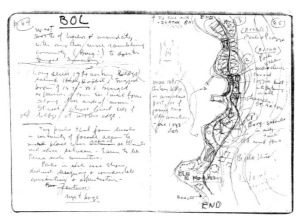

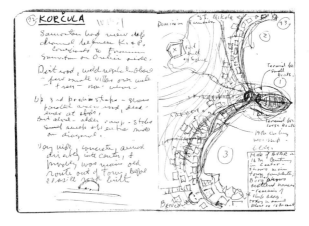

Mental mapping evolved from spontaneous responses to urban form recorded directly into field notes. Pučišća (top), Bol (middle), Korčula (bottom)

moved along with trial and error varied from place to place, though it developed into a sequence similar to this:

1. *Fully immersing myself* alone for three or four days in a chosen place free of any guidebooks, maps, or orientation beyond the knowledge I had already accumulated.

2. *Reconnoitering the place* in its entirety on foot, *looking with an*

Comparison of mental maps done from readings revealed contrasting urban forms. Pučišća (top), Bol (middle), Korčula (bottom)

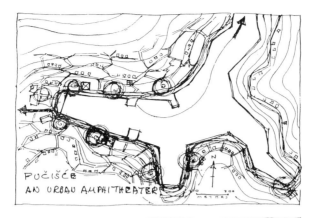

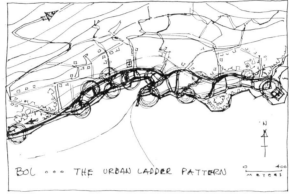

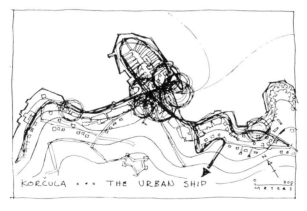

open mind for visual clues to the uniqueness in terms of its structural system, principal architectural and landscape features, and geographical origin.

3. *Intuitively tracing on my mind "mental maps"* with footsteps and visual alertness to monitor the changing directions of circulation routes and open spaces, thus creating a link between my mind and the reality of the urban structural system. Once I had refined these maps and sketched them on paper, I came to rely on them rather than on the abstract official maps, which, fortunately, were generally unavailable.

4. *Identifying each sector* with boundaries and features and inventing for each a name based on its geographical or functional origin and likely place in history. *Recording* my observations for each in a small, inconspicuous notebook in words and drawings spontaneously as they came to mind encouraged qualitative and perceptive responses. Avoiding the irrelevantly descriptive and factual, these notes came forth unexpectedly rich in symbolism, mood, scale, history, and human life as I expanded on them later.

5. *Reconstructing the natural environment* in its geographical form and its historical alterations to determine just how it shaped the built environment and its relation to the urban structural system.

6. In time, a *metaphor for the distinctive pattern* of the system came to mind that was useful in making specific the basic source of the uniqueness of each place.

7. *Observing and taking part in the interaction of year-round residents* with the patterns of circulation routes and gathering places to see how the urban structural system choreographed their daily lives.

8. *Interviewing* residents and visitors in a conversational way, as my sense of the uniqueness of the place became clear, to test out my conclusions. On occasion this included community leaders concerned with the issue of identity, suggesting their possible use in local policymaking.

9. Finding myself in time *changing from an "outsider" to an "insider"* equipped with a clear, comprehensive image of the uniqueness of the place and its source of identity.

10. On undertaking the next case study, *discovering that comparison of my responses to one town with my responses to another* is essential to clarifying the particular qualities of each.

This segmented, meditative observation was like peering into the "collective subconscious" of the generations of people who had participated in building each place, to draw on Paola Pignatelli's concept of the city as the physical expression of the values held by its centuries of builders.[28]

In "urban reading" it seemed as if the eye was being made to "think," as in reading a book. The more thought-provoking the eye finds the words on a page, the more one's intuitive sense is stimulated. Likewise, the more provocative the visual images in a given environment are, the more sensitive and perceptive one can become and thus the deeper one's identity with place. This Dalmatian experience proved to be so powerful that, after I had moved on to visit villages in Italy and Switzerland and finally returned to my hometown of Berkeley, these places took on a new and lasting reality and depth that has sharpened my perception in experiencing urban places ever since.[29]

Urban Reading in Italy and California

In 1987 I enjoyed a family visit of a few weeks in an old stone farm-house renovated as a retreat by a cousin from Dalmatia who works in Rome. The farmhouse stood just below the Umbrian hill town of Giove, some seventy miles north of Rome, one of a grand constellation of hill towns—some no more than hill *villages*. Knowing about my "urban reading," family members suggested I demonstrate my method in this mountainous geography. They asked how, using my intuition-generated approach, I would respond to places built on isolated hilltops far from the sea? Yet, like those of Dalmatia, they had been shaped over centuries by geography and the persistence of people constantly creating liveable environments. What would "urban reading" in a highly restricted yet in-spiringly airy and precipitous setting tell us about the sources of its own uniqueness as compared with that I found in those by the sea?[30]

I first discovered that, like the sea in Dalmatia, the awesome presence of far-reaching, unbroken space created a powerful source of uniqueness, augmented by constant visual exposure to the regional context of one companion hill town to another. Begun under the Etruscans, expanded under Rome, and brought to fruition from the Gothic to the Renaissance period, these places are now taking on new life in the late twenti-eth century through today's building technology and communications.

In seeking to understand Giove's structural system my wife and I began "reading" its "text" on the farm road along the north flank of the promontory on which Giove stands. Field workers' farmhouses and barns told us of the region's continued agricultural reason for being. From the prow of a ship that Giove's geography reflects we could see far up and down the Tiber River valley. Resting on the benches of the Belvedere, we joined the old men who daily gazed out toward Viterbo, beyond the mountains, and toward Orvieto, up the valley.

Turning inward to the town, we noted how its gently meandering main street stretches toward the readily accessible level plateau land to the east. To our right, the Porto Romano provided a narrow entrance to Old Giove, the town's birthplace in the eleventh century. Giove's chro-nology stands clearly evident in the Renaissance palazzo shaped by a massive addition to the old *castello* rising to greater heights above a clas-sic Belvedere. This abrupt change in urban form told us how the New Giove of the eighteenth century deliberately opened up the town to the world at large, as its symmetrical entry followed the revival of Rome's traditions of visual order and grandeur. Here, at the town's major point of arrival from the valley below, the rich facade of the parish church en-hances the terraced piazza and a park in the foreground.

We then walked eastward, down the Renaissance spine of the town. The narrow width of the Giove promontory allows space only for this

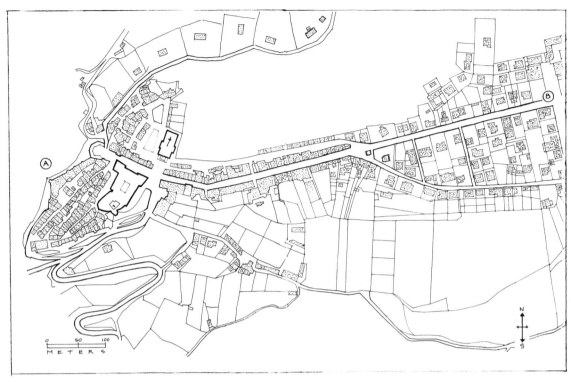

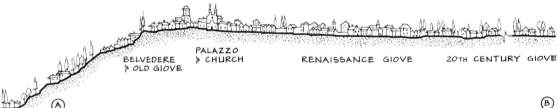

Giove: Plan and
section

single *corso,* flanked on each side by once-elegant dwellings. Its wall-to-
wall formality, broken only by occasional ground-floor shops, gives way
to Giove's third distinct sector, built entirely in the twentieth century. An
open piazza with a World War II monument boldly marks the onset of
a contrasting suburban subdivision of freestanding automobile-oriented
houses with front yards; Giove's incremental lineal system, rare in hill
towns, makes clearly legible the latest turning point in its thousand-
year history. Three periods of history stand side by side and reveal to us
Giove's essential uniqueness.

Our comparison of this "reading" of Giove with our readings of other
hill villages in the constellation revealed markedly different sources of
uniqueness. While Giove has two access routes, one arriving abruptly
from below and the other along the level highland, in nearby Lugnano
a single route curls upward along two-thirds of the perimeter of its
egg-shaped hilltop, just outside its medieval walls. Because of the way

Giove's palazzo, rising from the medieval *castello,* identifies its Renaissance origins. *Photo by author*

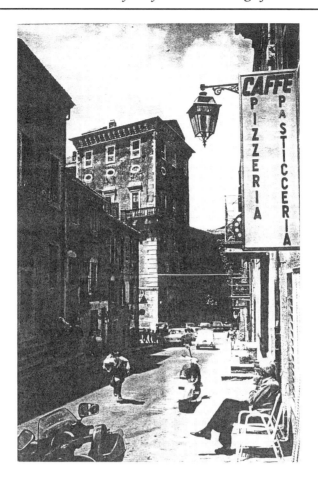

Lugnano's tiny articulated walks serve as hallways converging on piazzas of varying size, our experience was like intruding on the privacy of a stranger's home. The compressed urban texture completely blocks the outlook around the perimeter of the hill's relatively flat top. One loses completely the sense of being on a hilltop. Then suddenly a space opens up between the buildings and a panoramic cultivated landscape spills out before us and to Giove itself. The contrast between experiencing the confined built space and the sudden exposure to vast unbroken open space from this elevated, airy position instills a sense of wonder and awe at being thrust up above the world yet in such close and personal proximity to one's fellow human beings.

Porchiano's individuality stems from its pattern of two concentric circular streets and radial routes joined at the flat top of its hill. Walking Toscolano's convoluted circulation routes, we were choreographed into a spiral, upward movement around the hill shaped like a giant conch shell. The narrow ramped roadway unfolds step by step along with the residential facades. Amelia, by far the largest and a true hill town, holds an identity—so unlike Giove's—based on the hit-or-miss pattern of large-scale

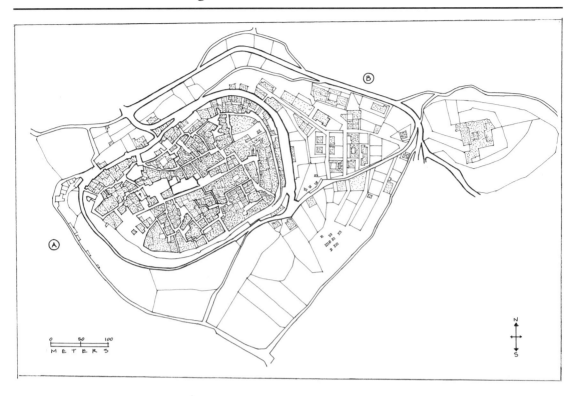

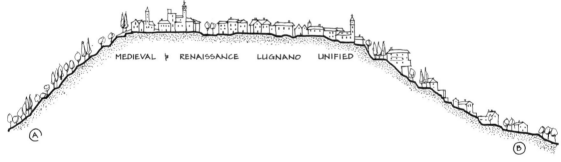

Lugnano: Plan and section (top and bottom).

Renaissance buildings scattered here and there on sites entirely within the extensive Etruscan-medieval street system. Here, the "reading" became complex and confusing but led to one opportunity after another for discovery and a challenge to find coherence in its baffling identity.

Compared with reading Italian hill towns, reading a city the size of Berkeley, on the eastern shore of San Francisco Bay, presents a challenge to put aside the fragmented images formed by the familiarity of daily life. I was required to drive each of the main elements slowly and to experience selected samples on foot, block by block, in the open spirit of phenomenological inquiry. Berkeley's strong east-west axis from the hills through the university's Beaux-Arts core, the city's downtown, down University Avenue to the Bay and out the Golden Gate came to dominate the "reading." First I drove the few straight downhill streets in order

Lugnano. A vast landscape spills out to Giove from the urban hilltop. *Photo by author*

to experience this visual connection to the Bay by movement and by time; then I drove the winding roads that serpentine along the hills from north to south. On these, dwellings and their landscaping have blocked and privatized the once open vistas of the superb landscape home of the Ohlone Indians. The few steeper downhill streets took on new value when I realized how they preserve public views of the Bay.

The popular image of social separateness of flatlands residents and hill dwellers held little truth in light of the fact of a gently sloping terrain all the way to the Bay. The grid system of the flatlands revealed a sense of the dynamics of development, from Victorians of the turn of the twentieth century to the workingman's bungalows of the 1920s, to the "shoebox stucco apartments slipped in by the speculative builders of the 1950s, prior to the community-generated downzoning to match the single-family pattern of existing land uses.

Out of this deliberate immersion into Berkeley's source of interplay between the natural and built environments grew my concept of the web. I had discovered the need to experience the third realm of environmental determinism—citizen participation—which was out of the ques-

Berkeley's urban structural system (top and bottom). The city and campus of Berkeley share an east-west axis marked by the Campanile, the downtown, University Avenue, and the Old Pier, which together are a strong source of collective identity (opposite). *Ansel Adams, U.C. Bancroft Library*

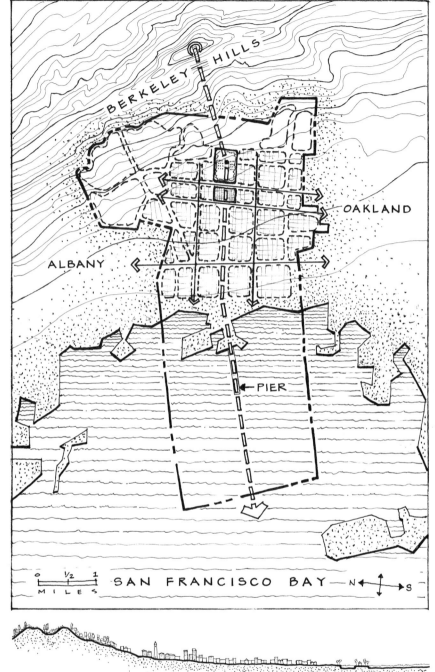

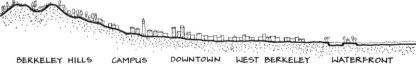

BERKELEY HILLS   CAMPUS   DOWNTOWN   WEST BERKELEY   WATERFRONT

tion in the Dalmatian towns. I turned to three planning issues that arose during the early 1980s: Codornices Park, near the campus and central to my own neighborhood; Berkeley's venerable downtown; and the threat of urban development on the waterfront. With virtually no previous involvement in the web, I found myself on its various strands centered on the politically sensitive city hall and radiating out to the activated neighborhoods. Within several years the intuitive judgments gained from my "reading" led me to a deeper sense of belonging to Berkeley.

In Codornices Park, a crisis of neglect and vandalism since the 1960s had led to complete deterioration and misuse of a fifteen-acre rugged site and its native, historical qualities. The landscape's creeks, oak woodlands, and meadows with views to the Bay were widely used by sur-

San Francisco's urban structural system (top and bottom). San Francisco's identity has grown out of its neighborhoods, formed by the unique geography of hilltops, ridges, and valleys (opposite). *Donald M. Nelson*

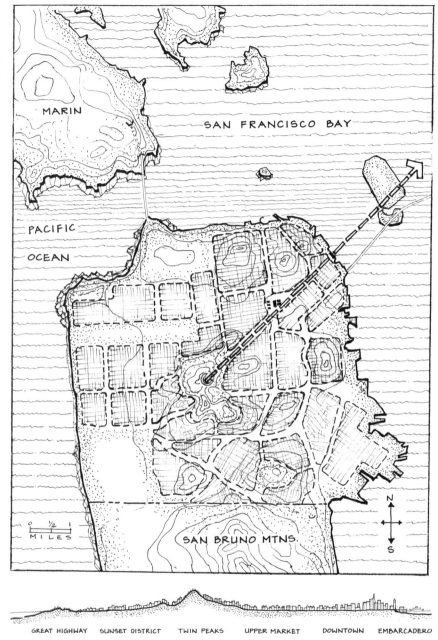

GREAT HIGHWAY   SUNSET DISTRICT   TWIN PEAKS   UPPER MARKET   DOWNTOWN   EMBARCADERO

rounding residents and the city as a whole. A grassroots initiative from our neighborhood brought together people who were largely unknown to one another and committed them to restoring the natural character of the park for more than a decade. Through taking firm positions on funded park rehabilitation and actually working in the park one Saturday each month, we developed a new awareness of the park's unique identity as well as a deep sense of community.

In Berkeley's downtown, a proposed block-long, ten-story office building on a site adjacent to a residential neighborhood triggered opposition by a group of community-oriented citizens. Over one year and in the absence of current planning by the city, a citizens' group of professionals, including myself, developed a comprehensive "Outline for a Downtown Plan." The group's pressure on city staff, commissions, and the city council generated a fresh awareness of the need for a renewed identity for downtown and related issues, goals, and opportunities. In a few years a participatory planning process was adopted, and by 1987 Berkeley had adopted a downtown plan. Our insistence on the issues of social well-being and a cultural presence in downtown, together with fresh economic approaches, led to a new and potentially significant sense of place for the downtown area.

On Berkeley's three-mile waterfront, the 172 acres of privately owned reclaimed land became the focal point of a major controversy over extensive development for profit versus open space for the public good. I took to the Bay itself by sailboat, along with with key activists and political leaders, to directly experience the nine square miles of water surface with which few Berkeley residents identify at all. Sailing close to the vacant shoreline of rocky bulwarks holding fill of nonorganic waste materials brought me a new, immediate awareness of the bayfront's dominance as a recreational and identity source. As an alternative to an entirely open shore, our community group with interdisciplinary environmental skills proposed to maximize Berkeley's diversity with a small marine-oriented strip of urban waterfront. After years of debate and negotiations against urban development, voters funded the purchase of the entire shoreline for public open space. By the early 1990s a portion of this landfill offered rolling grassy hills for kite flying and as a fresh way to experience Berkeley in its geographical wholeness by looking back to the hills from the sweep of the Bay.[31]

A "reading" of San Francisco, the place of my youth and first experience in urban planning, seemed redundant. Yet, my new method of place perception brought fresh insights when an artist friend from Dubrovnik chose to settle permanently in San Francisco, recognizing in it the magnetic quality of his own native town. For him, the two identities fused, and he asked me to write an essay relating the sources of identity of the two places in support of his visual work as an artist. Surrounded on three sides by the Pacific, the Golden Gate, and the Bay, the city was laid out in a grid pattern, as is Dubrovnik, though on a vastly larger scale. Also, like that city's Stradun, San Francisco's Market Street dominates and bisects the city from Bay to hills. The compact neighborhoods formed by the grids and hilltops make for intimate social life. San Francisco thus shares cultural and environmental qualities with Dubrovnik. The city also derives its regional identity as the centerpiece of a surrounding region of diverse landscapes.

San Francisco juts out boldly into the waters of the sea and the Bay on the very edge of the Western World, as Dubrovnik juts out toward the West. San Francisco's grid pattern of streets and blocks also terminates in a spectacular periphery, a source of identity of the adjacent neighborhoods. These shores plunge sharply into the sea to the north, as do the walls of Dubrovnik, and provide a clear-cut edge for the city. Likewise, mountains cut off the southern side of San Francisco, providing the populace with an individual sense of connectedness to their own urban places.

To this intersection of worldwide routes have come from Europe, the Atlantic Coast, across the Pacific, and the southern half of the hemisphere those seeking an outpost of human innovation and opportunity for self-fulfillment in a setting of strongly identifiable qualities. The city's frame of hills and valleys and shoreline accommodates each of these enriching cultures in a relatively self-contained manner favoring individual attachment and identity. Side by side, these local neighborhoods form a cultural mosaic that complements the natural and built environment: the Italians of North Beach, the Irish in the Mission, the Chinese in Chinatown, the French in the Richmond.

Fostering this self-containment of diverse cultural identities, San Francisco breaks down into a series of settlements established largely according to its pattern of hilltops, ridges, and valleys. These hills break the two-dimensional repetition of blocks, and the streets plunging downhill open the outlook to the marine periphery, ever-changing with its play of sunlight, clouds, and fog. As in Dubrovnik, this mesh of geometry spreading over the undulating land form results in a rare, three-dimensional variety, whether of hilltop looking outward, or valley floor bounded by hills. Yet, a ready identity is available among a wide range of choices for one's own place and a sense of connectedness on a scale easily understood, one's own Dubrovnik, Kuna, or Pučišća, as it were. Indeed, such was the case of my own forebears, which led to assimilation but also the stability that stems from having deeply experienced identity with place.

CHAPTER TWO

The Making of Dalmatia as a Regional Place

One of the most beautiful and interesting elements in Yugoslav history and culture is Dalmatia. . . . Harmony prevails there between the rugged mountains and azure sea; between the numerous layers of civilization, from the first Millennium BC and the manifestations of a Slavic culture in its western European form. — Cvito Fisković, 1962

THE URBAN SYSTEM AS A WHOLE

Dalmatia's strong environmental qualities and rich multicultural history have evolved a steadfastly collective sense of identity at the regional level. A diversity of natural landforms like no others in the Mediterranean interact with a wide variety of cultural influences on architectural and urban forms. These have produced a highly localized sense of connectedness for the people of each particular urban place and subregion. The sea, with its innumerable bays, inlets, and coves, has been a ready source of communication, giving the region social and economic integrity. A common geography facing the West contains a lineal chain of urban places that speak for centuries of turning points in history that shaped them, each in its own way. These characteristics have given the Croatian people of the Dalmatian coast a clearly identifiable culture.[1]

The region occupies the southern portion of the northeast side of the Adriatic Sea, itself just a major gulf running north five hundred miles off the Mediterranean Sea, dividing Dalmatia from Italy by only some hundred miles. Today the region of Dalmatia is about two hundred miles long and seventy miles wide in the north and as little as ten miles wide in the south. Covering much more territory under the Romans, its name was based on the Dalmatae peoples, a tribe of the Illyrians in the central coastal area. On the western edge of the Balkan Peninsula and looking toward that of southern Europe, this region has been impacted throughout its history by continental influences—from Rome, Venice, Italy—and by ties with the East through Byzantium and later through the coming of the Slavs and Ottomans. Thus, Dalmatian cities and people have been influenced by the sea, its waterways, and the world at large far more than those peoples subjected to the limited and divisive influences of inland mountain valleys and ridges have been.

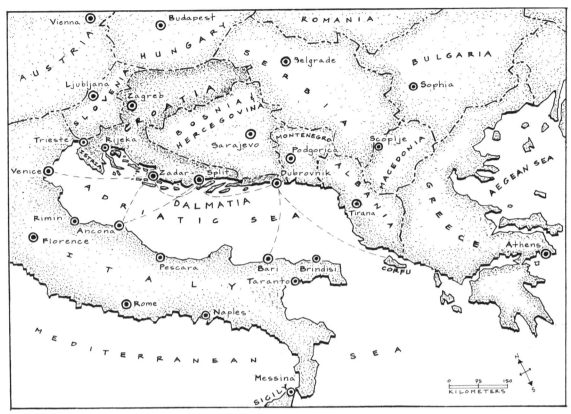

Dalmatia and the Mediterranean. A unique coastline developed as a part of the maritime life and culture of the Adriatic Sea

Throughout much of the past, maritime transport was superior to the mainland means of moving goods, people, and ideas. The Adriatic was more a link than an obstacle to interaction with other parts of the Mediterranean and Europe. At the same time, the mighty wall of the Dinaric Alps, running four hundred miles parallel to the Adriatic, with corduroylike ridges and peaks up to a thousand meters high, has been an obstacle to communication with the landlocked regions of Croatia, Slovenia, Bosnia, Herzegovina, and Serbia. These origins and sequences of the Dalmatians' cultural evolution within the larger context of these countries laid the foundation for ethnic and religious characteristics that differed markedly from those held in common by all Slavic peoples on their arrival in the seventh and eighth centuries. The Turkish occupation of the fifteenth century intensified these differences, which since then have been exploited for political purposes and have festered in spite of Tito's leadership to form a unified nation of *South* Slavs, that is, *Yugo*slavia. This led to the tragic wars of 1991–95, motivated both by aggression and by a desire for autonomy, over national boundaries and ethnic and religious identity versus tolerance in a shared environmental reality.

Dalmatia: Image and Reality

The image of Dalmatia carried abroad by seventeenth-century travelers focused on its archaeological context. A more distorted cultural image appeared in the writings of early western European travelers during the half-century between the coming of the steamship in the nineteenth century and the coming of the airplane during the twentieth. These writings portrayed the region as romantic and exotic, an image that had little to do with the geographical challenges the people had to face, the social and cultural realities of urban life, or the high levels of culture and urbanity in its history. Even then, as today, Dalmatia was viewed as an environment in which to escape from the drabness of European industrialized cities and to taste some of the flavor of the East in picturesque, often wholly intact medieval settings. Sensitivity to the critical development issues that had accumulated through centuries of neglect by "protective" powers was simply not a part of those times.

Between World War II and 1991, when the most recent war began, Dalmatia came to represent a microcosm of the forces at work transforming areas in many other parts of the world, made accessible today by communication and stimulated by world finance systems and the pressure for social mobility. Only decades ago international airports came to the medieval walls of Dubrovnik and the Palace of Diocletian at Split. The Adriatic Highway and the automobile replaced the sea itself as an internal means of universal circulation. The thirst for sun and romantic landscapes and town settings brought the industrialized northern Europeans, and with them new economic development. The desire for modern urban life lured young people from the villages to the larger cities. Now, late in the century, the impact of the war of 1991–95 on these places further threatens urban identity.

To respond to these changes in the interests of future generations, it is necessary to gain an understanding of the realities of the past that shaped urban Dalmatia and to look at the turning points over centuries that gave each place the particular qualities we knew before the war began in 1991. The right to be independent of foreign domination, finally bestowed on Croatia in the 1990s, provides an opportunity to link the past with a new sensitivity to local identity in the coming century.

The fragile physical and visual conditions of the natural Dalmatian environment throughout history have produced an urban culture that is unusually integrated within the region. In the twentieth century, modern development forces—large-scale tourism, mushroom growth—have violated the integrity and ecological balance that a decidedly urban culture has produced, in both the natural and the social sense. The free play of these forces on the rugged though beautiful coast, with its superbly intricate urban system, has diminished its cultural regenerative powers as well as its own inherent quality of land and sea. The failure to protect both the natural and the cultural environmental characteristics as prime conditioners to urban quality may mean the end of further evolution of Dalmatia's distinctive urban culture.

History as Collective Human Experience

My "reading" of urban Dalmatia showed me that "cities can be read, because they write their own history," as Henri LeFebre pointed out.[2] Therefore, using urban place as a "text," we can trace the record of the events that shaped them. I had explored this concept of turning points through my field and archival studies of the evolution of the cities of Spain. First, the processes of nature shaping the landscape itself and then Iberians, Greeks, Romans, Goths, Moslems, and Christians had left their marks in the design governing daily use of cities and towns today and the images the visitor's eye carries away. On a vast scale, these patterns came to influence much of Latin America's system of urban settlements.[3]

Though separated from us by centuries, these urban remains can in minutes connect us back in time to a "fourth dimension" of physical places. We can even reconstruct the reality of the "present" of centuries ago that produced today's environments and compare that time period with our own present. As Kevin Lynch pointed out in *What Time Is This Place?* these periods gone by stand permanently fixed as a visual reality in the incremental formation of any urban place over its lifetime.[4] The record can lie aboveground in single buildings, in confluences of structures, transformed for the use of each generation, belowground in archaeological diggings, or spread over vast acreages in the form of cities and towns. Together these inherited resources provide credible images to be lodged in the mind as reference points for identifying with our forebears. These have been well portrayed in the context of European urban history by Bruno Milić, my mentor of the University of Zagreb.[5] We can begin to understand that just as our forebears struggled over generations to accommodate themselves to environmental necessities and amenities, we are in the continuous collective act of giving form to the urban places of today.[6]

It is this linkage to history that we have lost in today's world of stepped-up communications, human mobility, and the dominance of building technology over the natural environment. Individuality in shaping environments, a fragmentation of regional character, a homogenization of urban quality, and a loss of collective expression have revealed on a global scale the fundamental weaknesses in what early in the twentieth century we called "progress." In this context, we begin our search for the meaning of place, by experiencing as best we can in the pages of a book, the unfolding, the evolution, of urban places in Dalmatia, first up to the invasion of Napoleon, then through the "modern period" up to the present.

SHAPING URBAN DALMATIA: FROM THE ILLYRIANS TO NAPOLEON

Illyrian Settlers, Greek Traders, Roman Colonizers

Dalmatia's earliest name, Illyria, invokes an image of a place with a poetic, or "lyrical," quality, an environment designed to inspire the human experience. Did the natives of that land create the lyre? Among the mosaic of peoples who dwelt in the rugged and varied terrain of the Balkans, the Illyrians—from the eighth century B.C. to the first century A.D.—remained unattached to their contemporaries, the Etruscans of Italy, the Iberians of Spain, and the Gauls of France. The multi-ridged wall of the Dinaric Alps shielded Dalmatia from the hinterland to the east, and the surging Adriatic Sea to the west acted as a vast moat; together the various tribes evolved a regional identity within a self-contained environment.[7] The rugged coastal terrain provided retreats for scattered settlements from Istria to Albania. The restrictions of the harsh land forced competition between three major tribal groups for the best sites. Between these settlements primitive routes were established. The Illyrians became such masters of the sea that for at least a century they were able to hold off the Romans, whose land-oriented armies could not cope with conquest on the sea, let alone readily gain footholds on the steep shores of Dalmatia.

The Adriatic provided a natural communication route for the Greeks, with their far more advanced culture, to venture northward by sea in the Mediterranean. Finding limited resources for strongly established settlements, they wisely established trade rather than military relations with the Illyrians. The small number of offshore islands along the Albanian coast made their early footholds to the north, including Korčula and Cavtat, called Epidaurus by them, and other settlements near Split, difficult to maintain.

Not until 168 B.C. were the Romans able, by means of a well-organized maritime strategy and a decisive sea battle, to take over the coast and hinterland, making Dalmatia a province of Roman Illyria. Clearly lacking resources for colonization, Dalmatia offered a foothold for the Romans to use to cross the Sava and Drava Rivers and settle the fertile plain of the Sava and Danube River valleys. Without Dalmatia, their frontier urban centers at Vienna and Budapest, famous for their Roman hot baths, might not have become the historic cities we know today. The highly integrated culture, fertile "civilization," and sophisti-

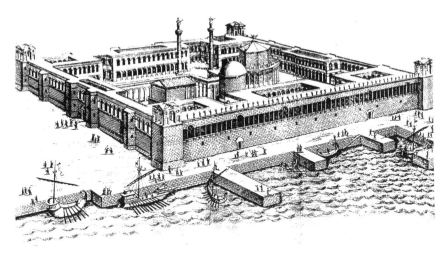

PALATIUM DIOCLETIANI EX QUO SPALATUM CIVITAS.

Diocletian's Palace, monument to Rome's urban heritage for all Dalmatia. Idealized version by Fischer von Erlach, 1656–1754. *Croatian Regional Archive, Split*

cated administrative system of Rome prevailed in Dalmatia. For more than five centuries this breed of people blending cultural qualities of both Rome and Illyria developed a new order of life focused heavily on cities, Pula on Istria, Zadar, Split, Trogir, and Starigrad on Hvar among them. This order came to an end when the Visigoths invaded the city of Rome in A.D. 410 and the empire fell in A.D. 476.

The architectural remains in these cities document the force of change that Rome brought to Dalmatia. J. J. Wilkes brought Roman Dalmatia to life in his volume researched entirely from engravings and monuments.[8] Yet, even Rome's political power and extensive city building served only to facilitate the permanent settlement of the Croatian newcomers.

Slavic Migrations and Croatian Dalmatia

By the fourth century A.D. the Roman Empire had overextended itself, resulting in the establishment of the separate eastern territory known as Byzantium. In the sixth and seventh centuries the tribes of Slavic peoples north and east of the Danube and the crescent-shaped wall of the Carpathian Mountains gradually moved south. Living a sedentary peasant life, people who had never seen lands as fertile as those of the Danube, Drava, and Sava River valleys nor the open sea were energized to seek more fruitful homelands. Diverse ethnic groups poured into the Balkans, some as plunderers and others who settled into fixed communities to cultivate the land. Ivo Banac points out that the first who came were nondifferentiated Slavs (with Avars), and during the reign of Heraclius

Slavic migration.
The Slavic peoples
migrated into Roman
Illyria in the early
seventh century A.D..
*Map reproduced from
school history atlas,
Školska Knjiga, Zagreb*

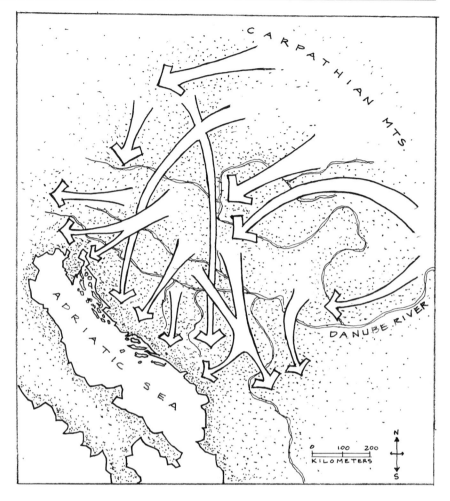

(A.D. 610–41), the Hrvati, or Croats, as their name evolved over the centuries, came, as did the Serbs.[9] Collectively they became known as the South Slavs of the former Yugoslavia.

The Croats, predominantly farming people from the Cracow area of southern Poland, established themselves in the former Roman provinces of Pannonia and Dalmatia and readily adopted Christianity. Strong ties developed with the West and Rome through the religious and cultural identity already in place rather than with the Byzantine Empire, which had survived the fall of the western portion of the Roman Empire and whose capital lay in distant Constantinople (Istanbul). Nin became their own bishopric. The Croats soon found the irregular shoreline ideal for independent settlements away from the old Roman towns. The moment of discovery of the vast plains of water and the infinity of a horizon never seen before surely would have stirred them deeply. This moment in Croatian history has been visualized by several nineteenth-century painters, including Celestin Medović, of whom we shall learn more since he was

The arrival of the Croatians at the Adriatic. *Painting by Oton Iveković*

a native of Kuna and a contemporary of my grandfather. As one generation learned from another, the Croats became settled crop growers, sailors, and fishermen. These new Dalmatians—though in a province of Byzantium—eventually became the first Slavs to become a maritime power on the Adriatic, on the threshold of the western non-Slav world.[10]

From the seventh century to the tenth a slow process of integration took place. In 879 the Croatian principality was recognized by the pope. Its status as kingdom became known when Tomislav was first named *ban* (governor), then crowned king of Croatia, as first recorded in 925. For the next two hundred years this kingdom of the Croats encompassed a territory between the rivers Drava and Sava, west to the Adriatic coast, stretching at times to the Neretva River. I saw plaques commemorating the one-thousandth anniversary of the crowning of Croatia's first king in many public squares of Dalmatian cities and towns. Šibenik became the first Croatian city, as the Croats gradually developed building skills, drawing on Roman models and using pieces of ruins, and made the urban areas their own. In the Great Schism, the splitting of the Christian Church into its Catholic and Orthodox segments in 1054, the dividing line ran north and south, unrelated to geographical and cultural patterns. This fostered a division among Croatia, Serbia, Bosnia and Herzegovina, and Montenegro that was a tragic source of conflict up to and including the war of the early 1990s.

Well established with its own parliament, coinage, and army by 1102 through dynastic continuity, Croatia entered into a union with the Hungarian kings. During the following centuries, local initiatives taken along the geographically fragmented coast helped evolve the many municipal statutes from these old Roman origins. These became a basis for future physical developments as well as developments in the areas of legislation and self-government. These stood Dalmatia in good stead when Venice

Šibenik in 1597.
Croatian State Archive

took possession of the region in 1409 on the basis of a contract of sale signed by Ladislaus, pretender to the Hungarian and Croatian crowns.

Venetian Culture, Turkish Threat, French Reform

As a result of sharing the maritime life of the Adriatic with Venice and other cities, Dalmatia became increasingly important as a regional place in its own right. The cultural and economic life of the towns became more lively, and the relationship between Venice and Dalmatia was more a useful partnership than one between colonizer and colonized. From the fourteenth century up to the fall of Venice at the end of the eighteenth century—the principal period of Venetian dominance in the Adriatic—shipping and trade generated urban development in the spirit of municipal autonomy. This, plus Byzantine influences, nurtured an identity for the Dalmatians built on their own well-identified accomplishments in the arts, architecture, urban design, literature, theater, and sciences, especially related to maritime life. In contrast to Venice, which was built mainly in brick, Dalmatian towns were built of stone cut from their own quarries. The major regional centers—Zadar, Šibenik, Split, and independent Dubrovnik—spread cultural influences to nearby settlements. The larger islands, with towns such as Hvar and Korčula, became necessary to the Venetians as places for preparing for long sea voyages and recuperating upon return.

One can make the case that Venice operated also as an exploiter and oppressor, especially with regard to the lower classes in Dalmatia, where a patriarchal structure had been inherited from Rome. Yet, since these

towns had developed their own collective stability and statutes for management of resources by the twelfth century, the Venetians had only
to negotiate allegiances based on mutually beneficial and established
regulations. Through this reciprocal relationship, the patriarchal system
conveyed Venetian culture to the upper classes, reinforcing the distinction between the people of the towns, who mirrored Venetian styles and
urban ways, and the rural populace, who maintained native Croatian
identities.

Of the cities, only Dubrovnik maintained the best of both these influences. For more than five hundred years Dubrovnik, blending Illyrian,
Roman, and Croatian cultures, remained an independent city-state with
a cosmopolitan life of its own. This gem of a city was in a sense a
little Venice, yet it had the self-reliant spirit of Dalmatia, mindful of the
threat of greater powers. This element of partnership between Venice
and Dalmatia provided the common moral motivation for keeping at
bay the newcomer to the Balkans: the Turkish Empire.

Looking back over the sweep of external events that shaped Dalmatia's collective identity, we see intrusions from all sides: the Greeks
from the south, the Romans from the west, the Slavs themselves from
the north, the Hungarians from the east, and the Venetians from the
northwest by sea. Now the Turks came from the east by land. They first
defeated the Serbs in 1389 at the decisive battle of Kosovo and destroyed
the state in 1459; Bosnia fell in 1463, Herzegovina in 1482, and Montenegro in 1496. In 1453 Constantinople gave way, and with it Byzantium.
Since the Neretva River valley provided the only access route through
the mountains, the Turks were held to a frontier at Počitelj within only
a few miles of the coast. This town of almost purely Turkish character, with its minarets and domes, had its own visual identity even after
the Ottoman exodus in the 1870s, when my grandparents left for California. These reflections of Turkish influence stood in sharp contrast to
the Venetian campanile, loggia, and piazza in urban Dalmatia. Inland,
Mostar, Sarajevo, and Jajce likewise reflected Turkish cultures in their
mosques, minarets, marketplaces, and walled-in hillside residences until
the disastrous military aggression of the 1990s.

The most momentous turning point of all comes from far to the West
in the person of Napoleon. With the French and American revolutions,
as well as others brewing in Europe, echoes of liberation accompanied
his arrival in Italy and the downfall of Venice in 1797. As a result, reform
movements were to resonate in Dalmatia throughout the nineteenth
century. Harriet Bijelovučić's highly readable book *The Ragusan Republic* presents the events leading up to and during the French occupation of
Dubrovnik in lively detail. Dubrovnik was the French beachhead for all
of Dalmatia from 1806 to 1813. The international intrigue between the
Austrians, the Russians, the Montenegrins, the Italians, the British, and

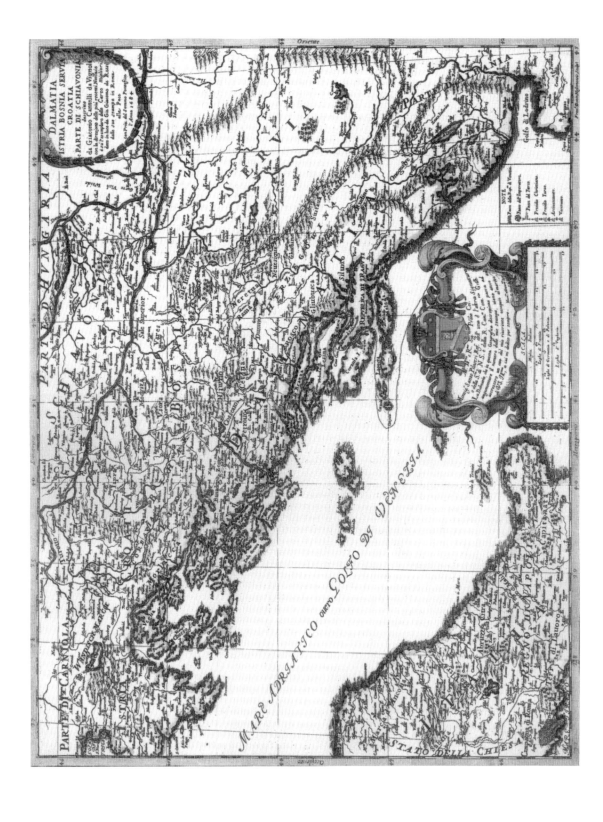

the Turks as each jockeyed to hold back the French and yet gain its own foothold in Dubrovnik, suggests many parallels to the conflicts there since 1991. Supported by well-selected quotations from archival sources, Bijelovučić's study of this short period vividly portrays the political and social turmoil as the yearnings for democracy surfaced. Cities became a focus of a lasting movement toward regional and national cultural identity.

One striking example of the drama taking place occurred when, after two years of French occupation, during which the French fended off foreign powers and played Dubrovnik nobility off against the middle and lower classes, the time came for the French to take command. Napoleon's general, August Marmont, ordered the Senate to meet. After berating them for their lack of loyalty, Marmont's representative "took another paper from his pocket and began to read . . . five articles." The first three speak for themselves: "The Government and Senate of Ragusa are dissolved"; "The existing civil and criminal tribunals are dissolved"; "M. Bruere, Consul of France, will be provisionally in charge of the administration of the country." The senators, stunned, "remained silent. . . . They asked for a copy of Delort's speech, but he answered he did not have any except the outline from which he read, but he promised them a copy. . . . but they received only the decrees. At this moment the army entered the palace. They locked all the doors and closed the Chancellery, treasury, customs house and all other offices and placed seals on them." A few days later, the French consul reported that "only the nobility show any anger at the change. The bourgeoisie are very happy to become the subject of the first and most powerful rulers in the world. They have shown their satisfaction with a magnificent ball which they gave the General on his return."[11] Thus, it took no more than the presence of Napoleon's troops in Dubrovnik and the stroke of the pen of his general, August Marmont, to wipe out the centuries of independence and self-made local identity. This set new directions for the "Illyrian Provinces"—a new, yet very old, name for Dalmatia and its hinterland.

THE URBAN HERITAGE: SOCIETAL AND ENVIRONMENTAL QUALITIES

Well-Defined Social Characteristics from Land and Sea

In the social fabric of Dalmatia the consistent and strongly asserted group identity with settled places is significant.[12] The Illyrians' prolonged warfare against the power of Rome marks its earliest expression; the Romans built their own monumental identity into the first cities, an identity adopted by occupants up to our times. Later, the urban character of the Dalmatian cities, identified with piazzas and campaniles, evolved during the Middle Ages out of the urban culture evolving on

Dalmatia in 1684, at the height of Venetian influence (opposite). *Croatian State Archive* the east and west coasts of the Adriatic. But the Croatian tribal organization—in contrast to the structured Roman patterns—fostered the development of the compact village and hamlet, or *zadruga,* clustered houses of an extended family, usually on slopes above the precious arable land. In my experience at both ends of The Bridge, Dalmatians readily express feelings of attachment and affection for their own particular home place and maintain its superiority over other places. The vigorous sense of commitment to a physical place has reinforced a cohesion among members of families and larger social groups, an important human resource in the many periods of stress in Dalmatia's rugged history. This affection for native villages was given voice dramatically in response to their being shelled and even destroyed in the attacks of 1991 and 1992.

In spite of migrations within Dalmatia over the centuries, the scale and distribution of urban population varied little from the early Middle Ages to the beginning of the nineteenth century. As a result, a stable urban system has nurtured both the collective identity of the region and the relatively uninterrupted evolution of well-defined social characteristics distinguishing one place from another. This continuity of the urban system facilitated the population's mobility. For example, the scattered Greek settlements intended only for trade were replaced by new Roman towns, whose settlers eventually absorbed the native Illyrians. After the fall of Rome the Croatian tribes occupied the agricultural lands outside the Roman towns, and only after many centuries did they become urbanites in these established centers. The old population dwindled to a minority through the ravages of time, plague, and earthquakes. By the late Middle Ages the continuity of social institutions was maintained even though virtually all of the urban system became predominantly Croatian.

The maintenance of the Roman model of class structure, even after the Croatian population had taken over, contributed to the continuity of this pattern of settlements. This patrician system tended to be softened by the vigor of the lower classes and their ability to bargain and negotiate for their rights and demands. The limited resources of the land put the upper and the lower classes in somewhat similar positions in the struggle for survival. In the case of the city-state of Dubrovnik, this dominance by the elite held sway entirely through the nonmilitary force of its own self-government and the maximizing of economic advantages for all carried out by its ruling class. This style of government provided opportunities for advancements in science, the arts, and culture, general marks of distinction between urban and rural life anywhere.

Dalmatia's harsh physical geography and lack of readily usable land and resources would not appear to foster social and cultural development. Yet, the social evolution of this region reveals a high level of creativity and self-reliance. Minimum resources brought maximum social development. Larger urban centers—Dubrovnik, Hvar, Korčula, Split, Trogir, and Zadar—produced their own writers, poets, historians, philosophers, architects, artisans, and even statesmen. More importantly, Dalmatia's urban history demonstrates that a population developed that had the stamina to withstand the constant pressure of external forces and a commitment to build and manage its own environment. They became a particularly sturdy people, sensitive to both the forces of the natural environment and the requirements for social organization.

Religion, family life, and community structure also contributed to the stability of Dalmatia's urban system. During the Middle Ages Christianity linked people to nearby Rome rather than to faraway Constantinople. For example, under the guidance of the Benedictines on Mljet as early as the twelfth century, then the Franciscans on Pelješac, and later the Dominicans on Brač, a consistent social basis for the urban system evolved through education, agriculture, and the arts. The advantages of strong family and community structures stemmed from Slavic tribal systems. Such institutions as *kumstvo* (godparenthood), for example, added to the already firmly accepted family traditions related to rearing the young. At the same time, the *zadruga* system brought extended-family members together into cooperative arrangements generally for purposes of production. *Bratinstvo,* the concept of brotherhood organization for welfare and religious purposes, lent strength to patterns of community organization; the tradition played an important role even among Dalmatian newcomers to California.

Economic Resources and Maximized Social Qualities

Throughout the centuries, Dalmatia's economic life grew out of the imbalance between the poverty of the land and the richness of the sea. This imbalance also brought human development. The land had little to offer: a basic rugged limestone, or *karst,* broke forth in jagged ridges from areas of red and stony soil. Arable land for crops and grazing was formed by removing stones that could be lifted and thrown into enormous mounds known as *gomile,* which typically form a pattern of giant polka dots on the hillsides of Hvar, Brač, and the mainland. Miles of ancient olive and grape terraces, some still in use, were built of stone. The small, family-oriented farms thus created required constant human care and watchfulness. These factors produced a genuine, lasting rootedness to each particular element of the environment. Nature's most generous contribution to the urban system has been the wide variety of stone out of which the cities have been built and because of which they endured so well, even during the severe shelling of 1991. Stone ranged from the creamy white marble of Pučišća, used in constructing the Palace of Diocletian, to the slate that roofed that village and many others.

Fortunately, the sea offered far greater compensation. In this sense, the Adriatic has been both the lifeblood and a source of learning for Dalmatia, carrying its products to distant places and bringing back goods and knowledge from foreign places. In its fertile depths urban Dalmatia found its daily food, and in its innumerable coves, havens from storm. These maritime resources have governed the location and growth of the towns and cities that provide the link between sea and land. The sea, more than any other force, has shaped the urban system.

Poor land and rich sea together reinforced the lesson that human resource development, in addition to transportation, is basic to economic development, and both provided a source of connectedness to the natural environment. In these formative centuries of the early Middle Ages the human capacity for skill, discipline, self-reliance, and inventiveness was critical in advancing Dalmatia's economic life; out of scarcity and adversity grew individual human strength and collective growth.

This process of economic development based on human perseverance succeeded in direct ratio to the freedom from foreign domination or interference. For example, Dubrovnik's handling of its own affairs for more than six hundred years created a highly advanced economy, social order, and a workable urban and regional spatial system. In Korčula, Hvar, Brač, and, in a much more troublesome way, Zadar, the limitations on the economic enterprises set by their Venetian governors prevented these places from making the most of their own resources. Local self-assertion and initiative, however, did reduce interference and led to a certain level of economic maturity. For example, while competitive trade with Venice was not allowed, small port towns on Brač and Hvar prospered in the seventeenth and eighteenth centuries. Although the size of ships built on Korčula was restricted by Venice, shipbuilding contributed to urban growth there and in Orebić on nearby Pelješac.

Because of Dalmatia's fragmented regional geography and its distance from major European populated places, high economic achievement in individual places tended to foster local cultural expression and thus nurtured sources of identity with place. Through contact with other cultures via the sea, Dubrovnik, Hvar, Zadar, and Split ranked high in creativity in painting, sculpture, literature, architecture, and urban development when they prospered via trade and export during the Middle Ages and the Renaissance. As the size of urban centers in Dalmatia increased, the range of regional economic activities increased. The larger centers—Zadar, Split, Dubrovnik—played a dominant role in the economic life of Dalmatia because of their strategic position linking extended inland routes with the sea. Smaller towns played a smaller economic role because they primarily served the nearby hinterlands.

In the formative centuries when there were few communities, each city, town, village, and hamlet could aspire to physical growth only commensurate with adjacent zones of production and trade routes. For example, when the Romans built Salona, they chose a site on a protected bay linking the sea and natural land transportation routes inland. The larger, well-located cities have thus become sufficiently mature to adapt to modern development and serve as anchors to Dalmatia's strong collective identity within the Croatian state.

Distinctive Urban Places Derived from the Natural Environment

The very nature of the geological formation, exposed at the surface and virtually without soil, challenged the most rugged families and the better-organized social groups. Dalmatia's supplies of fresh water were always precarious, with underground rivers frequently flowing directly into the sea rather than into valleys. The limestone formations worked against alluvial deposits and flows of surface water. To this day, many offshore islands that are drenched in winter are still dependent in summer on water collected from cisterns on rooftops.

Summer heat seared the crops and livestock, while the *bura* (the chill winter north wind) and occasional snows, even at sea level, drove families to the hearth, a response so essential that it became a prime element in social formation. Given a predominantly vertical topography, settlements sought scarce valleys, inlets, and flat uplands. These factors still limit the intensity and patterns of urbanization. This fragile environment lost much of its forest cover when the Venetians cut timber for shipbuilding.

Dalmatia's geography forged a system of compact urban settlements at key points of access to the sea, a ready-made system of communication to all parts of the region. This gave Dalmatia a coherent urban character that distinguishes the region from nearby inland and mountainous regions. The close proximity of people and their collective dependence on limited resources brought about a spirit of community. The sea provided not only food and nourishment but also lessons in science and direct contact with other societies, their cultures, and economics. The Dalmatians, who otherwise might have been merely marginal farmers, became expert sailors, seamen, traders, and fishermen.

Although Dalmatia's economic opportunities have been limited, its natural environment has always been attractive for rest, relaxation, beauty, and climate. The Romans, who knew the Mediterranean well, clearly were drawn to the varied shoreline settings to build permanent settlements and country villas; Diocletian spurned the luxuries of Rome to build his seaside palace at Split. The sweeping power of the coast made it inevitable that the Croatian tribes would make this their permanent home. From the sixteenth century on, Hvar became a logical resting place for Venetian sailors to recuperate from long sailing journeys, and providing accommodations for them required special building programs.

The accumulation of human-made elements has evolved a heritage of architectural and urban design elements that attract universal interest because of their compatibility with the natural environment and their superior state of preservation. As Andre Mohorovićić has comprehensively documented, on the level of individual buildings, hamlets, villages, towns, and cities this rich assembly of urban environments speaks for a broad range of historical periods and cultural features that is unique in rapidly modernizing northern Europe.[13] To experience these places affords the contrast and sense of perspective we need to evaluate the scales and patterns of our contemporary urban centers.[14]

The built environments that this history shaped according to the particular geography of Dalmatia generally became known for their own sake in Europe through a series of travelers drawn there mainly by the archaeological remains. This field took hold through the seventeenth and eighteenth centuries, especially in Italy and Greece. The Palace of Diocletian in Split served as a particular focus. Beginning with George Wheler in 1689 and Robert Adams in 1764, both from England, published reports and writings appeared into the twentieth century, increasing in number and breadth of subject matter.[15] These published works were superficial and introduced images and writings that encouraged a style of tourism that was out of touch with Dalmatia's true identity. Today, in the aftermath of the 1991–95 war, with greatly increased mobility and a level of communications unimaginable in past centuries, Dalmatia is faced with the challenge of correcting that image through the cultural power of its own genuine identity with place.

Local Government Shaped by a Desire for Self-Determination

The two characteristics in Dalmatia's urban history that show the most continuity are the urban places the centuries produced and the nature of the governmental institutions that evolved to shape them. The Roman administrative structure for municipalities was introduced full-blown in a number of widely separated places—mainland, peninsula, and islands. In the year 945 Constantine Porphyrogenitus, the Byzantine emperor, identified six such city-states as provinces, or themes, of Byzantium, in his book *De Administrando Imperio,* the only written account of these times. The principles of Rome's legislation survived up to the early Middle Ages.

By the late Middle Ages most of these entities continued to operate under their own municipal charters; even under the Venetian "protectorate," Korčula, Vis, Hvar, Brač, and Zadar had well-established codes. Free of Venice, Dubrovnik was virtually independent from 1358 on. Its institutional system was highly developed, administered consistently in a formal way, as demonstrated by records of all municipal affairs contained intact in the archives of that city over seven centuries, from 1278 until the present day. Although these localities were separate entities, often competing against each other and under differing degrees of autonomy, the overall system fostered regional cohesion. The heavy hand of Venice somewhat weakened local autonomy in these states, and the remainder was all but eliminated by Napoleon's invasion at the beginning of the nineteenth century.

The tradition of sustained efforts to maintain local autonomy and collective identity made it possible for these Dalmatian entities to survive during centuries of foreign domination. The self-contained island character of Korčula, Hvar, Vis, and Brač especially stimulated the early evolution of independent administrative statutes. Dubrovnik was a symbol of self-sufficiency for the other centers. Although all the statutes were based on Roman models, they acquired Croatian elements and varied considerably. For example, most of the municipal statutes covered urban dwellers as well as rural residents, but those of Dubrovnik and Zadar contained separate laws for each. Both of these places gave greater emphasis to maritime law.

With resources limited on this sparse coastline, foreign powers saw the region as a buffer against other aggressors rather than one to be colonized. Venice held off the Turks to the south and the Hungarians to the east by obtaining naval and related support from the Dalmatians in time of war. The Venetians continually exacted all possible tribute and curbed local or regional initiatives that would lead to alliances among the separate Croatian entities.

Thus, the Dalmatian towns were left alone to a degree. Without such well-respected and effective municipal laws, they could have been engulfed by Venice or even by Dubrovnik, which did expand to the north by buying Pelješac from the Serbian king Dusan in 1333. For this reason this elongated peninsula, almost as complete a geographical entity as nearby Korčula, became the only major subregion that had not developed laws of its own. However, when it was finally freed of Dubrovnik in the nineteenth century, Pelješac's citizenry participated actively in the regional government under Austria.

The tradition of self-government established by the Dalmatian towns in the early Middle Ages also enabled them to protect and make the best use of the limited resources in their jurisdictions. Motivated by a desire to effectively manage the use of land and to meet local needs for economic development and social well-being, these statutes dealt in large part with matters related to the physical environment. These activist attitudes toward self-management demonstrate the depth of the people's identity with and care for their places of living.

For example, as early as the thirteenth century, Korčula's municipal charter regulated in considerable detail the cutting of timber, since its pine woods, valued even by the early Greek settlers, were of critical importance to shipbuilding. These already well-founded laws were used to prevent Venice from stripping the island of its forests. Such practical considerations frequently became the basis for negotiations with the Venetian governors. Several paragraphs of Korčula's statute also deal with the protection of viticulture, one of the mainstays of the economy, and severe punishment was to be dealt to those guilty of damaging vineyards.

The statute of Brač, from the twelfth century, dealt particularly with the relationships between those who made their living primarily by growing grapes or olives, breeding cattle or raising sheep, fishing, producing grains, or other limited forms of agriculture. The statute restricted grain exports and wine imports to protect local production. The regulations dealt specifically with the relations between the peasants and landowners and between the shepherds and the owners of livestock, as well as with the use of extensive lands for cultivation or grazing that were held by the municipality. The Brač statute also established regulations to control public health, particularly relating to the spread of infectious diseases at ports of entry.

As the maritime Dalmatians, especially those of Dubrovnik, increased their contact with other areas of the Mediterranean, they acquired knowledge of other forms of government and legislative systems related to social and economic betterment. For example, they became expert in maritime law. They not only realized the sophistication and strength of their own relatively high level of self-government but also absorbed new concepts. Among the lower classes the awareness of the revolutionary struggle taking place in eighteenth-century France grew from sailing to foreign places and paved the way for locally based identities. Ironically, however, the immediate effect of Napoleon's occupation was its tacit and blunt destruction of many of the self-directed municipal institutional systems in favor of centralized regional authority.

TURNING POINTS IN THE NINETEENTH AND TWENTIETH CENTURIES

Napoleon Plants Seeds of Social and Political Change

In the nineteenth century, the slow pace of Dalmatia's urban history boldly accelerated with an unprecedented momentum. Like a stone suddenly cast into calm waters, the French administrative and military takeover of 1808 caused changes to ripple out through the century, events directly experienced by my great-grandparents. Cutting off ties to a weakened Venice gave the Croatian population of Dalmatia the opportunity to deepen their basic sociocultural and ethnic ties with those from the interior. The concepts of *liberté* and *égalité,* on which the French Revolution had been built, induced these changes. The Dalmatians, for centuries marine traders with Marseilles, learned of the French political upheaval and its intellectual leadership through literature. A vision of social betterment reached the nonelite peoples of the towns.

Recent historians have pointed out that the revolutionary democratic movement in Europe of the late 1700s, together with the static condition of Dubrovnik's own medieval institutions, facilitated the French influence's reaching Dalmatia in such a short time. Nor is it believed that the French intended a full democratic revolution. Two similar cases were the old city-states of Genoa and Venice, likewise on the brink of major change. Soldiers recruited from Dalmatia to assist the Republic of Venice in 1797 returned and spread discord and revolutionary ideas.[16]

With the French headquarters in Dubrovnik, Napoleon's appointed governor, General August Marmont, developed a markedly sympathetic identity with Dalmatia. This was exemplified by the French revival of the ancient name Illyria for Slovenia, Croatia south of the Sava River, and parts of Bosnia, for administrative purposes known as the Illyrian Provinces. Marmont had a tendency toward romanticism, and his letters to Napoleon reveal how the character of Dalmatia made him sensitive to the region's uniqueness. The landscape and ancient settlements, the environmental imagery, and the history clearly preserved in its cities, towns, and villages engaged him. Although the military purposes of the Dalmatian occupation were acknowledged by Marmont, the physical and cultural environment held a particular attraction for him, and that led to an array of improvements that ultimately benefited the environment as a whole.

Rebecca West writes warmly of Marshall Marmont:

In 1806 Napoleon had still much of his youthful genius. It made him take over this territory . . . and found the two provinces of High and Low Illyria that comprised Croatia, Dalmatia, and Slovenia. . . . He made Marshal Marmont the Governor of these Illyrian provinces, and it was an excellent appointment. . . . he was an extremely competent and honorable man, and he loved Dalmatia. His passion was so great that in his memoirs, his style by nature . . . pompous, romps along like a boy, when he writes of his Illyria.

He fell in love with the Slavs; defended them against their Western critics. They were not lazy . . . they were hungry. He fed them and set them to build magnificent roads along the Adriatic.[17]

Austria's Heavy Hand Fosters Croatian Identity

Austria's role in Dalmatia endured from the departure of the French in 1814 to the close of World War I, yet the first half of the nineteenth century, up to 1848, differed greatly from the second. After the Congress of Vienna in 1815, because of Austria's relatively static government there were no major changes in urban evolution. This was a period of gestation, fertilized by the new ideas of the French, a period of push and pull for power between Italy and Austria and for dominating the new technologies that were to lead to railroads, steamships, and mechanized industry. However, the Austrians became well aware of the competing international forces touched off by Napoleon's aggressions. They used their experienced engineering and architectural skills to form an integrated defense system for the vulnerable Dalmatian coast even up to World War I. They brought nineteenth-century advances to the old medieval fortifications and city walls and built new facilities at Pula to the north and at Zadar, Split, Dubrovnik, and Kotor to the south. These had the indirect effect of reinforcing these places as focal points of development within the urban system as a whole.[18]

These external political forces indirectly provoked a revival of Croatian self-assertion. In Croatia and its Dalmatian coastlands enlightenment and popular movements took the form of a search for identity, both national and local or regional. Self-awareness and individuality in localized social expression began to loosen the Dalmatian city dwellers from the hold of Venetian culture and language as well as from the hold of the even more alien Austrian culture. The continuing contact with the sea and other countries gave the Dalmatians an advantage over those living inland. The New World and the American Revolution were known to these people; indeed, early in the 1800s the first Dalmatians began to take passage from Marseilles to the French territorial city of New Orleans. It was via this route that they first found their way overland to California.

The second half of the century brought the flowering of these seeds of change. The year 1848, when the young Franz Josef succeeded to the throne of an expanded Austrian Empire, marked a decisive turning point for Dalmatia. Carl Schorske describes his pragmatic yet broad vision when, in 1857, he determined to convert the open belt of land outside the old city walls from military to civilian use.[19] At this same time peasant revolts swept many of the rural villages of Europe as people living in the countryside became aware of improved living conditions in cities.

Under the general orchestration of Franz Josef, Vienna flowered as a center of music, education, science, art, architecture, and urban modernization, some of which filtered down to Dalmatia. The replacement of the walls of Vienna by the Ringstrassen, those tree-lined boulevards encircling medieval Vienna, created a vast array of public buildings designed to bring a degree of civic and cultural benefits to the rising middle class—courts of justice, municipal administration buildings, concert halls and opera houses, art schools and art museums, universities and public gardens, and palazzo-like apartment houses. These examples of civic embellishment and new styles of living in cities provided models for the new sections of cities like Zagreb, Ljubljana, Sarajevo, and, in Dalmatia, Zadar, Split, and to a lesser degree Dubrovnik.

These urban improvements were based on well-engineered and comprehensive survey maps of urban places drawn from 1823 to 1837 and updated in the 1870s. The Venetians had made cadastres, inventories of real estate, as a basis for taxation as early as the fifteenth century and with greater accuracy in the mid-eighteenth century. The Austrians applied a far more exacting method of triangulation by use of the geodesic table, covering 744 communes and producing 6,725 cadastral plans. These contributed to a more equitable local administration of each commune and a focus on urban improvements, and achieved a greater community awareness of the particular sources of identity with each place. The extensive information on the uses of land and types of vegetation provides historical background for environmental planning today.[20]

These refinements became readily identifiable expressions of Austria's coordinated architecture and urban design, yet they enriched identity with place by adding new styles to the styles from previous centuries of city building by Rome, the Croatian state, and Venice. In spite of Austria's foreign and authoritarian imprint, a degree of autonomy and political identity was given to Dalmatia in 1861 in the form of a Parliament located in Zadar.

Austria was a positive force in Dalmatia's physical development, but it operated in a cultural void that inhibited needed social and urban reform. This failure generated a lasting movement toward Croatian autonomy and identity reaching into the twentieth century. New technologies in the form of the railroad and the steamship in the nineteenth century and the airplane and electrical power in the twentieth century brought changes that were far more penetrating than the changes brought by either Napoleon or Franz Josef. Modernization, the watchword of the opening of the twentieth century, became the driving cause of change in the form and scale of Dalmatia's urban places. It was inevitable that this period of external stability would break down, as it did with the murder of Archduke Francis Ferdinand in Sarajevo in 1914, the subsequent outbreak of World War I, and the death of Franz Josef himself in 1916.

From Royalist to Communist Yugoslavia and the Republic of Croatia

With 1918 and the Treaty of Versailles we come to the next major turning point for Dalmatia: the establishment of the Kingdom of Serbs, Croats, and Slovenes, later named the Kingdom of Yugoslavia. For the first time the Slavic Balkan states became free of all foreign political interests

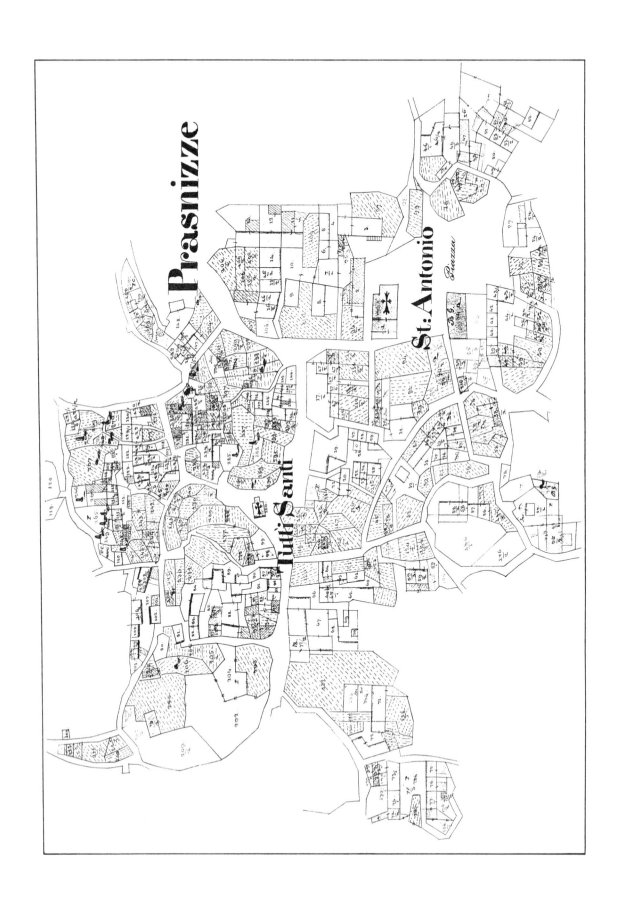

Prasnizze

St: Antonio

Tutti Santi

Piazza

(though not economic ones). The underlying problem—so well revealed by the medieval-mindedness of the tormenting war of the 1990s—lay in their attitudes toward ethnic differences, religion, and language, as well as in divisive patterns of geography and, until modern times, primitive transportation system. Ivo Banac sees the chief source of Yugoslavia's failure to unify to be the ethnic groups' lack of a clear collective identity within the interior of the country: "the total lack of a common state concept made manifest in the conflict between Serbian supremacists and Yugoslav integralists, on one hand, and non-Serb federalists, notably Croats, on the other."[21]

From the outset Dalmatia's environmental assets and its Mediterranean culture were a matter of contention as a part of the new Yugoslavia that should be shared by all. The brevity of this period between the two wars and the economic failure of the Belgrade administration led to the neglect of this resource and promoted development damaging to Dalmatia's urban system and its environment.

After only twenty years, however, history stepped in with World War II, and all of Europe became involved. The problems of social and economic advancement that had festered all through the nineteenth century had their catharsis in Tito's revolution. The revolt was against both royalty and capitalism. Dalmatia, with its maritime frontier directly available to the West and its relative ethnic and cultural integrity, played a significant role. The island of Vis sheltered Tito in 1944 and provided liaison for the Allies (U.S. and European). From that small, most westerly Dalmatian island Tito led the successful drive to bring together the highly differentiated ethnic groups into a tenuously held—and now discredited—Marxist federation.

The extremely difficult *karst* terrain played a role in avoiding an invasion by the Allies to occupy Yugoslavia. This would have resulted in a greater number of casualties and the disruption of the urban places. This same situation also contributed to Western European countries' reluctance to intervene militarily in the 1991–95 war.

And now, in the 1990s, we find history's accelerated pace discarding all that has gone before and Yugoslavia breaking up into the pieces that have sought their independence for centuries, though at an enormous cost. Dalmatia's own symbols of cultural identity in the built environment have been attacked, damaged, or destroyed. Now, in order to restore these and make a fresh start for the future, it is critical to understand the accumulated urban and environmental problems of the nineteenth century, yet to be resolved to give Dalmatia's regional system the integrity needed for the coming century.

Cadastre map of Pražnica from the 1830s, when my grandfather was born there (opposite). *Croatian State Mapping Office, Split*

A CENTURY OF FRUSTRATED SOCIETAL AND URBAN REFORM

The Need for Social Reform

A marked increase in population underlay the social condition of urban Dalmatia in the nineteenth century.[22] Yet, no population census was taken by which to evaluate welfare until 1857. In the second half of the century the population of Dalmatia increased by some 50 percent, from 415,628 in 1857 to 645,604 in 1910. This increase of 230,000 people—equal to twenty new settlements of 11,500 each—gives a realistic idea of the pressures on the existing urban settlement system and the limited agricultural resources, the overcrowding of urban facilities, and the consequent social neglect.

In the budding modern industrialized society of nineteenth-century Europe, technical skills and knowledge became critical as alternate opportunities were made possible by the greater freedom of the working class. Dalmatia's urban attitudes and taste for learning to meet the demands of a more complex world distinguished the people of that coastal region from those of the relatively nearby inland mountain regions. However, the majority of those on the coast were in less advantageous positions, and schooling remained a basic, neglected need. The Franciscan and Dominican religious orders had taken the lion's share of responsibility for schooling during the previous centuries. Reports made at the time of Napoleon's occupation showed over 90 percent illiteracy. By 1880 the figure was 89 percent, and in 1910 it was only 78 percent. Over the half-century from 1862 to 1912 the number of students in *gimnazijama* (public high schools) increased from 552 to only 1,233. The number of primary-school students, however, increased from 8,084 to 63,662 in the same period, reflecting the partial response to pressures the people of Dalmatia put on the Austrian government for schools. Sources of illiteracy were not removed until the mid-twentieth century, when social-development policies were made a cornerstone of the new Yugoslavia.

Beginning in 1861, when the Dalmatian Parliament was established and representatives from the local governments brought their concerns to Zadar, communities consistently pressured the government to build schools. Since my own grandfather represented Kuna, I took a particular interest in reviewing in 1979 the verbatim records (in Italian and German) of the Parliament's meetings held in the Croatian State Archives in Zadar. These social conditions came to life for me as I noted the formal requests for such urban improvements as hospitals and street paving. The records document Austria's failure to respond with building programs. The strongest demands were for schools, which were seen as a vehicle for advancing social and economic levels and breaking down the remnants of medieval serfdom.

A deeper motivation for education beyond learning was the desire for Croatian-speaking teachers, rather than Italians, who had dominated the education field for centuries. In 1862 only 49 of the 192 public primary schools used the Croatian language entirely; all others used either Italian alone or both languages. The fundamental issue here was to put cultural identity clearly in the context of identity with an established locality.

The shortage of adequate health facilities and services was widespread. As late as 1910, as Dinko Foretić emphasizes, among 600 Dalmatian villages fewer than half had drinking water supplies other than unreliable wells. As we learned from Rudy Palihnich in the introduction, *kupus, blitva,* and undersized potatoes were the main items of food. Only these vegetables could endure the heat of summer and the cold of winter. Again, I noted in the archives of Zadar that one of the urban improvements most frequently requested to the Dalmatian Parliament was water supply systems.

Primitive urban facilities, inherited from centuries of incremental building, often resulted in grossly substandard sanitation. Travelers in Dalmatia in the nineteenth century experienced the smell of gasses arising from sewers in towns such as Split, Korčula, and Dubrovnik. I found this still to be the case on my 1937 visit, though it had been resolved by 1968. Among the more prevalent diseases were malaria, tuberculosis, and pneumonia, as well as those of the digestive system caused by poor sanitation. As is common in maritime towns, venereal disease reached even the healthful offshore islands in spite of the primitive preventive methods then used. However, means of prevention were kept on hand by all sea captains in Dalmatia, out of a sense of responsibility to their sailors, as I learned from examining memorabilia of retired sea captains in Cvito Fisković's home in Orebić. Because of their centuries of taking precaution in time of plague, the people of the outlying islands became known for their concern for sanitation. Their natural isolation made it possible to control contagious disease at ships' point of arrival. Clinics and hospitals were located only in the larger cities. The ratio of hospitals to population in 1912 was one for every 680 inhabitants of Dalmatia and one doctor for every 500 patients in the hospitals. The records I reviewed of the Dalmatian Parliament from 1861 on showed frequent demands for these facilities.

Community leadership was almost entirely concentrated in the upper classes. Foretić has documented the rise of popular community organizations in Dalmatia that gained remarkable momentum with increased urbanization after the establishment of the Dalmatian Parliament in 1861. This gave impetus to collective identity free of class distinction, largely through the benefit societies of *bratinstvo* (see the introduction).

By 1872 there were ninety-five nonreligious organizations, some of which demonstrated the ability of the Dalmatians to organize themselves in their own localities. For example, the Maritime Society of Pelješac at Orebić developed the shipping industry of that region to an unprecedented level to protect the established sailing vessels against the new steamships built under Austrian initiative at Trieste. Other local initiatives included a Vintners' Society in Split, savings-and-loan groups, and scientific and artistic societies. The setting up of *čitaonici* supplied with newspapers, journals, and a limited number of books in the Croatian language represented a further example of community formation indirectly to promote self-government. A few socialist workers' organizations made their first appearance about the turn of the twentieth century.

Another significant indication of the growth of collective identity was the rapid rise of communication media in the Croatian language as part of the nationalist political movement's effort to replace German and Italian for official uses and in the schools. Between 1859 and 1866 twelve newspapers and journals appeared in the native tongue; by 1914 there were twenty-eight in Croatian and only three in Italian. Even during the 1870s newspapers in Zadar were printed in Italian and Croatian; official communiqués from Vienna, even those dealing with critical development problems in outlying areas of Dalmatia, were in German. Under those circumstances little progress in solving social problems could be expected. Thus, community organization leading to cultural integration and political unity became a critical resource for urban and regional development. Collective identity with place had gained new significance.

Urban Economic Development Restricted by Repression

The shifting of economic systems from agricultural to urban contexts in Europe in the late nineteenth century seemed to offer a potential for new vitality in Dalmatia's urban system. Protected sea routes and ports and sites for railroad connections were available. Nevertheless, Dalmatia's urban centers remained in a state of stagnation, in contrast to the economic development concentrated in the northern Adriatic region, where more accessible urban areas of primary interest to Austria—such as Rijeka and the Istrian peninsula—were located. Franz Josef's empire viewed Dalmatia as a useless and impoverished appendage, of interest more for its exotic history and scenic landscapes than for capital investment.

Some of Foretić's research indicates the enormity of the task of building the Dalmatian economy after World War I. He surveyed the occupations in the six Austrian censuses taken between 1857 and 1910, during which time the population increased by 50 percent. In comparison with the increase in the population, he found very little industrial or financial modernization over the fifty-three-year period. In 1910 the ratio of the population working in agriculture (limited by tenant-farming practices), forestry, and fishing, on the one hand, to the population employed in industry, mining, and trade, on the other, was still 17 to 1.

Although viticulture, a dominant agricultural activity after 1850, of- fered Dalmatia a transitory period of prosperity, the sweeping vineyard blight—phylloxera—in France and later in Italy stimulated the market for Dalmatian winemaking and export and thus gave new impetus to growth in both the inland villages and the island towns, Brač, Korčula, and others. So much land, however, was put to grape production that when the international market was no longer favorable, the Dalmatians had nothing to fall back on. This became a major reason for emigration to the New World and the loss of needed labor, and the situation was even worse when the grape plague hit the Dalmatian vineyards them- selves in the 1880s. Finally, the Austrian government leveled the fatal blow in 1894, when import tariffs on Italian wines were deliberately re- duced at the expense of wine products from Austria's own province of Dalmatia.

Only during the final twenty years of Austrian rule did industrial de- velopment begin to take root in Dalmatia, with its first rail line to Split in 1877. In 1880 agricultural pursuits were still predominant in the areas of Split and Šibenik, while civil-service and military activities predomi- nated in Zadar. Commerce and banking were hindered by the lack of industrial growth and by the shortage of domestic capital available from the upper classes, who traditionally followed the Mediterranean practice of investment in rentable property rather than in productive ventures. Only toward the end of the nineteenth century were hydroelectric plants built, and these were financed by foreign groups, whose profits flowed out of the region for reinvestment elsewhere. The first, built on the Krka River near Šibenik in 1895, provided both public and private lighting. Celebrations held when electricity was introduced to Dalmatia's towns and villages illustrate how highly valued this was as a social improve- ment, let alone for its use for industry. Indeed, at the time of my first visit to Dalmatia, in 1937, people in all of the villages had to read, dine, play cards, sing, or walk about at night by the light of candles or kero- sene. By my next visit, in 1968, electric lines, with all their social benefits, had just arrived, together with the blessings of indoor plumbing, auto- matic washers and dryers, and modern kitchens.

The New Spatial Scale of Urban Development

The visit of Emperor Franz Josef in 1875 sparked recognition of both the social and the economic benefits of improving the physical environ- ment. Since the emperor's name and foreboding image had been the focus of controversy in my youth, when I came upon a memorial bro- chure illustrated with towns he visited, as well as an article published in Zagreb in 1906, I took a special interest in the showmanship of the event.[23] These sources suggested that he may have been motivated more by the need to strengthen military security on the borders of Bosnia-Hercegovina against the Turkish domination than by an interest in the welfare of Dalmatians.

Although he made his tour of several weeks by ship rather than on land, he saw and was seen in a large number of coastal urban places, both large and small, a contrast to landlocked Austria. These towns called attention to their problems, achievements, and scenic and economic attractions. Records of the emperor's visit describe which permanent improvements were to be made to docks where his vessel would land, to the public buildings that he might visit, and to the roads and streets and public areas where parades and welcoming events would take place. Temporary decorations, more theatrical than truly civic, at his predetermined points of contact gave an exaggerated image of urban well-being to these backward towns, for they had never experienced such an illustrious visitor nor international attention. This pretentious public display had its negative side: it tended to polarize the pro-Croatian forces among the populace away from those in power who leaned toward Italy and Austria. A long-term outcome of this visit was the recognition abroad of Dalmatia as a unique focus of tourism, which attracted a considerable number of more intrepid tourists and British writers of rather romantic travel books around the turn of the twentieth century.

An immediate effect of Franz Josef's visit was the sending of a Viennese merchant and industrialist to Dalmatia to study means for increasing productivity. The resulting Stockhammer Report proposed a remarkably progressive, ten-point program of development based on the urban modernization experience in Vienna that could stand beside the comprehensive concept of urban planning today. The report dealt mainly with the physical environment as a key to social and economic improvement. The proposals covered (1) abolition of serflike practices of land tenure; (2) compilation of land uses and registration of land ownership; (3) building a network of highways; (4) connecting Dalmatia's railroads with the main lines to the interior; (5) establishment of a state-subsidized interisland steamship system; (6) construction of schools and establishment of scholarships; (7) customs reforms; (8) establishment of banks favoring trade; (9) tax incentives for new enterprises; and (10) stemming emigration to the United States with skilled employment. Little of this program, however, was taken to heart or implemented as positive city and regional development policy.

When, in the 1860s, the first large steamships traveled the length of the Adriatic, they bypassed key Dalmatian ports. The local sailing-ship activities of Dubrovnik, Korčula, and Orebić could not compete with the new vessels. Also, it was in Austria's interests to locate the new steamship terminals at Trieste and Rijeka, which were linked by rail to European cities. Toward the end of the nineteenth century, small steamships took over the communication of people and goods between the port towns and villages along the coast, eliminating arduous all-day walks on foot or *mazga*. In spite of the high cost of capital investment, railroad lines connecting either Zadar or Split with the interior would have been technologically feasible. The main deterrent to inland railroad lines was the political reality of Austria's interests in keeping the Slavic territories of Dalmatia, Croatia, and the interior regions disunited.

A major obstacle to the modernization of the towns and villages in the interior of the islands and on the mainland was the lack of a road system even for horse-drawn vehicles. Although the French undertook vehicular transportation systems that were ultimately of value for civilian needs, they were essentially in response to military purposes. In ten years they built some 580 kilometers of roads. The Dalmatian communities depended on the graded trails paved with stones that typified the entire circulation system in Dalmatia for centuries. Still intact in 1937, these roadways stepped up the steep slopes in stone tiers designed to prevent erosion and to fit the gait of the *mazga*. Routes of this type, built under the French for the movement of troops on Korčula, for example, continued to serve the local needs even up to recent times. On my own walks from village to village, I could identify by improved construction and gradient those routes built by the French.

During the hundred-year rule of the Austrians only 440 kilometers of roads were built, primarily for military purposes. In many parts of Dalmatia modern graded highways did not appear until the 1950s. In the 1960s, the construction of the Adriatic Highway, financed in part by the United Nations, finally opened up all of mainland Dalmatia to the automobile, bringing with it an unprecedented regional scale of urbanization.

Marked Deterioration of Municipal Autonomy

The autocratic policies of the French and the strong administrative leanings of the Austrians resulted in the near elimination of the relative level of municipal autonomy, evolved over the centuries to protect the marginal economic life of towns and villages and their collective sense of identity to place. Dalmatia was characterized by an ever-increasing local cultural unity and economic potential, but it lacked a governmental system of its own to guide the irregularly shaped coast into becoming a functioning urban system matching the new regional scale of communication.

In 1823 Austria created a so-called government of Dalmatia, with Zadar as its capital, with responsibilities for civil and military administration. Four districts were established: Zadar, Split, Dubrovnik, and

Kotor. These were subdivided into *kotari* (counties) and, within them, municipalities. In 1854 there were thirty-one *kotari* and eighty-nine municipalities. During this time the number and boundaries of these changed very little. The populations of the municipalities varied from large—ten thousand or more—to small—one thousand or less.

Under this *kotar* system two-thirds of the councilmen were to be elected from among the one hundred most important landowners, and the remaining third from among the owners of the secondary commercial or industrial businesses. Jurors, judges, and village heads were appointed by the district administration, and the Austrian emperor appointed the municipal heads for the four large cities—Zadar, Split, Dubrovnik, and Kotor. Thus, from 1822 to 1865 virtually no autonomy prevailed since a single social class and the central government ran the affairs of the municipality. Furthermore, this ruling group was essentially the Italianized class, the vestige of feudal nobility, which had no feeling for the increasingly popular political movement for *narodni,* or national, Croatian identity.

The 1864 basic laws of municipalities were more progressive and democratic since they opened the door to wider participation. The most important reform, one that sparked the first full expression of social aspirations, was that every taxpayer was given the full right to vote at both the municipal and provincial levels of government. This measure intensified identity with place and reflected a general trend in the increasingly liberal ideology of Europe at that time. Due to the backwardness of the masses and the benevolent though firm hold of the provincial governments over the municipalities, however, these codes did not go far toward achieving environmental betterment or social and economic development.

The location of the new provincial government at Zadar contributed to the inadequacies of the new municipal code. Zadar represented the periphery rather than the heart of Dalmatia both culturally and functionally. Virtually all of Dalmatia lay considerably to the south, and the journey by sea—one my grandfather made many times in the 1880s from Pelješac—was a deterrent for the municipal representatives who came to speak for their localities.

Through the process of newly electing municipal governments, Croatian, as opposed to Italian, identity gained increasing headway in the 1870s. During Franz Josef's visit in 1875, competition ran high between the nationalist and autonomist forces. The former demonstrated in civic displays the use of their native language and the Croatian character of Dalmatian towns; the latter emphasized the old Italian influences. Both, however, portrayed the social and economic backwardness. Without question, the pro-Croatian forces succeeded: on the emperor's departure from Vis, the mayors of fifty-nine nationalist-dominated municipalities were represented, in contrast to only twenty-one mayors of autonomist-dominated municipalities.

After 1875 it became clear that the divisiveness of the issue of nationalism kept local authorities from tackling the accumulating problems of urban development and working on common objectives for the local environment that both groups shared. In this climate of individualism and the resulting lack of cultural cohesion, the actual operations of many of the municipalities remained in the hands of the autonomists.

By 1900, however, the nationalists had won over the municipalities one by one. Foretić has identified the municipal leaders for the period 1857–1900. In each *kotar* the conflict was essentially between the people, who desired practical betterment of urban and rural life, and reactionary forces, who wanted to continue to control both public affairs and economic activities, in spite of new, liberalized municipal laws and the new constitution.

The greatest step toward maximizing participation came about in 1887, when Croatian finally gained full acceptance as the official language. After twenty years of struggle, Croatian, the language of 80 percent of the people, became the official language in all municipal administration. From that time on, the central government increasingly responded favorably to the Parliament's requests for schools, roads, hospitals, and other urban improvements. Yet, once more the response was inadequate because of the increasing population, accumulated needs, and rising social standards. Any progress made in this direction ended abruptly with World War I.

FOREIGN TRAVELERS' IMAGES OF DALMATIA, 1689–1941

As already mentioned, from the late seventeenth century through the nineteenth Dalmatia was "discovered" by foreigners; its history became a subject of fresh archaeological interest thanks to travelers, mainly from England. Few, however, attempted to relate urban development directly to the continuity of cultural and political change. By the twentieth century the Dalmatians themselves had become their own historical archaeologists. Their writings suggest the sense of discovery they experienced in reporting on cultural qualities and customs over some three centuries.[24] The firsthand descriptions of these travelers offer us a sense of their present in any given period to link up with their past, which created the urban forms of Dalmatia that we now experience.

The Seventeenth and Eighteenth Centuries

In 1689 George Wheler became the first European to report on the attractions of the Dalmatian coast and its archaeological remains. Almost a century later, Robert Adams was sent by the king of England and a "host of subscribers" among the aristocracy to document in writings and drawings the classical heritage of the Palace of Diocletian and its environment. Like Goethe two decades later, he had already studied in Italy and sought a more accurate picture of Roman architecture on the eastern shore of the Adriatic. In five weeks in 1757, with the help of a French artist and two draftsmen, Adams completed a set of measured—and somewhat romanticized—drawings that have become part of history. He made numerous references to Vitruvius, whose principles of design he recognized as being well applied in the design of the palace. In his report accompanying the famous drawings and addressed to his royal patron, the king, Adams clearly hoped to encourage him to act as Diocletian had done in support of inspired urban design:

To the King, I beg leave today before your majesty, the Ruins of Spalato, once the favorite residence of a great Emperor who, by his munificence and example revived the study of architecture and excited the Masters of the Art to emulate the elegance and purity of a better Age. . . . Your majesty's early application to the study of the Art, the extensive knowledge you have acquired of its principles, encourages every lover of his Protection . . . not only a powerful patron, but a skillful judge.

Adams signed his report "Your Majesty's Most Dutiful Servant and Faithful Subject."

Shortly after Adams, in 1774 Alberto Fortis provided a reliable account of his Dalmatian tour in *Viaggo in Dalmazia,* a work still valuable and republished in 1971. L. F. Cassas and J. Lavallee, from France, made drawings of the palace in 1802 that showed little change since those of Adams. Between 1806 and 1813, Marmont wrote of the beauties of the coast and its historic urban places.

From January to July 1847, A. A. Paton, a member of the Royal Geographic Society of London, traveled by horse-drawn coach in the Balkans to compare conditions with those at the time of his previous visit, in 1839. His broad knowledge of cultures, history, and languages make his observations vital and realistic. For example, crossing from Zagreb to Zadar (on the road built by the Austrians to connect the coast to Vienna), readily identifiable, the Dalmatian coast first unfolded to him at dawn at the summit of the Velebit Mountains: "Dalmatia in all her peculiarity, laying stretched out before me—like seeing from a tower, Zara (Zadar), plain and mountain, city and sea" (vol. 2, ch. 2). Paton saw the geography of Dalmatia as the source of its social uniqueness in the Balkans: "From numerous land-locked anchorages, from productive fisheries and milder climate, from reciprocal wants of Highlander and Islander has risen skill for maritime enterprises, which makes Dalmatians perhaps the best sailors in the Mediterranean" (vol. 2, ch. 4).

Paton noted an increase in the population from 343,000 in 1843 to 428,000 in 1847. Traveling from city to city, he portrayed well the human life of Dalmatia and its varying local identities at that time. For example, he described a carriage companion as "a man of tall stature, boldly chiseled features, sunburned complexion, independent bearing . . . a true Dalmatian . . . intelligent and communicative." His companion told him: "Dalmatia, sir, has the best air and water, but is rather deficient in corn and vegetables" (vol. 2, ch. 3).

The Nineteenth Century

Whereas Paton spoke for the natural environment, Sir J. Gardiner Wilkinson in 1848 turned more toward the quality of urban places. He found Dalmatia, broadly identified by the name Illyrian Kingdom, instituted forty years earlier by Napoleon, divided by Austria into three *circoli* (departments): Zadar, Split, and Dubrovnik. These three, together with Kotor, held some 403,000 people in 1844. Wilkinson, however, confirmed Austria's failure to improve urban conditions and was impressed with the strong imprint of Venetian influence in Dalmatia's buildings, towers, and streets, "narrow and badly paved, though cleaner than many in Italy and the people civil and courteous" (ch. 3). Wilkinson noted that the French had built the military road from the Neretva River to Split and thence to Trogir, which had been extended by the Austrians to Zadar for access to Vienna. He recorded the capacity of a Turkish caravan terminal near Split as 621 horses and 272 people, and he detailed the variety of goods.

Like other early writers on Dalmatia, he made little mention of Pelješac, so inaccessible in those times by sea and even more so by land. In Dubrovnik Wilkinson found people in a depressed condition, "the streets deserted and grass between the paving stones, but no beggars" (ch. 5). Seeing the palatial villas at Ombla, he sensed a marked division between the upper classes and the lower. To visit a caravan terminal three miles south of Ragusa, he had to be escorted by an armed guard. The city had no carriages or truck horses; goods had to be carried by porters, and even a few sedan chairs for nobility were to be seen.

E. A. Freeman, of Trinity College, Oxford, visited Dalmatia in 1875, 1877, and 1881, and his records of those visits provide a sense of the changes that took place. However, the travels over several years covered in three volumes of the British architect Thomas G. Jackson give us the most complete account of the historical background of Dalmatia as of 1887, especially of urban places. Written at the time when European travel was being facilitated by ship and train, the work brought widespread attention to Dalmatia as an exotic place to visit. Differing from other works by travelers, Jackson's writing includes more history than firsthand observation and shows how the regional pattern of concentrated settlements we know today was established. Jackson pointed out that "the history of Dalmatia is in the history of its towns" and recognized how the Roman institutions of local government carried over into the Middle Ages.

Jackson also noted how the tribal nature of village formation, a cultural resource that came with the Slavic culture, has allowed villages to maintain their collective identity and independence in the face of a history of external authorities who wished to dominate the strategically located region. Each place considered its ancient charter to stand as evidence of its territorial right not to be interfered with. Jackson makes clear Dalmatia's dual cultural origins — Latin and Slavic — and dismisses "those who believe the Latin fringe to be derived from Venice. . . . Nothing could be further from the truth . . . for Zara, Spalato, Trogir and Ragusa were Latin cities when Venice was non-existent."

The writings of the next visitor, in 1894, have relevance today. Robert Munro, secretary of the Society of Antiquarians of Scotland, entitled his work *Rambles and Studies in Bosnia-Herzegovina, with an account of the Congress of Archaeologists and Anthropologists held in Sarajevo in August 1894.* His purpose was to popularize the results of the congress and to publish his own "sketch" of his visit to Bosnia. His experiences ranged from exploring the numerous prehistoric sites of the region and traveling the "Roman Road" from the Adriatic along the Neretva River to seeing towns of Ottoman character that today are in ruins. Munro could never have imagined the tragic events that would take place just one century later.

The Twentieth Century

In 1908 another Jackson — Hamilton — wrote with greater emphasis than had Thomas Jackson on the picturesque people with their colorful costumes, the scenery, and antiquities but little on the life of the people in their urban settings. Rather, he praised the earlier Jackson's work and echoed his defense of maintaining the Italian culture that the Croatian renaissance movement was eroding. The freshness of discovery in the writings of Adams, Paton, and Wilkinson was absent, and Hamilton Jackson's work became a prototype for others to follow into the 1920s.

These later works, devoid of social content and framed in oversimplified history, were the forerunners of the typical "travel books" at times financed by steamship lines to promote tourism. A number of these books — Holbach's (1910), Moque's (1914), and Brown's (1925) — occupied my family's bookcase. The colorful Dalmatia they portrayed had little to do with the images gained through communication between California and Dalmatia via The Bridge. I experienced firsthand the difference between the Dalmatia portrayed in these books and the reality of family and village life on Brač and Pelješac in 1937. This awareness stimulated me to understand more fully the nature of the people of Dalmatia and to gain a genuine sense of their identity with the environment.

Rebecca West's two-volume opus of 1941 on the former Yugoslavia contained a major section on Dalmatia and thus was the first work about Dalmatia to both provide comprehensive coverage and have an engaging literary quality. History, art, architecture, social life, personalities, culture, and ethnic differences, landscape and sea coast, combine to serve as a setting for the wide array of people with whom she interacted. Her personal responses to them and to the places visited are useful as we trace our own paths and responses to Dalmatia's cities, towns, and villages.

THE QUEST FOR URBAN BETTERMENT CONTINUES

The Early Decades of the Revolution

The nineteenth century stands forth in Dalmatia's urban history as the turning point in which the people gained the right to establish their own identity, free of foreign domination, in adapting the environment to human needs. In 1937, crossing The Bridge from California, where we took our urban advances of the 1930s for granted, allowed me to discover firsthand how Dalmatia's cities, towns, and villages had been kept from moving into the modern era. The damaging effects of the push and pull of history and of World War I were still seen and experienced daily, and the threat of World War II hung heavy in the air. Within two years, Hitler's invasion of inland Croatia and Mussolini's attempts to annex Dalmatia—which blocked my return there—reinforced the old pattern of obstacles to independent identity and set off the explosive Marxist revolution in all of former Yugoslavia.

After communism was firmly established, many of these issues were boldly tackled as modernization was launched in the 1950s; by the 1960s a marked degree of local self-determination had been achieved among the *općinas* (municipalities). The new Yugoslavia's break with the Soviet Union over highly centralized authority brought a style of socialism based on the opposite belief in decentralized responsibilities, local initiatives, and self-management.

The response to the thirst for local identity generated large amounts of collective energy and human resources for urban and regional development. For example, even during the wartime 1940s, seminars were being held by Croatians concerned with the loss of cultural identity as a result of damage done. Professionals in architecture, history of art and culture, and urban planning promptly established Croatia's first Town Planning Institute, with a particularly active branch in Dalmatia. Although the federal government adopted broad policies for the protection of cultural monuments that occupied a critical place in the nation's identity, the responsibility for the implementation of programming and financing was local. Out of this initiative grew the revenue-producing reuse of Split's Palace of Diocletian, reconstruction in Zadar, and the preparation of master plans for new growth in the 1950s.

On my study trips between 1968 and 1972 I could share the optimism generated by the U.N. assistance with the planning and environmental reforms undertaken by the Dalmatian leaders in order to achieve historic preservation. It became apparent that in the early 1960s the decentralized economic system had stimulated initiatives for development of local resources. For example, on a visit to Vis, I met an agronomist who had come there to select palm trees for landscape improvements at the Split airport. His father, a wartime pilot, had been one of the founders of the Split airport and a leader in a grassroots movement by aviation followers. In the absence of a central agency for planning and building, they sought funding on their own from the municipality and local banks and ultimately built, managed, and owned the new airport, even defending it against potential invasion at the time of the Soviet march into Czechoslovakia in 1968.

On the other hand, the backward smaller places on the islands, lacking social facilities and economic resources, lost a younger population to the mainland cities. Even as tourism shifted from access by ship to cars and buses on the Adriatic Highway, urban scale increased. Yet, goals based on social needs were not incorporated into the U.N. urban planning of the 1960s. An emphasis on the infrastructure needed to support tourism rather than on the infrastructure needed to support residents eventually resulted in an imbalance in the entire urban system. The lack of secondary schools, community clinics, communication technology and facilities, and water systems encouraged further movement from traditional settlements on the islands — from Brač to Split, for example, or from Pelješac to Dubrovnik. This and employment in the summer tourist season tended to break important and socially productive ties to the heritage of local identity and social cohesiveness.

By the 1970s, vacationers from northern Europe and the rising middle class from the hinterlands of Yugoslavia had discovered the sun and clear water of the Adriatic. Airports increased accessibility, and hotel construction began. The economic vitality in the early decades provided promise of income for all of the former Yugoslavia and support for the Communist form of government. Originally, tourist facilities were in keeping with the scale and cultural character of the existing environment, and policies of respect for the historical monuments and the natural environment were followed. It appeared that tourism on a modest scale, commensurate with the natural environment and focusing on the cultural assets of Dalmatia, would both enhance the indigenous identity that Dalmatia's urban dwellers hold for their places and offer a unique, noncommercial experience for visitors.

From the 1970s to the 1990s

The promise of integrating tourism into the pristine landscape was not to be. Rather, Dalmatia's unique urban settlements became enmeshed in the intricacies of international travel and the world economy. The region's traditional identity was invaded by the same forces that are eroding the individuality and rootedness to local cultures that provided places with human meaning in other parts of the modernized world.[25] Through an increasing centralization of management and economic resources, the quality of urban development deteriorated.

By the 1970s, incongruities had begun to appear in the native quality of the cities, towns, villages, and still pristine natural environments. The political monopoly saw the growing tourist industry as a ready-made economic base for socialist reform and moved priorities toward large-scale complexes. These were oriented more to modern resort life than to the rich cultural past and the fragile environmental heritage, which earlier official policies had required. The "workers' " collective enterprises for hotels—of modest scale at first—took on a corporate, "big business" character. The funds that became readily available for expanded hotels and restaurants, access roads, and water supply might better have gone to bringing modern services to smaller towns and villages seeking stability. Furthermore, competing socialist building enterprises and *općinas* themselves increased pressure for new lucrative construction, as chambers of commerce do in the United States.

By the 1980s the golden egg of tourism rested in the hands of a highly centralized political authority. As a result, the potential of the early days of conservatism and environmental policymaking, reflecting deeply rooted values, was usurped by the political aims of the Communist Party. This had the effect of weakening the institutes dedicated to the protection and preservation of cultural monuments from incompatible new construction. In other cases, long-term plans for new urban extensions balancing community facilities with participation of the citizenry were revised in favor of housing alone and at higher densities.

Several examples from my 1990 visit, expanded on in chapter 3, demonstrate this loss of responsibility to the environment. A major controversy took place over the building of a new highway that cut through the ruins of Salonae, the original Roman town nearby Split, an area of great archaeological, cultural, and tourist interest. Traveling by bus from Zadar, as I neared Split, I saw that demolition and waste from the highway construction had replaced my customary view over the Salonae ruins to the ancient Roman aqueduct that continues to supply Split with water. Beyond, the relentless three-mile Manhattan-like wall of Split's skyscrapers, built since my previous visits in the early 1980s, underscored the politically derived rupture with the past.

The "New Skyline," as I have named it, housing some 160,000 people, was allowed to take on a life of its own. The authorities had not learned from the planning principles of high professional quality used in the earlier "model" district of 10,000 called Split III, first built in the 1970s, nor had they followed the well-considered master plan for Split as a whole, prepared in the 1960s with the assistance of the United Nations. Here we have an example of how a supposedly noncapitalist system, with access to political control of investment capital, could produce an environment that makes almost a fantasy of modernization. By contrast, the human appeal, diversity, and flexibility of the highly popular old urban core make traditional Split the epitome of what a humanized form of modernization, with all of its technological resources, should strive for. Split was responding to a longstanding need for modern housing for permanent residents, augmented by the newcomers from the nearby islands, who were linked by ethnic and family origins and sought schools, health services, and employment. In the 1970s an unpredicted influx of people of a nonmaritime culture decided to trade their life in the backward mountain villages of the adjacent inland regions for the new dynamics of city life. Political leaders of a failing economic system gained credibility by supporting the maximum housing construction, even without the greatly needed social infrastructure. "The more housing that was built, the more people came," was said too frequently.[26]

In addition to the issue of the traditional versus the newly built environment, Dalmatia's natural environment has been put at higher risk. The limited capacity of its agricultural resources to sustain such population growth has become an established fact. The development of tourism, industry, settlements, and the accompanying infrastructure reached threatening levels just prior to the war. Most types of environmental degradation occurred in areas that enjoyed the greatest growth in the 1970s and 1980s, such as Split, Zadar, Šibenik, Makarska, and Dubrovnik. Environmental impacts came from air pollution due to automobile traffic and industrial plants, pollution of the soil due to poorly treated effluents, which lowered the quality of the sea and resulted, along with overexploitation, in a subsequent loss of marine life.

Other forms of environmental degradation included erosion of fertile soil and loss of native vegetation along the increasingly commercialized Adriatic Highway. This resulted from residential and tourism-related development. Perhaps the worst impact on open land was from forest fires. Thousands of square kilometers of woodland were burned, and the bare landscape remained for years to slow recovery processes. These visible types of damage to both the natural and the cultural landscape and the consequent loss of ecological, self-sustaining integrity could ultimately have serious economic consequences, because the principal resource base of tourism is slowly eroding away. At the same time, the destruction of the cultural, or human-made, landscape or nonurban areas leads to the eradication of traditional patterns of environmental management.

In 1986 these types of deterioration of the environment were carefully accounted for within a comprehensive range of categories in a study the Yugoslav government requested from the Organization for Economic Cooperation and Development (OECD). The recommendations came too late for implementation, as Tito's Marxist union began to fragment, leading to war. On the other hand, the excellence of the study suggests that it could become a valuable guide for reconstruction within the separate governments emerging with the peace.[27]

It could be said that more damage was done to Dalmatia's environment during these decades of overdevelopment than by the war itself. The findings and recommended policies can give directions to enable postwar and post-Marxist Dalmatia to avoid returning to the upper limits of exploitative development. As Dalmatia approaches a new century, this fragile coastal region calls for a political philosophy of public versus private interests in the context of local participation and ecologically based environmental management.

CHAPTER THREE

Cities of the Mainland

*The urban places of Dalmatia . . . represent a higher quality than
their individual monuments alone, when considered as whole entities
preserved within a complete historical context. . . . They document the
entire life span of a place from the first colonization of our coast to the
present. Their architectural works of the past woven into harmonious
urban patterns, offer us a unique quality of critical importance in
our times.* — Tomislav Marasović, 1958

FROM REGIONAL TO LOCAL IDENTITY

Throughout my youth, Dalmatia stood forth as a single geographical
entity, identified, like California, by its lineal borders of both mountains and sea. Though distant, that coastline became the prime source
of ancestral identity for our family and other immigrants attracted by
environmental similarities. In 1937 I learned that Croatia represented a
higher level of historical and political identity, as a frame for Dalmatia; that region, in turn, included a variety of identity sources arising
from the marked geographical differences between subregions. And each
subregion, whether island or coastal, had a dominant urban center. My
awareness of these various levels of placeness led me to focus my inquiries
not only on cities but also on small towns and coastal villages and to try
to understand them in their own geographical and historical context.

As we undertake our "urban reading" of the mainland cities Zadar,
Split, and Dubrovnik, rather than perceiving history as a chronological
sequence of events, we shall try to grasp these events as simultaneously
spread out before us in a single city, as a visual accumulation of the environmental outcomes of history. In this visual experiencing of place we
shall allow our intuitive responses free rein in interpreting what we see
and feel. We shall watch for the continual interplay of building styles,
open spaces, people's use of them, and materials from different centuries,
letting our own individuality enter fully the process of perception. This
reality offers the fertile opportunity of creating closer bonds between
ourselves as "readers" of a selected place than would the superficiality
afforded by descriptive responses to a given place.

A balance between systematized and intuitive responses can reveal a
city's identifiable form and its nonphysical "spirit," just as individuality

in people grows from their "life" experiences from childhood. In our reading of Dalmatia's three main cities, we will discover how each stands in its own right, just as we might discover individuals among parents, family, and friends through the experiences of growing up. Just as a person's uniqueness and self-identity are often shaped through personal will and recognition of resources at hand, so a city's uniqueness and self-identity are often shaped by deliberate planning and management by city government.

This parallel between people and cities has been used to striking advantage by authors from both sides of the Adriatic. In Ivo Andrić's *Bridge on the Drina,* it is almost as if the Bosnian city of Višegrad is telling its own story of five centuries of growth, from before the Turkish occupation in 1463 to World War I. More recently, Ismail Kadare, in his *Chronicle in Stone,* described a city in his native Albania as a living organism, a place whose particular physical qualities shape the lives of characters in the story, who are deeply intertwined with the city as a physical place. Similarly, in the novel *Illyrian Spring,* by Ann Bridge, the intrinsic qualities of Split and Dubrovnik as distinctive places before World War II become the integrating fabric for the characters, each in search of his or her own identity. Another example is Ivo Vojnovi'c's *Dubrovačka trilogija* (The Dubrovnik trilogy), a classic of nineteenth-century Croatia that takes place in and typifies the spirit of Dubrovnik. From Italy we have Italo Calvino's *Invisible Cities,* in which "Marco Polo" describes distinctively different city forms, imagined with deliberate inventiveness from his expeditions; these tales demonstrate how human yearnings for individuality in today's cities are suppressed by the weight of technological complexities.[1]

Some essential differences in identity emerged from my own reading of the three principal Dalmatian cities. Zadar's central street, the Kalelarga (in Croatian, *široka ulica,* or "wide street"), was designed by the Romans for pedestrians and their social interaction and has served that purpose ever since. Turning points in history came and went, and new generations of human beings living in changing cultures and using different languages, were choreographed into the same movements, determined by this narrow route and its fronting shops, churches, and the like. Even today residents of the suburbs come by bus to this traditional heart of the city for the intimate contact the restricted scale fosters. The Kalelarga is Zadar's strongest identifying element and the main focus of its community experience.

Whereas Zadar stands on a site surrounded on three sides by water, Split grew within the formally structured summer palace of Emperor Diocletian, which was not intended to become an urban center. Thus, the city expanded gradually, with no planning, outside the walls during the quiet centuries of limited urban growth under Byzantine rule.

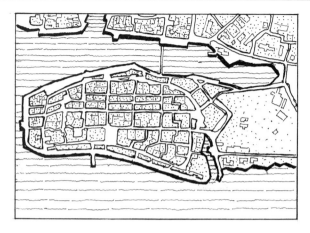

Distinctive urban forms: The basic source of identity with place. Zadar. Rome's ancient grid plan predominates in today's identity (top), Split. A formal Roman villa contrasts with the incremental growth of the Middle Ages (middle), Dubrovnik. Nature's bold forms combine with the rational urban geography (bottom)

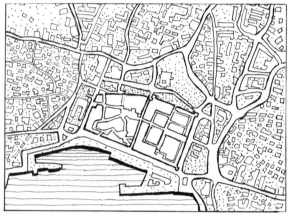

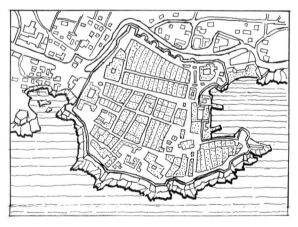

In time the Venetian influence prevailed, giving the western addition a strongly contrasting character. Its informal main piazza serves a similar function to that of Zadar's Kalelarga. Split is permeated with the juxta-positions of styles over a long sequence of centuries. In Dubrovnik, the counterpart to these examples is its Main Street, the Stradun, "street" in medieval Latin, the official language in Dalmatian cities in the Middle

Ages. For centuries Dubrovnik's growth has been determined by its topography, bringing far larger numbers of people to live within or close to the old city in densities higher than in Zadar or Split. Some of these differences between Split and Dubrovnik in the period before World War II have been described vividly by Rebecca West.[2]

We can search for these kinds of uniqueness of urban form for each city, first by tracing briefly the turning points that shape each city, then by walking through the urban structural system, linking the built environment's form with the daily human experience. Our aim is to replicate the firsthand experience of reading a city. Indeed, all three cities were built to be experienced by walking, standing, and sitting in public and generally accessed on foot from docks. Taking this as a phenomenon in itself, so completely different from the present long-distance travel by car, bus, or transit, we can, in our readings, bridge the gap in centuries.

ZADAR: CITY OF DIVERSE CULTURAL ORIGINS

Urban Identity through Cultural Change

From the Earliest Settlement to the Coming of the Croatian People

Zadar has a varied and balanced geography, something rare on the rugged Dalmatian coast. This range of landforms has allowed the city to maintain a certain spatial integrity in spite of the many cultural changes over time. Out of this wholeness of place has grown the identity we experience today. To the west, several tiers of parallel, attenuated islands, some with their own settlements and many more mere fragments of rock, protect the city's frontage and channel from the open sea. Back of the city, the terrain rolls gently inland, creating a well-defined and potentially self-contained subregion of the upper end of Dalmatia.

The prehistory of Zadar reveals no permanent, constructed settlement on this small, flat peninsula. Rather, the population of the region, although considerable in number, lived in small, scattered settlements on nearby mountain sites, chosen for the protection and access to fresh water, game, and fish they provided. Artifacts from the ninth century A.D. on indicate that the Liburnians, one of the many Illyrian tribes, first occupied the city's site. In the fourth century B.C. the Greeks settled in the area, though only for the purpose of trading, as evidenced by coins and pottery used for shipments of goods. Zadar's name was first recorded by them as Idassa, and a half-millennium later the Romans gave the name Jader to a newly planned town of permanent regional importance.

After being held at bay for centuries by the well-established Illyrian tribes, Rome succeeded in building six cities from Istria to Montenegro. Zadar, established as a Roman *municipium* in 59 B.C., was the most

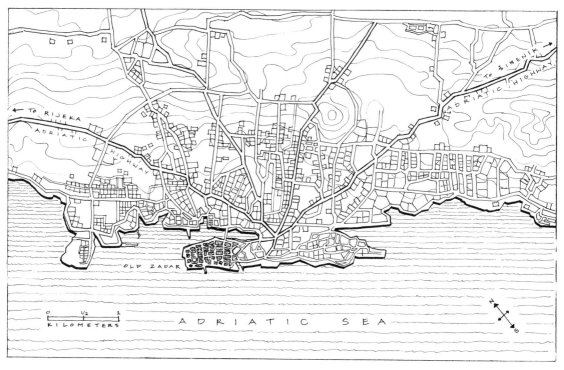

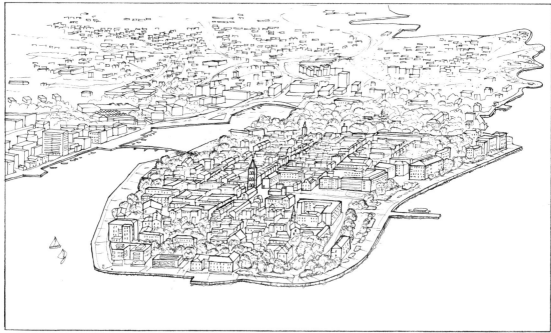

Old Zadar: Focal
point of a growing
urban region

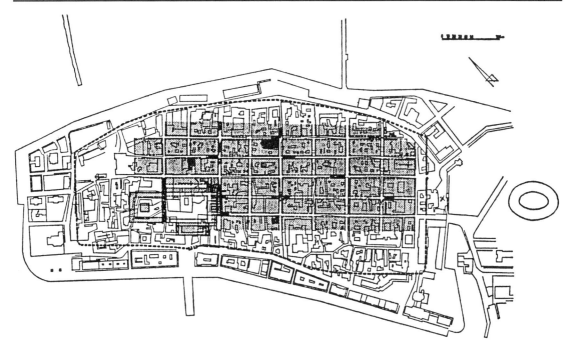

Zadar: Faithful to its Roman planning. From *The Ancient City on the Eastern Adriatic Coast. By M. Suić*

completely built. The Romans created a system of streets, defenses, and access points so well suited to the site and its surroundings that for two thousand years changing populations used much of it, especially the grid pattern, as the basis for their own land uses. Even though medieval Zadar built over the old Forum, that area remained a focal point of activities until the air raids of World War II revealed it to be one of the more elaborate of those built by Rome.[3] Zadar's central position opposite Italy's port city of Ancona, from which it was only some 150 miles inland to Rome via a pass through the Apennines, made it a strategic base for Rome's imperial operations to the east. This centrality attests to the soundness of Roman urban planning and Zadar's ready adaptation to the uses of later occupants.

When the nomadic Croatian people arrived in the region of Zadar in the seventh and eighth centuries, they found ample land on which to settle. The city itself remained in use by the native-born Roman Dalmatians, though in the absence of Roman rule and vitality the city gradually fell into decay. As time passed, the newcomers became the permanent settlers, learning from the Roman plan laid out for urban life and coming to identify with it. Although there were major ethnic, cultural, and political changes throughout the ninth and tenth centuries, the physical city maintained its original Roman purposes and became the core of the Croatian state, with its capital at nearby Nin. In addition, the Croatians learned from a close relationship with the Roman Christian Church.

From 1160 to 1414, Zadar, a city-state in its own right, lived under a sequence of rulers—the well-established Venetians and the Hungarian kings—transforming itself from a Roman city to an early medieval city. Through this contact with the West, the Croatian culture evolved and spread to the surrounding region. This period brought new architectural forms adapted from the Roman precedents. For example, in our urban reading we can see how the monumental church of Sveti Donat was built in the Forum in the ninth century of stones that had been cut by the Romans in the second, then left as rubble through the Byzantine centuries. Laid by the Croatian people, who were of rural rather than urban background, they were inventively adapted to a boldly circular form securely buttressed by semicircular apses, creating a new cultural identity that blended Byzantine and Frankish influences.

Nada Klaić and Ivo Petricioli's history of Zadar stresses the long evolution from Illyrian times through the Middle Ages.[4] It accurately describes how the city's multicultural origins were welded to each other by the forces of history. Uniting both time and space, the work is more comprehensive than those on other Dalmatian cities since it presents a balance between the three nonphysical components of a city—a sociocultural structure, an economic system, and an institutional-political dynamic—all set in and dependent on a spatial framework of land and buildings. In this sense the work gives us an image that we can understand based on our perspective of cities today. The special coverage of Zadar's leading role in the Byzantine period suggests that the city regained the prominence and regional responsibility it had in Roman times, its original grid of streets and open spaces giving a common denominator for all periods up to the present.

Venice, Austria, Italy, and Croatia

Thomas Jackson points out that in 1202 Venice gained a lasting foothold in Zadar, five days distant, by taking advantage of the Fourth Crusade to occupy Zadar by force with its fleet of fifty galleys and tens of thousands of knights and troops.[5] By 1409, Venice, desperate to have Zadar as a strategic point for its overseas trade, bought the city from Ladilas of Naples, pretender to the Croatian-Hungarian kings. Its first act was to restrict all shipping activities, sending its potentially strong economy into a decline and causing it to greatly expand its fortifications over the centuries. It also marked the beginning of a four-hundred-year-long transformation of the architectural qualities of structures built by the Dalmatian people themselves. Although Zadar's economy suffered under the Turkish threat, its dwellings increased in number and density, and churches and public facilities were built blending Zadar's own traditional forms with innovations in style from the larger Adriatic region. This influence brought campaniles, intimate piazzas, Renaissance

Zadar's massive
fortifications, 1695.
Croatian State Archive

churches, and some neighborhood streets meandering over the Roman grid. This gracious architecture and urban form, densely confined to its geometric site, became the basis for for the distinctive identity of Zadar.

In 1797 the Austrian army entered the city and Venetian soldiers became soldiers of Austria; in 1805 Zadar came under the French. The citizens of Zadar, who generally belonged to the urban upper class, were opposed to the French democratic reforms and suppressed all alliances with Croatia that encouraged French anti-aristocratic ideas. Napoleon's administration built the first major access roads to Zadar. With Zadar virtually turned into a military garrison, the short period of French rule was socially stimulating for the deprived classes, who were eager for equitable reforms. In 1813 the British took control of the sea, and the Austrians, after bombarding by land, marched into Zadar and took full charge.

A. A. Paton gives a sense of the diverse cultural life of Zadar in 1847, when he spent six weeks there, in his *Researches on the Danube and the Adriatic*. Some six thousand people lived within Venice's sixteenth-century walls, and twenty thousand lived in the surrounding rural region. His descriptions of the very urban activities seen from the promenade atop the walls that still ringed the town are insightful. He saw "Austrian officers reading German newspapers . . . native nobles reading French literary reviews . . . people playing dominoes." He attended carnivals and plays and noted the prevalence of "sailors and porters, broad shouldered with brawny legs and sun-burned faces." Visiting the islands across from Zadar, he observed how different the people there were from those of the mainland: "All fields are fenced, venerable trees at pleasant spots cast welcome shade to invite repose. The Islander is provident from an hereditary consciousness that all he saves he can keep"[6]

Austria's highly organized bureaucracy focused on modernizing Dalmatia's urban places in the image of Vienna. This sharp break with the Venetian influence left little room for expression of Croatian culture, especially since Zadar was made the administrative center for all of Dalmatia as a semi-autonomous province of Austrian in 1861. Public and private buildings on land gained by building new sea walls and landfills gave Zadar a monumental nineteenth-century image that did not match its medieval past: it seemed to be more a part of the Adriatic than of the Balkans.

In 1922, after the fall of the Austro-Hungarian Empire, Zadar was separated from Dalmatia and given to Italy as a reward for its cooperation in World War I. Horatio Brown describes the Italian architectural character that dominated the internal districts of "Zara" in spite of the century of Austrian rule. From his descriptions of his visits there in 1883, 1910, and 1924, we can see the reality of drastic environmental changes that are taken for granted today: the automobile first making travel possible by land; urban dwellers of Zadar distinguishable in town from rural folk by their particular regional form of dress; an Adriatic Highway to bring Dalmatian cities into direct contact with each other; and airplanes to bring tourists. Each change threatened the accumulation of historical identities.[7]

This "Italian" status of Zadar fitted in well with Mussolini's ambitions in the 1930s for territorial expansion around the Mediterranean, inspired by Roman imperialism. Heavy bombing of the city's medieval quarters of Venetian character by the British in 1943 marked the end of Zadar's Italian cultural foothold. By 1944 the Partisan resistance had occupied Zadar and put the city, half-destroyed for the first time since the early Middle Ages, under Croatian self-government, and Croatian became the common language. New populations took the place of those who had left during the war, and Zadar made a fresh start. Its Croatian people were now able to create a physical environment that gave fuller expression to their own cultural identity. Energy and funds to rebuild the bombed areas seemed to come out of nowhere. Rather than replicate

Old Zadar, the progressive young architects of Zadar took the opportunity to follow the "modern architecture" movement that had swept postwar Europe.

The architectural historians, however, remained faithful to the multicultural past, making sure to leave exposed the Roman remains, even in their ruined condition, and to restore fully St. Donat's Church. This principal example of the Croats' version of the early Romanesque style now stood as undeniable evidence of their claim on Zadar back to the very early Middle Ages. Related artifacts were gathered into the new Museum of Archaeology. The urban grid system of Roman streets was preserved, and the historical monuments in the Forum area were restored to deliver a strong cultural message from the past and thus fortify the present population's identity with Zadar. In their daily lives, residents and visitors can clearly experience in a single place the entire evolution of Zadar's Croatian identity. This heritage came to bolster the sense of belonging for Zadar's citizens as the vengeful shelling that began in 1991 continued to threaten these monuments even into 1995.

An Urban Reading of Zadar

Entry at the Square of the Five Wells

The flat, shiplike simplicity of Zadar invites you (as reader of the book) and me (as author) to begin our urban reading together, but the question is, where? Unlike a book, a city has no "page one." Thus, we need to look for quick clues to how the street system is organized. After a little reconnaissance, we discover intuitively from the partly "legible" grid pattern that the Kalelarga could be the city's "binding." You may sense a certain "pull" in the social vitality of the morning shoppers, giving this *cardus,* the north-south axis of Roman city planning, a commanding presence over other streets. Furthermore, we note the Kalelarga's starting point at the spacious Trg Pet Bunara (Square of the Five Wells), occupying a break in the medieval walls we stand before.[8]

With a fresh sense of discovery and orientation, we face westward down the central axis of the island from the point where Zadar attaches to the mainland. Here we seem to have found the city's "front door" and the right place to begin our reading. In the pleasantly irregular open space of the Trg Pet Bunara, now pause and gaze around at the surrounding elements that appear separated by centuries in a space that brings together not only people bound for the Kalelarga but also a record of the city's history. The scattered, disorganized quality of the place gives it an identity all its own, a "lobby" with its strong sense of entry, an "overture" to our reading of Zadar as a whole. Note the flight of five very wide steps connecting the Trg Pet Bunara to the lower Trg

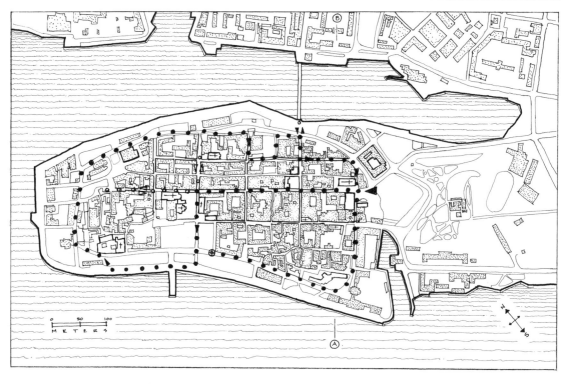

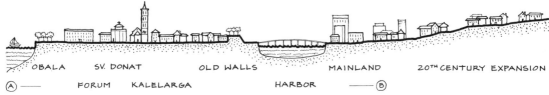

OBALA   SV. DONAT   OLD WALLS   MAINLAND   20ᵀᴴ CENTURY EXPANSION

(A) ——   FORUM   KALELARGA   HARBOR   —— (B)

Zadar: Urban structure
and reading paths

Oslobodjenja. These steps provide a sitting place for young people and
separation from the parked cars that have invaded this world traditionally belonging to the pedestrian.

See how the upper *trg,* connected to the old Roman sea gate by a
constricted, tunneled walkway, is closed in on three sides. Behind us the
wall of the medieval fortifications slants upward some thirty feet; opposite, a lower wall and its hexagonal Bablja Kula (Old Woman's Tower)
from the eleventh century rise. And to our left, joining the two walls,
is a two-story building labeled as the headquarters for the Zadar Communist Party, which by 1990 had become one more remnant of the past
and a symbol of the latest turning point in Zadar's history. Little could
we imagine that the tower itself would be shelled within another year.
Now a feature of the upper *trg* catches our eye: the line of the five stone
wellheads that rise from the stone paving, complete with handsomely
carved basins and iron framework. They were the city's source of fresh
water, Rome's gift to all its cities—an aqueduct used up to recent times.

Stepping down to the lower *trg,* we feel as if we are on a theater stage;

Square of the Five Wells

we become background players, resting on the broad steps just as performers do on those warm summer evenings. The overlapping buildings of varying pastel colors spaced out from the Middle Ages irregularly enclose the far side. In contrast, a second, smaller square displaying Renaissance symmetry asserts itself as the anteroom to the corridor of the Kalelarga. Note on the left a nineteenth-century building, perhaps an old hotel, backed by another of weathered stone. A massive structure of a rich yellow-ochre stucco from which Austrians ruled Dalmatia juts forth to form the west end of the square and announces the beginning of the Kalelarga. A monumental balcony overlooks the architectural open space, as if at any moment a resident governor sent by Franz Josef would step forth to greet us, or to issue an edict to the populace below.

To our right, a long stucco building of a deep rust red provides a backdrop for a minute sunken park that in turn protects ancient ruins of the Roman entrance to Zadar, some exposed at the original Roman grade of the old city. A freestanding Corinthian column left from Roman times rises at the intersection from the cross street, moved there to mark the beginning of the Kalelarga, further enhances the sense of theater. The green of trees dapples moving shadows on paving and benches, and a tiny cafe suggests that we deepen our experience of place and linger over strong coffee. The long facade of Sveti Šimun, handsomely restored, and yet another ancient freestanding column, together spanning the entire history of Zadar, are there for us to contemplate. Absorbing the gemlike quality of this architectural assembly, we intuitively recognize a frame for the long, narrow Kalelarga ahead, waiting to direct us to the Forum, the powerful symbol of Rome.

This quality of chance in the placement of these bits and pieces calls upon us to examine a sequence of small, irregular spaces in which all cannot be seen at once. Yet, focusing on each part in terms on both time and space simultaneously and assembling them into a coherent mental image, we can experience visual delight and grasp our own foothold on identity with Zadar.

The Kalelarga

After the Trg Pet Bunara, with its open, unstructured quality, we experience a feeling of confinement in the narrow corridor of the Kalelarga. The rigidly parallel and converging lines of cornices and eaves compel us to shift from the easy looseness of the *trg* into the straight-line conformity imposed by the Roman planners. This realization impels us to move along, knowing that further down its length the majestic shaft of the Campanile awaits us.

Between us and the Campanile lies a grid pattern of long blocks parallel to the Kalelarga; these are the *insulae* of Roman town planning, clearly visible on maps of Zadar in those times. After the bombing of World War II, Zadar's planners widened three of the cross streets to form "superblocks" in the contemporary sense of Le Corbusier's site planning. At first glance the consistency of shops on the ground floor and residences above seems monotonous. But when we focus closely on the variety of shops, the remnants of the past alongside contemporary design, the rich detail, the breaks in facades, and the small courtyards, the Kalelarga begins to read as a place with its own identity. When I visited in May 1990, Zadar's old Roman "Main Street" was alive with human activity, filled with the new vitality, enterprise, and political posters and banners. We now experience the new turning point, the holding of free elections in Croatia and the eroding of communism. We sample in visual sequence the new shops, which verify dramatically the resulting change in the environment since my first reading, in 1981: a smartly designed cafe and bar with outdoor terrace, a boutique, a travel agency, shops selling videos and records, flowers, shoes, women's clothing, and books. These are run largely by self-assured young people who previously had no such outlets for enterprise. At the distant end of this first "superblock" we see the Campanile rising into the blue sky, marking the spacious opening to the Roman Forum. A lacelike scaffolding speaks for the eagerness of the citizens' restoration movement to maintain Zadar's rich and compact accumulation of history.

Here the Kalelarga opens up on our right to display the splendidly civic Narodni Trg (People's Square), which has served as the equivalent of Rome's Forum from the Middle Ages to the present. On the corner, the Loggia has been handsomely restored, with a glass front facing the Clock Tower, both from the sixteenth century. Together the two build-

The Kalelarga at
Narodni Trg

ings anchor the square to the Kalelarga and frame a third at the far end
housing the Općina. Here our experience abruptly changes. After moving in a predetermined straight line along with a mass of humanity, we
stand and gather with others who hold forth daily as a ritual, conversing
on the state of affairs in Zadar. In the presence of the autonomous government for which the Croatian people fought so long, we can feel ourselves a part of the throbbing pulse of this city at its new turning point. I
sensed this in 1981, when I entered the Općina for an interview with the
then mayor of Zadar. His liberal views toward local autonomy through
a democratic form of socialism and a more humanistic future fore-
shadowed changes taking place in 1990. The position of the Općina ties
Narodni Trg to the surrounding metropolitan area. To the left, the town's
principal cross street stretches toward the mainland and pierces the giant
berm, once part of the fortifications, with a tunnel and crosses the harbor
on the only pedestrian bridge connecting Zadar to its sprawling suburbs.

We return and move along the Kalelarga's second "superblock." Down
one of its surviving culs-de-sac, housewives have hung their washing
in perfect order from wall to wall on each of five stories. In contrast,
we find a variety of new enterprises—shops for clocks, shoes, lingerie,

The Kalelarga at the
Campanile

beauty products, yardage, dishes, and glassware—all occupying the first
story of older buildings. Overhead, the pattern of lined-up cornices and
eaves tells us we have arrived at New Zadar, reconstructed after World
War II, when airborne bombings demolished some two-thirds of the
city. Had we done our reading in the 1930s, we would have seen Italianate Victorian facades of hotels from the nineteenth century up to
four stories high facing stucco apartment buildings with shuttered windows allowing a narrow view of the Campanile. Their rich architectural
variety, texture, and once fashionable period styles would contrast with
the austerity of the now widened Kalelarga. The sterility of its uniform,
five-story facades reflects the influence of Le Corbusier, who inspired
the new generation of architects in Yugoslavia after World War II.

Moving on to the third "superblock," we see that the entire frontage is even more uniform, with the stark surfaces and severe lines of the
postwar modern architecture movement. Here we find shops the size
of small department stores: shops for clothing, yardage, and rugs and
a supermarket, all with large glass fronts for window shopping during
the evening *korzo* (stroll), when the Kalelarga reaches its peak capacity.
Arcades of plain square columns on both sides of the street mean that
we must view the merchandise at close range. We lose the sense of con-

The Forum and Sveti Donat

tinuity and intimacy of experience we enjoyed throughout the length of the Kalelarga. The new width, however, broadens the framing of the Campanile and prepares us for the exhilarating experience of passing from the Kalelarga into the vast open space of the Forum.

Now, we must turn our minds to the coming of the Romans, two thousand years ago, and their permanent successors, the Croatians, and then to the influence of the Venetians. Superimposed on the open space of the ancient Forum rise before us the Byzantine-inspired circular mass of Sveti Donat, the nearby cathedral of Sveta Stosija (Saint Anastasia), and the Campanile. See how they stand there with an air of dignity, as if grateful for their faithful restoration from the bombings of World War II, which destroyed the imposing Bishop's Palace and other buildings, laying bare the Roman Forum.

As if released from the confines of a ship's corridor, we respond to the Forum's openness to the sky and to the sea at the far end. Only two blocks away, along the *decumanus,* the Roman east-west axis, lies the open sea. You can sense a vastness that contrasts with the intensity of Zadar's core. The Forum itself, with the fluted drums of Roman columns lying about speechless, has epic stories to tell us, for we are standing at the ancient heart of Zadar. Here all of the turning points in Zadar's history appear to us in rapid sequence, and we feel that Zadar's identity is within our grasp.

Now the Kalelarga pulls us north past the Forum to its remaining two blocks. Here we do not find the strong commercial character of earlier blocks, though one does see, tucked in below residences, numerous tiny bars, which are more gathering places for the gregarious young people than places for heavy drinking. In fact, the predominant quality of this area at the end of Zadar is sacred rather than commercial, for here are clustered the Orthodox Sveti Ilija, the Glagolithic Seminary, and the elegant Romanesque Cathedral. We reflect on the role these numerous large buildings must have played during the centuries of struggle for survival by sustaining spiritual beliefs, art, dance, and music in a public setting.

Nearby, the vacant site of the Opera House, bombed so senselessly by the British, speaks for the role music played in urban life in Dalmatia. At the far end of the Kalelarga we reach the Trg Tri Bunara (Square of the Three Wells). Close to this ancient source of water, at the end of the Kalelarga, is a small chapel of a pale orange hue that suggests a spiritual relation to the wells—as a basic life source. Fittingly, this Renaissance rotunda personifies Gospa od Zdravlja (Our Lady of Health), and we read human well-being into what appears to be a mere urban embellishment.

The *Obala*

We return to the Forum to rest and gather our thoughts on Zadar's interior before experiencing the city from its waterfront perimeter, the *obala*. There we give in to the pull of the open sea that tempted us earlier. Buildings give way to the shoreline park the people of Zadar developed along their seafront on the sites of apartment blocks created by the Austrians and destroyed by the bombings of World War II. Passing through a leafy green setting, we see people sitting on benches on the *obala*, gazing on the calm of the waters and the breathtaking expanse of the sea that opens to us, that powerful force that turned inland nomads a millennium ago into settled maritime peoples. Beyond us stretches the archipelago of low and lyrical islands—among them Ugljan, Dugi Otok, and Kornati—that are the major regional source of Zadar's environmental identity.

After the dense core of Zadar, this great open stretch of sea generates a moment of reflection—a potential of all places of first arrival, as I have learned—on the ever-evolving nature of the built environment in comparison with the unchanging quality of the sea. Each of Zadar's epochs speaks for its own identity and documents it with the changes we have witnessed. Yet, these basic elements of sea and nature remain just as they were for the people at each of the turning points we have established. This phenomenon of the sea provides an infinite and eternal "menu" of perceptions: wind, tide, current, texture, color, sound, and the motion of water itself, its reflection of sun and moon, and its relation to the stars that guide sailors away from and back to Zadar.

The broad, sweeping *obala*, sparsely populated on a cool windy day, thronged on a warm summer evening, functions very much like the deck of a ship. Note how it defines Zadar as an irregular rectangle, stretching east and west some one thousand meters along the sea. To walk this two-

Children dancing the *kolo* on the promenade. *Photo by author*

and-a-half-kilometer "rim" around the entire island allows you to perceive Zadar as an integrated whole, which is rarely possible in the case of most cities. While our experience of reading the Kalelarga gives us an internal view of Zadar, the *obala*'s uniqueness stems from its consistently external outlook to the world at large. The contrast between these two experiences becomes a major clue to the identify one feels with Zadar.

As we move westward along this city edge composed of sea, promenade, and green parkland, note how the pure line of its smooth white stone paving and sea wall dominates our vision. Its generous width boldly divides the built environment from the natural environment, and now we feel cut off from Zadar, more a part of the sea than of the town. Down the shore, the Hotel Zagreb and the adjacent apartment buildings that survived the World War II bombings mark the line where, under Austrian rule, the new sea wall was built. Beyond stand the remains of the medieval walls. The water laps softly against the sea wall; below, a sailboat goes by and seagulls coast in the breeze—a picture of peace and environmental quality of a high order.

Let us now walk toward the west end, where the roadway rises to some twenty feet on to a great berm, once part of the fortification system. A grand baroque stairway takes us to that level. There we find Zadar's terminal for large ships, giving the residents of Old Zadar direct access to ships traveling to the major Adriatic cities, from Dubrovnik to Rijeka, Trieste, and Venice. What is the huge mansion with extensive grounds richly landscaped and enclosed within a spiked fence just behind us at the higher elevation of the old walls? We assume that it was built for the Austrian army and might have been used by Franz Josef himself. It turns

out that it served as the headquarters and residence for the Jugoslavenska Narodna Armija (Yugoslav People's Army), the same JNA that only
a year after our reading participated in the shelling of Zadar! Abruptly
turning to an environment of peace, we visit the tranquil courtyard of
the church and monastery of the Franciscans atop the old walls and
ponder its privileged position alongside the military mansion, with its
commanding views out to sea and its future in the new democratic Croatia anticipated the year before.

From this elevated position we can experience Zadar as a whole from
a new perspective. Stretching all the way back to the Trg Pet Bunara,
the red rooftops define the street system we have walked. We wonder
how so much history and so much contemporary living can be fitted
into the small space so skillfully. Walking further toward the mainland,
we note the revolutionary nature of street names: Ulica Karla Marxa,
Bulevar V. I. Lenjina, Obala Marsala Tita, all promptly changed when
Croatia gained its independence. Just below we find Ulica L. Adamica,
named for my mentor, who inspired my youthful interest in my Dalmatian roots.

Rounding the corner to the long north side, we see the suburbs of
New Zadar, across the channel, stretched out before us, providing a
striking contrast to the open sea and the outlying islands to the south of
the *obala*. Zadar no longer seems to be a small, isolated, city surrounded
by water; instead we see it as part of a vibrant, growing regional center
set on an ample and buildable terrain. As the fresh panorama unfolds,
we slow down and wonder at the range of experiences Zadar has offered
us: our entry at the Five Wells, the tight urban Kalelarga, and the *obala*,
transporting us out onto the sea. The question presents itself: In these
troubled times, in the face of an uncertain future, can we bring into unison the very different identities of such superb urban places as Zadar?

We rest on a bench on the grassy, tree-lined top of the ancient earth
wall. Below, on the harbor side, the formerly busy regional bus station
on the Radnička Obala (Workers' Waterfront, yet another revolutionary
name) now serves only the local residents. Many people from outlying
areas are still magnetically attracted to Old Zadar. This local maritime
scene has the traditional appeal of a small harbor alive with people enjoying their fishing and pleasure boats, which they care for themselves in
the shipyard. Across the harbor, along the shoreline of this mainland district called Vostarnica, only two large buildings of prewar grandeur break
the low skyline. Leaving this commanding site we see a low monument
with a plaque placed on it in 1981 and now outmoded by the swing to
democracy in 1990. The plaque commemorates the fortieth anniversary
of the founding of Zadar's "First Cell of the Communist Party," perhaps
in a vain attempt to reaffirm its failing vigor. This is an illustration of
how the great urban works of Zadar's past have become permanent and

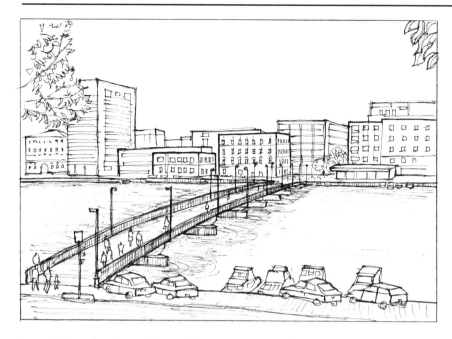

The pedestrian bridge
to the mainland

how their value is unaffected by momentary changes in outlook. Pedestrians from the mainland throng the low-lying bridge across the harbor. They have come by bus to take advantage of the markets and services of Old Zadar. It becomes clear that the old city has been maintained as the focal point of the region as a whole.

Crossing the old embattlement, we face Zadar's rooftops and look down into the large open-air market and the cross streets leading to the Kalelarga. Down the steps, the principal street leads us through a tunnel, handsomely framed with a stone portal, to sample this busy footbridge across the harbor. Returning via the tunnel, we shortly find ourselves on the same street leading to the Kalelarga and the Narodni Trg. We return to the Trg Pet Bunara, not via the Kalelarga but by the minor cross streets teeming with decidedly modern-looking people. Most are walking fast, as if with purpose; you sense it in the firm clicking rhythm of the high heels of the women's stylish, nonproletarian shoes. Even on the side streets, we see an explosion of new shops of smart modern design with goods of high quality, sparked by the rise of private enterprise. Old Zadar has become a downtown shopping area for all metropolitan Zadar.

This urban core functions like an unplanned shopping mall—but closed to the automobile—where a two-thousand-year-old pedestrian tradition works very well: tables are placed outside the cafes; the compact streets are arranged in diverse patterns; and the cream-colored stone, polished by years of walkers, reflects the bright sun. Consumer goods of great variety have replaced those of a decade before: shops featuring televisions, stereos, flowers, pets, hobby supplies, and services indicate the fading away of Marxist economics. One sign reads: "Čeka vas Evropski Business—Marketing Agencija Colombo" [European Business Awaits You—Colombo Marketing Agency].

Continuing our tour of Zadar's rim, we cross the Trg Pet Bunara, where the medieval walls pick up again, and pass through the tunnel to the old eastern Land Gate, an arch built by the famous Venetian architect Sanmichelli in the seventeenth century, and again arrive at the water's edge. Here the sea is no more than a quiet inlet, a harbor for small fishing and recreational boats from which I have sailed to explore the Kornati Islands with cousins. The sheer walls of early times rise high above on one side; below, a narrow walkway takes us around to the windy corner of the peninsula. There, as we turn west where the wall comes to an abrupt end, the entire kilometer of the broad south *obala* sweeps away before us. The vast sea and lyrical skyline of the islands offer us the image of Zadar's regional identity.

To our right the scale and landscape setting of monumental public buildings allow us to recall the grandeur of Franz Josef's Vienna. Among them are the University of Zadar and the Archives, where my grandfather represented Kuna at the Dalmatian Parliament in the 1880s and where I—ninety years later, in the 1970s—explored the records of requests for schools, clinics, and roads under the new Dalmatian autonomy. Now, given the mosaic of visual images and historical insights gained from our multifaceted reading of Zadar, these buildings contrast strikingly with the harmony, intimacy, and variety of human input we experienced in Zadar's internalized realm. They symbolize the need in our times to build anew in a context of what has gone before.

What better place to end this reading than in the ruins of the old Roman Forum, where Zadar's urban life began. We now follow the route of the old waterfront roadway, past the Hotel Zagreb, through the verdant parklands. Turning our backs on the sea, we step on paving stones of the Romans and sit on the weathered remnants of fluted columns. We wonder at the Basilica built by the early Croats, at the force of Venetian culture in the Campanile rising above, and, in sharp contrast, at the contemporary architecture of the post–World War II period. With each speaking to us for its own epoch in spite of the centuries that separate them, we feel the articulation of a firm amalgam of time and space on which our sense of identity with Zadar is built. Again the question: What direction will identity with place take in the future? Will it be a focus on metropolitan growth within ecological and humanitarian guidelines made possible by the new democracy in the making? Never could I have imagined that within a year of my visit Zadar's renewed sense of identity gained through self-determination would be the victim of a war blatantly out of tune with the world order.

The New Environs

Perhaps the answer lies beyond the lively shoreline and harbor we saw from the heights above the north *obala,* the suburban development extending up the gently sloping land, virtually non-existent at the time of my first reading, in 1981. New Zadar spreads out before us in a pattern of clustered housing that appears to have been guided by enlightened professionals of the rather progressive municipal planning agency. But its extended coverage, geared to the automobile, marks a sharp change from the highly structured patterns of Old Zadar, fitted so well to the pedestrian. A tour in the car of my cousin, an architect with the municipality, revealed little evidence of centers to encourage community life—shops, schools, play areas—accessible by foot. Clearly, the old core remains the single center of community life. In fact, the vitality of the old core stems in part from the new growth in the surrounding area: Zadar's population tripled between 1970 and 1990. The new residents, housed largely in low-rise apartment buildings, depend on the diversity of urban activities in Old Zadar. In 1990 some seventy thousand people occupied the mainland slopes, compared with some eight thousand in Old Zadar. Buildings in New Zadar are kept to a reasonable height and set in ample open space where future community needs might be met. Thus, the medieval tradition of compact, self-contained pedestrian life that thrives in Old Zadar could well serve as a model for local centers for community life, each with its own identity, in the metropolitan Zadar of the twenty-first century.

SPLIT: FROM ROMAN PALACE
TO SKYSCRAPER SKYLINE

Urban Rome and Venice Live on through Millennia

In order to search for Split's historical sources of identity with an open mind, we must set aside the insights and clues that revealed the identity of Zadar. Although Split's historical turning points conform generally to Dalmatia's, it has a different environmental heritage. Zadar hides its history under layers of building, devastation, and rebuilding; in Split we cannot escape directly experiencing of the city's essence, both past and present. Furthermore, Old Zadar, almost surrounded by water, is separated from adjacent modern growth, whereas Split has spilled far beyond the confines of its historical core to form an extended metropolitan area that breaks harshly with its heritage.

The environmental assets that made Split the largest and leading city of Dalmatia were twofold: its central position in the region, which gave it strategic advantages at all the turning points in its history, and its site

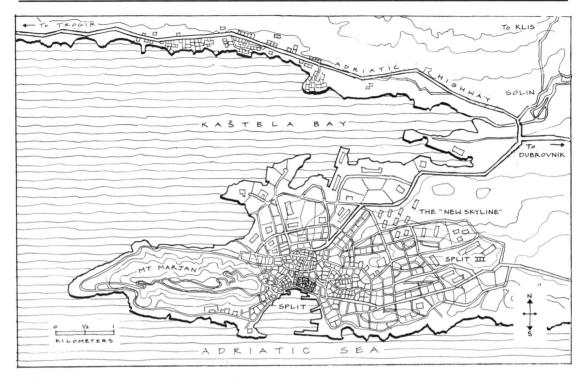

Split: An urban area
stretching inland from
the Roman palace to
the massive high-rise
skyline (top)

General view of Split
(bottom)

on the mountainous peninsula of Marjan, fronting some twelve kilometers on the sea and an equivalent amount on the great Bay of Kaštela. This impressive jut of land tapers down to a sharp point to form a protected entry to the bay and is topped by Mount Marjan, rising high above the sea. From its craggy heights settlers could watch for potential invaders from either sea or land. Visible from great distances, Mount Marjan provided a strong identifying image for the city.

A natural deepwater harbor ultimately made this location far more suitable for long-term regional development than Rome, so the Romans built Salonae on the inner bay. Furthermore, two major islands, Šolta and Brač, some fifteen kilometers off-shore, provided almost continuous protection from the open sea. Brač's potential for settlement contributed to rounding out Split's geographically varied maritime subregion. In addition, a short distance beyond lies the tip of the island of Hvar, a critical part of Dalmatia's urban system.

Illyrian and Roman Beginnings

The Illyrians left nothing on the site of Split with which to identify their times but their own beautiful ethnic name for the entire east coast of the Adriatic. The name of the tribe of the Split region, the Dalmatae, however, was given to Dalmatia as a whole. The tenacity of the Illyrians in holding off Rome for centuries stemmed from their firm occupation of the Bay of Kaštela. The Greeks found outposts for trading on the site, then known as Aspalathos (later to become Split), as early as the fourth century B.C. The reality of Grecian settlements in Dalmatia as modest, nonaggressive trading posts became clear to me when I saw the difference between the meager remains at Stobrec, south of Split, and the full-blown temples of colonies at Agrigente in Sicily and Paestum on Italy's west coast.

Aware of these outstanding advantages for city building, the Romans planned and built Salonae on the inland side of the protected bay in 119 B.C.. Located at the mouth of the Jadro River, Salonae's system of urban facilities covered an area roughly four times the size of the original Zadar. The space between Salonae's widely spaced facilities, which included theater, a temple, a forum, an arena, walls and gates, public baths, and an aqueduct used up to the twentieth century, is evidence that the Romans intended Salonae to be a very large city. The provincial capital, Salonae became a vibrant city and served as a base for Rome's penetration inland for more than three centuries, dominating by far the little settlement of Spalatum (as the Romans called it) on the sea. Beginning in A.D. 295, Emperor Diocletian built a vast summer palace in the form of a fortified Roman *castrum* (encampment) covering some seven acres about ten miles south, on the Adriatic. This served as a place of retreat from his imperial responsibilities, with all the luxury and comforts of Rome in a maritime setting that surpassed those of other imperial palaces throughout the Mediterranean. These qualities may well have influenced Diocletian's abdication in A.D. 305; however, he used it only briefly, for he died in A.D. 313. J. J. Wilkes describes this period in sufficient detail to make real the vast scope of the Romans' creation of new

Evolution of Split, 1675–1990. By Jerko Marasović. By 1675 Split's ancient beginnings stood protected by walls against the Turks (top left). Spontaneous settlement had covered the western hillside by the 1880s (bottom left). Before 1914 unplanned suburban growth continued (top right). By the 1990s urban sprawl had taken over on a metropolitan scale (bottom right)

cities and new societies of great coherence, a scope unmatched in recent times.[9] It has been announced that the seventeen hundredth anniversary of the building of the Palace will be commemorated during the decade from 1995 to 2005 under the auspices of the Mediterranean Center for the Built Heritage in Split.

For the next three hundred years the Palace served as the waterfront showplace of Salonae and as the home of exiled emperors during the gradual disintegration of the Roman Empire. Diocletian unknowingly made an even greater contribution to future generations than the Palace ever would have been as an emperor's villa: by A.D. 614 Salonae had fallen to the Avars, and its residents, fleeing from their demolished city,

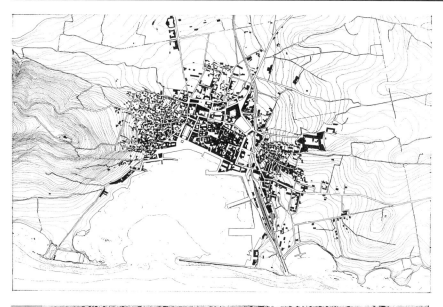

became permanent refugees within the solid stone walls of the Palace. Salonae was never rebuilt, since Spalatum had the advantage of the deep-water natural harbor.

Extended historical research on the unusual evolution of Split's urban form has been carried out by Jerko Marasović, of the Mediterranean Center for the Built Heritage in Split. Working with archival documents in Zadar, Venice, Vienna, and Paris gaining further information locally,, he has shown the type and heights of buildings in graphic form, first in plan, then in three-dimensional perspective from six points of view. These drawings show the site in Illyrian times, when the site comprised three wells, a rock, and a creek. In the first century A.D., the Romans built a road from Salonae and used the outcropping rock as a dock. This became a logical site for the emperor's palace in the late third century. By the tenth century scattered settlements had grown up in the countryside, often around churches. In the eleventh century the waterfront began to be filled in, and in the twelfth century wooden buildings were built just off the west wall to form an intrinsic part of the central core of the city we know today. Under the Venetians in the following centuries more permanent construction was added, and many of the architectural features can be seen today. Marasović's study includes the modern docks and related railroad lines and industry of the nineteenth century and continues up to the present.

Split's distance from Venice, its proximity to Zadar, and the adaptability of Diocletian's Palace, which allowed it to be transformed into a city, eventually made Split the stronger foothold for settlement by Croats. The very urban character of the complex, large enough to hold thousands of residents, facilitated the transition from rural to more settled urban life that the Croats had already begun through their organized state and aristocracy. This was a gradual process through which urban communities could absorb new settlers from the vicinity. Becoming participants in the development, the new settlers transformed the Palace into their own cultural space. Over the centuries, the newcomers built dwellings between the existing columns of the central peristyle, repaired ruined walls according to the original masonry systems, and used fallen stones for new construction. A clear circulation pattern provided direct access to common public areas and to the three gateways leading to the open countryside. Diocletian's Temple was converted into a Christian church, Sveti Duje, which today is Split's cathedral. Because of its continued use, it has been preserved intact up to our times.

During the early Middle Ages Salonae (or Solin) became recognized as a major Croatian locality when the coronation of King Zvonimir took place there in 1075. Meanwhile, the refugee encampment in the former palace at Spalatum developed into a new town that experienced rapid social and economic growth. For several centuries Byzantium held the city, while the nonurban land beyond its walls remained under Croatian rulers. By 1069 the increase in the Croatian population in the region and the possibility of using the palace-turned-city as an anchor reinforced the leadership position of Split and its subregions as a viable element in the evolution of Dalmatia's urban system.

As the rural Croatian people gradually became urbanized, they also learned the art and science of sailing and thereby gained contact with other cultures in the Adriatic. The adjacent islands of Brač and Šolta shared in this economic and cultural growth as settlements linked to Split evolved there, which placed them in a strong position for bargain-

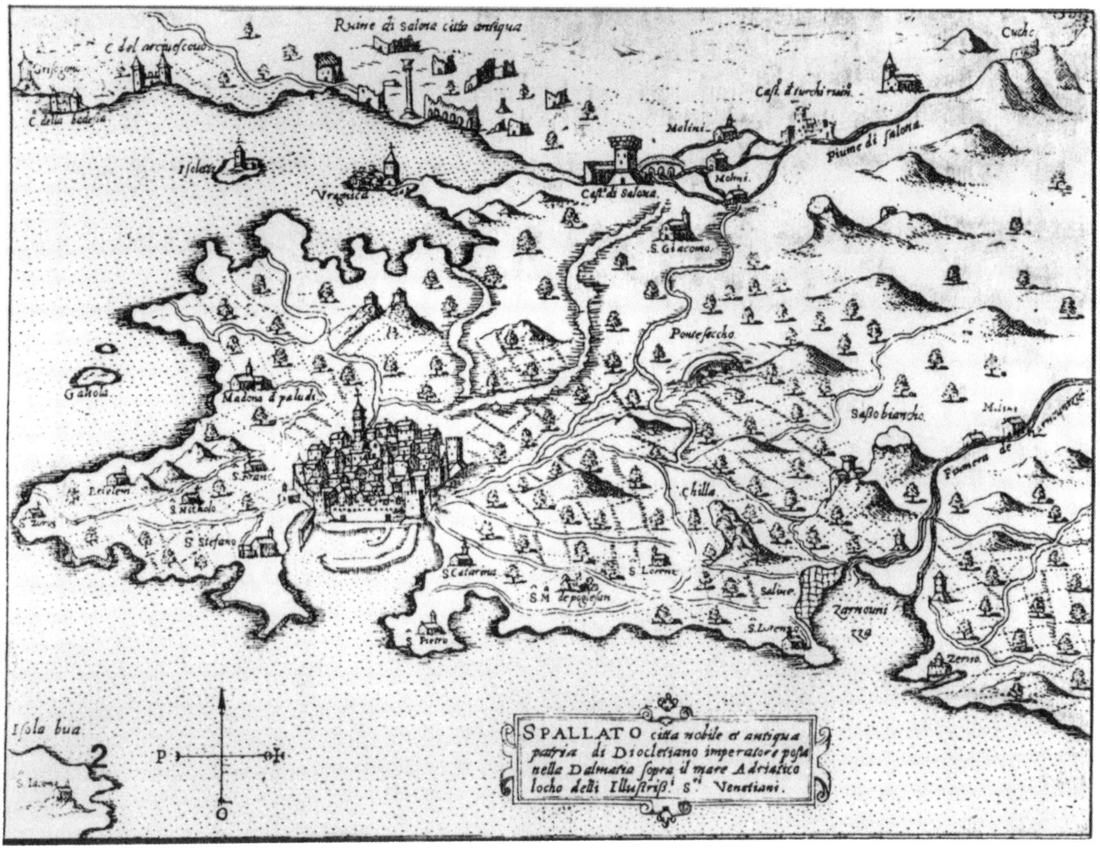

"Spalato: Noble city from the ancient land of Emperor Diocletian in Dalmatia on the Adriatic Sea under the Venetians," 1570. *Croatian Archeological Museum, Split*

ing with the Venetians based on the "protection" they offered in the later Middle Ages.

From Venice to Austria to Metropolis

In 1420 Split accepted Venice's offer, and in 1537 Klis, just above and in sight of Split on the mountain pass to the north, was lost to the Turks, who held it until 1647. A new threat came along in the form of plagues, spread by the concentration of population in the confined quarters of Diocletian's Palace. In 1527 some 6,000 people died and 250 homes had to be burned; even by 1553 Split's population numbered only 3,073. The need for a place of quarantine resulted in the establishment of Split's first hospital In 1607 a second major plague struck the city; 2,700 of the city's population of 4,223 died.[10]

The neighborhood that had grown throughout the Middle Ages immediately outside the West Gate of the old Palace evolved essentially in response to needs at varying times, in sharp contrast to the fixed plan of the Palace. By collaborating with Venice throughout the Renaissance, Croatian Split flourished and brought a new architectural richness to the

Roman framework of Split, dual sources of its distinctive imagery and identity today.

Because of the constant threat of Turkish attack, new walls for defense were built around both the Palace and the Venetian addition. The star-shaped pattern of these massive embankments built in 1667 altered the severe rectangular form of Split. Grga Novak gives us a sense of the reality of this enormous undertaking, which necessitated the demolition of a Franciscan monastery and some 150 homes of poor people. One result was reduced crowding in the old city, which stimulated new building and began the extension of Split out onto the adjacent flatlands, highly suitable for building the large city we have today.[11] Robert Adams's drawings of Split in 1757 show virtually vacant space between the Palace and these ramparts.

As the Turks' power in the Adriatic diminished toward the end of the seventeenth century, peace with the Ottoman Empire and Venice brought Split greater security and economic gains. The eighteenth century brought new public works, outstanding buildings, harbor development, a new sea wall and landfill that cut the Palace off from direct access to the sea, and the beginning of the broad *obala,* a major element of Split's identity today.[12]

After a brief Austrian takeover in 1797, the French occupation of Split had an effect similar to that of Zadar; that is, the antidemocratic upper class resisted French reformist policies. However, with their tradition of urban embellishments during the Renaissance, the French, under General Marmont's authority, undertook a number of urbanistic projects. Land access to Split from Trogir along the eastern shore of the Bay of Kaštela was improved; parks and public gardens were created; for public-health reasons, cemeteries were no longer allowed within the city limits; and in 1809 the first general hospital was built as part of immediate public-health reforms that the French considered critical to their own welfare.[13]

Austria paid less attention to Split than to Zadar, which was its Dalmatian administrative center in Croatia. Making Split a major seaport would only have brought it into competition with Rijeka and Trieste, both of which were well connected to Vienna. Furthermore, rail connections to the interior were hampered by the Turkish occupation. The Austrians were interested in architectural and engineering advancement, however, and therefore took a special interest in Diocletian's Palace. In line with their Ringstrassen projects, they added public buildings, parks, opera house, and new avenues, the first formal extension of the city outside the walls of the Palace and the medieval walls. These improvements form a belt around the historical core. Austrian engineers also widened the *obala,* built new docks, enlarged the Roman aqueduct, and installed

The palace on Split's activated waterfront, 1782. *By L. F. Cassas*

a new water-distribution system installed in anticipation of population growth.

Unlike Zadar, Split came through the two world wars and the Revolution virtually untouched and with no loss of its population and cultural continuity. A strong feeling for the unique heritage of two thousand years was expressed by leaders in architectural restoration and urban history. Fueled by the sense that now, at long last, the Croatians were fully in charge of Split, these civic-minded leaders began the ambitious task of excavating and restoring for contemporary uses Diocletian's Palace, the central focus of our reading. Training for this task was based on scholarly research and teaching in architectural history by urban historians at the University of Zagreb, such as Frane Bulić and Vicko Andrić, who had begun the excavations of Salonae in the nineteenth century. The improvements not only contributed to the sophisticated, urban quality of life but also contributed to the cultural image and identity of Split, making it known throughout the world.

Paton's description of his firsthand experiences in Split at mid-century conveys a personal image of this identity: "From the ridge of Mt. Marjan, one sees the Gulf of Salona, the shores of which form the noblest part of the whole land. Far beyond, smooth wide waters of the bay, a rich land of smiling villages and gardens, dotted at regular distances with Venetian castles, beyond which rise the rugged mountain chain. This locking of water in the embrace of land has produced such beauty to be found nowhere else in Dalmatia."

Arriving in Split at the height of the Carnival season, Paton attended costume balls of both the upper and lower classes. The "Spalatinos" were "kind, cordial, and witty" and responded to his singing with "thunderous applause for him." They followed the latest Italian operas. Paton enjoyed lamb roasted whole at Easter, as well as "the dancing of the Kolo, the national Illyrian amusement." He found "monotonous" the music of the ancient two-stringed *gusle* and its Homeric storytelling, just as I did in 1937.[14]

An Urban Reading of Split

As we embark on our second reading experience, you and I again have to decide where to begin. A Dalmatian colleague rooted in Split's identity and origins advised me that "to best approach Split for the first time, one must come by sea, for this was the way the Romans did, when they first arrived, and all others until the building of the Jadranska Magistrala [Adriatic Highway] in the 1960s."[15] So, stand with me aboard an imaginary ship moving into the deepwater harbor as if arriving from Brač. Split's spacious *obala* and docks stretch out before us: to the left is the wooded peak of Mount Marjan; in the center are Diocletian's Palace and the towers of the Venetian addition; the low-rise residential and commercial areas of the nineteenth century form a belt outside the walls; and finally, the skyline of jagged spires of the high-rise apartment district stands at odds with the ancient mountains beyond. Built since 1980, this building mass, the first unit of which was named Split III, constitutes an intrusion that would probably drive the Romans back to Rome should they return today.

The irresistible pull of Diocletian's Palace and the power of the geometry laid down by Rome for our use today emanates out across the whole waterfront facade with deliberate finality. Although in Diocletian's times the main entrance gate was on the north side, in the direction of Salonae, we shall enter by what was the "back door," opening directly to the sea. Centuries later this became the "front door," cut off from the sea by landfill for the vast *obala*, popularly called the *riva*. We begin there, guided by the image and identity established two thousand years ago, overshadowed now by the land-based system of highways and airports, the mode of arrival for the majority of today's visitors.

Modern Vitality in a Roman Palace

With our back to the sea and the spacious, tree-lined *riva,* we face the massive three-story walls of the Palace. A colonnade of sixty arches at the second-story level breaks the mass of its 600-foot facade. Dwellings built atop the walls and between the columns, as well as tiny shops attached at the street level, tell us of the passage of centuries and the mixture of uses that have turned the Palace into an organic urban core with a contemporary life of its own.

To compare what we see with the reality of the Palace as it was in

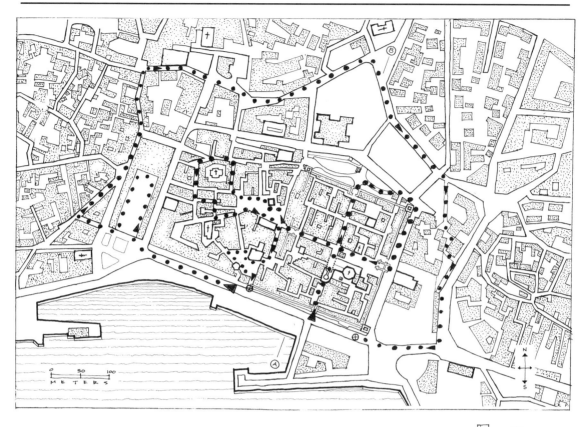

HARBOR   OBALA   DIOCLETIAN'S   NORTH   OLD   19ᵀᴴ CENTURY EXPANSION   20ᵀᴴ CENTURY SKYLINE
Ⓐ   PALACE   GATE   WALLS
Ⓑ

Split: Urban structure
and reading paths

Roman times, we must remove from our vision much of today's Split: the landfill of the *riva* in front of the walls, together with its docks, buildings, promenades, trees, cars, and buses. Gone is the shallow facade of two- and three-story buildings plastered over the face of the almost perfectly preserved walls. Instead of this bustling urban waterfront, picture ships unloading supplies into the old "back-door" entrance to the Palace, standing serenely at the edge of the sea. The vast open space, the southern exposure, and especially the *bura,* the winter north winds, in the twentieth century became—as I learned in 1937—the basis for the popular nickname *obala,* "Poor Man's Overcoat."

To our right, the rounded fortified tower rising above the walls exemplifies the program of preservation through reuse begun in the 1950s. In the 1970s, the self-managed office for architectural restoration convinced the Croatian highway agency to locate its new administrative quarters in the tower rather than in the suburbs. The interior was vacated of slum dwellers, gutted, reinforced, and rebuilt with up-to-date offices and meeting rooms with wall-to-wall carpeting and electronic facilities. Overhead, atop the walls, on what was the main living level during Roman times, the colonnade that joined the private rooms of the emperor and his court has been filled in with poorly built dwellings. Tawdry living places built during recent centuries lessen our sense of wonder at this remarkable element of Split's present-day identity and frustrate our impulse to bridge the centuries with an accurate, full perception of this phenomenon. Several of the spaces between the columns that have been opened spark in our minds an image of Romans strolling there in the evening, enjoying the sight, sound, and smell of the open sea, unaware of us standing below on the *riva,* nonexistent at that time.

We enter in the center of this commanding facade through a narrow, almost secretive passageway to the underground quarters where supplies were brought by ship for storage in the vast, dark, cool chambers ahead of us. In the late 1960s I followed closely the excavations of refuse and leavings of centuries to recycle the space for modern uses. Handsome buttress foundations soar above us and join structural walls of stone. Some twenty thousand square feet of space was created for exhibitions, performances, and tourist uses. One vast chamber leading to another gives us an idea of the enormity of the undertaking, the scale of population served by the supplies stored, and the intent to maintain the same level of luxury here in the province as in Rome. Local handicrafts and artwork occupy well-lighted stands that line the wide main passageway to daylight beyond.

An inordinately steep staircase with very little headroom, designed for rugged workers of antiquity, takes us out of the cool darkness of the purely Roman depths and up to the glorious, sun-filled Peristyle, alive with arched colonnades, people, and pigeons. Reading now in daylight at ground level suddenly shifts the quality of our experience from the utilitarianism of the Roman underground storage chambers to the sophistication of the Roman architecture at its flowering, topped off with additions made a thousand years later under Venetian influence. On our right we see seven arches — the classic number; their rusty brown Corinthian columns still stand free. On our left seven more arches are filled in with four-hundred-year-old Venetian palazzos that became slums in the nineteenth century but are now restored handsomely for public use. Converted in a fashion highly faithful to their historical diversity and integrity, these buildings became headquarters for the United Nations Regional Activity Center of the Mediterranean Action Plan, a program initiated by Split's highly experienced conservationists. On the ground floor, rich elements of an old palazzo have been converted to a

From the Vestibule
through the Peristyle
to the *cardus. Photo by
author*

cafe, with outdoor tables that give life around the clock to the Peristyle. The Peristyle readily allows us to see how history's turning points have shaped Split's visual identity today.

We turn back in the direction of the sea to mount a half-flight of stone steps and enter the Vestibule, a circular court joining the east and west wings of the emperor's living quarters, which open out to the Colonnade and the sea. An opening in the perfectly preserved dome of alternating brick and masonry high above casts a circle of sunlight on us and the stone floor. The thought that Romans—and Diocletian himself—stood here experiencing this same circle of warm sunlight brings on reflections endemic to urban reading: wonder at the continuity of nature and architecture interacting in building design, together handing down a daily, identical experience for countless generations to enrich our own today.

Our reverie is interrupted by boys who have turned the classic space into a handball court. Vines reaching for the sun from a roothold high up on a nearby wall join in an ecological readaptation of the space over the years. Four empty niches set into the curving wall evoke visions of the former occupants, an honored family and their friends long lost

The Peristyle:
Vestibule, Colonnade,
and Campanile.
Drawing by author

since the fall of empire. An inner court area, also taken over at the moment by energetic boys, provides access to the offices of the Urban Planning Institute of Dalmatia, where the United Nations Development Plan for the Dalmatian Coast housed its international consultants in the late 1960s and extended assistance for my studies in the following years.

We retrace our steps to the Peristyle. On our right, beyond the end of the open Colonnade, Diocletian's octagonal Temple—converted to Split's cathedral—stands in perfect condition thanks to its having been used as a Christian church since the fall of Rome. Seeing it together with the Campanile, added during the Romanesque period of the early Middle Ages, we experience directly an architectural amalgamation of centuries of changing times, peoples, and traditions. History fades away as pigeons in midair catch bits of bread thrown by a lady on a balcony and people stand, talk, and move on with passive acceptance of this scene that fills us with wonder. The remains of Diocletian's Palace breathe meaning into the place, though those people who spend their daily lives here may be only subconsciously aware of it. The geometry and the purity of line of the majestic Colonnade built in the fourth century A.D. provide a solid base for the soaring Venetian Campanile. As we ascend its height step by step to look down on the Palace's cluttered array of rooftops and far out to sea, we feel that Diocletian continues to dominate this collaboration in urban design even now.

My own earlier images impair this vision. When I saw the Peristyle for the first time, in 1937, it was Croatia that rose above either Venice or Diocletian, in the form of Ivan Meštrović's gigantic sculpture of Bishop Grgur Ninski (Gregor of Nin) looming high in the center of this ancient space. Placed there in 1929, this great work of art then carried a powerful message to me, namely, that the people of Split had now been validated fully as the leading group among a triad of contributors over twenty centuries. For it was this bishop who, in the tenth century, defended the right of the Croats to use their own language. This message of history was dimmed in the early 1940s, when the bronze figure was moved. According to some scholars, the Italians were eager to suppress the Croatian language, and later the Communists had no love for the Church. Eventually this stirring work of Meštrović's was relegated to a wooded park outside the North Gate and thus lost much of its symbolic meaning.

This first experience here, at the very core of Split, imbues us with a sense of eternity, a reminder that stone will outlast people, yet new people will come to take over the long process of creating the uniqueness of this place called Split. Eternity speaks through the walls, cut from the stone quarry of my grandmother's Pučišća, across the channel on Brač. The classic columns stand streaked with centuries of stain and weathering, bleached white to a smoky gray in the damp shadows. They absorb us as we gaze, and the regular spacing of the Colonnade sets a rhythm to our feet as we move to the north end of the Peristyle.

There we reach the late twentieth century in the form of the Splitska Banka, a creative example of revenue-producing historical restoration typical of the 1960s and 1970s. The corner wall of glass exposes the original Roman street, the base of one corner column, and a portion of another, marking the continuation of the original Colonnade to the North Gate. Even the street paving, a stone catch-basin, drain, and curbing have all been put to twentieth-century use.

This intersection of the *cardus* and the *decumanus* is as wide as when it was built between the two sides of the Colonnade; then the street narrows to a mere six feet in width, the next segment of our reading. Picture the dramatic effect of that arched Colonnade continuing to the northern main entrance, Zlatna Vrata (Golden Gate), and on to Salonae! To the east, that is, to our right, Srebrena Vrata (Silver Gate) and to the west Željezna Vrata (Iron Gate) still serve as reminders of the total sense of enclosure the Palace impacts us with and of the inflexible geometry that now so fully contains and directs us.

Moving down this narrow Dioklecijana, if we could enter the minds of the men and women of Rome and reconstruct what they saw here, we would experience a double row of some twenty columns framing the route, identical to those of the Peristyle. They give a majestic air to this approach to the emperor's quarters through the housing areas for the members of his court. But today the narrow route severely restricts our view of the once grandiose Venetian facades of the homes of Split's elite, later subdivided into apartments. One with Gothic arches bridges the street. The largest of these once stately urban palaces was restored for use by an adult education program of the University of Split. We pass through the graciously detailed courtyard of the finest and best preserved of these palaces to the handsome Municipal History Museum. In 1968, bent on seeing the interior of one of the old palazzos, I entered and found a treasure of material on the history of Split. This brought me into conversation with an official, who I discovered, to my delight, was the nephew of my grandmother from Pučišća who had helped look after me thirty years before. Our meeting, charged with emotion on both sides, made real my metaphor of The Bridge and my exploration of meaning in environments. On subsequent visits, I found that he had aged with the palazzo itself, and by the time of my 1990 reading of Split he was gone, but the museum was undergoing an inspired restoration and renewal as a result of its recent reassertion of Croatian cultural heritage.

At the end of the *cardus* we reach the inner gate, where the street returns to its original width. Passing through the outer gate, we sense relief from the physical confinement and from the weight of the past. Sunshine defines the great space outside the walls that was preserved as a park in the nineteenth century, verdant and pungent with pines extending along the length of the ancient walls. At the east and west corners rise the two massive and most fully preserved towers. The interior of the West Tower has been handsomely restored to provide an office for the Center for Historic Preservation, a branch of the University of Zagreb's architecture school that has supported the studies of Split's great urban heritage. The park, with its villas on the far side, recalls Split's Austrian period.

Soaring above the leafy verdure and out of place against that backdrop stands Meštrović's soaring figure of Bishop Grgur Ninski, the savior of the Croatian language, which I had seen in the Peristyle in 1937. There he had appeared to rise above Roman antiquity; now the bishop looks down intimidatingly, with forefinger pointing to the sky as if to remind present-day generations of the ultimate identity of the Croatian people independent of outside rulers—truly a message for the coming twenty-first century. Here, outside the walls, the street with its cars, the pace of the people, and the blocks of buildings from the turn of the century together make us mindful of our inability to find urban patterns as clear and coherent for our times as as the patterns the Romans found for theirs.

In that frame of mind, we walk around the East Tower of the Palace and sense what it was like during the Middle Ages to be unprotected outside the walls; this feeling is possible only on the north and east sides, where the great walls stand unencumbered by old buildings. The broad space flanking the east wall, filled with the market's sun-shaded vendors, shoppers, and pedestrians, may have had just the same character centuries ago. Vendors from Bosnia and Macedonia extend the fantasy of the past with their hand-crafted products: woven goods, leather belts and sandals, and carved wooden trays and utensils. The East Gate, a kaleidoscope of architectural elements from earlier periods, stands well above the original grade of Roman times. From our high vantage point as we enter, a rich view of the Peristyle calls for thoughtful attention. We now see the other four sides of the Temple, the Colonnade behind it this time, stores and kiosks, the bank, a market with brightly hued flowers, the last segment of the *decumanus.*

How different is the experience of this welcoming entry to the heart of the Palace compared with our rising up from the darkness of the underground route into the brilliant Peristyle. Walking, we can feel beneath the soles of our feet the smoothness of the original paving stones and identify with the innumerable human beings who have experienced this same grand entry immersed in morning sunshine in the fresh, fragrant air. As we again cross the end of the Peristyle, the *decumanus* narrows down to six feet, as the *cardus* did earlier in our reading. Opposite the

The flower market at
the East Gate. *Mindy
Dulčić Hurlbut*

corner bank we saw earlier, one of the many duty-free shops in Dalmatia
(with 1990 price tags in German Deutschemarks) is a symbol of modern
international economics invading ancient Rome.

A four-story building envelops us in shade. The experience of openness, sun, structures of the past, and people today shifts to the imme-
diacy of merchandise in shop windows. We see at close range faces of
people intent on moving along this busy commercial passageway. The
narrow, shaded walkway crowded with people leads us toward the West,
or Iron, Gate. Shop windows brim with signs of new competitive vitality
since my visits of the early 1980s. The gloomy, nineteenth-century look
of those years and the heavy hand of bureaucratic socialism are gone,
along with the sparse shelves, the shabby interiors, and the halfhearted
salespeople. Instead, the stores have been redesigned—they appear to
be the work of young, private architects—in the fresh spirit of modern
times, yet they still reflect Split's history. Marble floors and displays of
goods of high quality and great variety are looked after by alert young
people. Small private owners were bringing new life into the traditional

stone buildings in 1990, an alternate and practical way of maintaining the urban spatial and social patterns that have worked so well for centuries.

The Contrast of the Unplanned Piazza

The storefronts stop short of the West Gate's monumental Roman arch, which rises the full three-story height of the remaining portions of the Palace wall. To the right and left, narrow routes lined with tiny shops follow the inner face of the old wall. Once we pass this powerful reminder of Roman urban design, the geometry and commanding order that have shaped our experience until now vanish. In but a single step we span a thousand years, from ancient Rome to medieval Venice, as the Narodni Trg opens before us. Also called the Piazza, its angular and irregular form evolved incrementally and became the focal point of today's Split. In contrast to the feeling we had of Rome's rigidity and authority choreographing our movement, we feel free to choose from a variety of ways to move about. Groups can gather to define their own turf—a shortcut to regaining one's own personal identity with place.

As in the Palace, many of the shops and cafes here are alive with new enterprise. Meandering, diagonal side streets bring people from the surrounding neighborhoods to shop and socialize in the cafes, even more so when the winter *bura* sweeps the popular seafront *riva*. The fine Loggia and Clock Tower symbolize the communal autonomy Split gained during the Middle Ages. The air is filled with the humming sound of pigeons and the chatter of people in endless conversation over coffee and chance encounters with friends. Here you can witness the social traditions of the past living on.

Three large buildings, one with an inlaid plaque dated 1886, bound the square on the left and superimpose a nineteenth-century quality onto the medieval urban form. At the far end of the Narodni Trg, looking back to the West Gate we see the Piazza's bold Clock Tower, built snugly against the walls, its hands telling us the hour. To its right, a building also dated 1886 displays on its third story a balcony well recessed for privacy with a double row of delicate columns that discreetly expresses social status. In 1981 this facade had just been opened to enhance the historical message of the Piazza. Archival research by the Mediterranean Center for the Built Heritage showed that this graceful architectural element had been walled over during the Austrian era. A spacious cafe on the corner opposite the West Gate, several bars, and more modern shops all give the Narodni Trg a fuller social identity than does the Peristyle.

A short detour behind the Loggia reveals a piazzeta appropriately called Iza Lože (Behind the Loggia), a turf where older men exchange the latest news on sports and politics while patronizing the popular GaGa's coffee bar. Nearby, another foray off the *trg* yields us one more

Narodni Trg with the
Clock Tower at the
West Gate. *Mindy
Dulčić Hurlbut*

small open space, a turf where boys play ball. These are retreats that only the "insiders," the people of Split, would know, especially at the peak of the tourist season, even in the very center of the city. Returning, we pass under a building spanning the street that creates a dramatic entry to the loosely shaped piazza, so typically medieval. As we walk from the quiet, deep shade of the back street into the bright sun and urban life, the contrast between that entry and entry into the formality of the Peristyle becomes striking.

Turning south now, we leave the *trg* via the diagonal route toward the *riva* and the point where we began our reading, at the *obala*. We are choreographed into a zigzag movement following the narrow shop-lined streets to arrive at the Braće Radića Trg (Radić Brothers Square). Less a focal point than the piazza, its graceful medieval irregularity make it just as appealing. Actually, two piazzas join here: Radice, the smaller, and Preporoda. Together, they express the essence of medieval organic urban form. Above them, standing freely, rises the one remaining fortified tower of the fifteenth century, and the tallest, the Hrvojeva Tvrdjava. Here also stands the Milesi Palace, now the Maritime Museum.

By noon the paved open space of the *trg* is dotted with tables pro-

Marulić statue by Ivan
Meštrović in Brače
Radića Trg

tected by new yellow sunshades, and the storefronts also have a fresh look. In the center stands another Meštrović sculpture, that of Marko Marulić. In the fifteenth century Marulić became famous among the intelligentsia in Europe and in Split for his contributions to Croatian identity by means of his poetry in the native language. In perfect scale with this most gracious urban gathering place, the statue is silhouetted against the small, almost hidden opening to the vast space of the *obala* and the sea. We move through it to the waterfront, turn left, and walk along the base of the Palace facade to the Sea Gate, where we complete our reading of Split's historical core, sometimes called Split I, the "first Split." Surrounded by nineteenth-century urban growth, Split II took form, and beyond, in our times, high-rise Split III has become an incongruous backdrop.

The processes by which these irregularities in street direction and open space evolved through human use over centuries take on new meaning after we have experienced them side by side with the rigid pattern of the Palace, planned and built by a central authority in a single decade. We become aware of the richness of urban form shaped by people participating over time. In the evolutionary style, we are allowed our own route; in the latter style, we must conform to fixed patterns imposed by a single comprehensive decision, whether a thousand years ago or today.

We ponder over the collective experiences of differing peoples having produced across centuries an environment of such magnetic and fulfilling sense of place. Three questions arise: Is this phenomenon to be at the heart of the problem of building meaning into urban places? How can we establish processes of new growth in our time that will generate the flexibility, diversity, and richness of human assembly we have experienced? And how can we bring forth from our present planning practice places that contrast with both the rigid planning of the Palace and the loosely structured pattern evolved by the Dalmatians under Venice as a stimulating format for the ordinary everyday life of our cities?

A Belt of Nineteenth-Century Diversity

A reading of Split must include a sample reading of Split II, the broad belt of mixed development that wraps in a crescent around the Palace and its medieval addition. In this area, with a population of some twenty thousand, automobiles and pedestrians together create an engaging vitality. Old villages of diverse historical origins and new urban forms mingling in an unstructured pattern provide a rich array of unexpected visual and social experiences.

We begin at the west end of the Venetian addition. Overlooking the *riva,* a nineteenth-century complex that includes shops, offices, and a small hotel faces out to the sea across a large formal piazza, the Trg Republike, designed to emulate Venice itself, even in its faded reddish color. Close by, the unplanned hillside Borgo of Varoš has been lived in continuously since the fifteenth century. We lose our way in its maze of meandering streets, the result of the coalescing of separate villages, which occupy an area more than twice the size of Split's entire core. Its constant use has provided minimum housing for newcomers from the interior, attracted by Split's growing economy.

Moving north along an old street closed to cars, the Bulatov Trg, dominated by the handsomely restored Opera House, brings us to the Austrian period. There, a large monastery from the Middle Ages has been skillfully restored in a contemporary spirit. The Prima Department Store, Split's only large-scale retail business, has been completely modernized to meet First World standards. The medieval image returns in the form of two star-pointed bastions in the vast scale of the seventeenth-century walls that frame Split's north side. They form a stabilizing backdrop for this bustling scene of cars, buses, pedestrians, and highly mixed uses. All appear to thrive in a well-orchestrated urban design growing out of life as it is lived rather than based on preconceived concepts. The green, leafy open space lined with turn-of-the-century villas, which we saw from the North Gate of the Palace, is grandly anchored between two of these massive bastions.

We now swing around to the east side of the Palace walls, where we find an example of inventive collaboration between tradition and modernity. The depressed railroad line that cuts across this mellowed mix of modern uses to reach the docks of Split has been covered over for a distance of a mile or more to provide a wide pedestrian shortcut to

The diverse urban core outside the palace walls. *Photo by author*

neighborhoods. On one side it is lined with open-air pavilions to serve a variety of new small businesses. Benches, snack bars, a fountain or two, and landscaping encourage the accustomed meeting of friends and watching of people passing by. This is a pleasurable access route to the transfer point for buses, the open-air market, and the broad, tree-lined *riva,* where we first entered the Palace walls.

We complete this segment of our reading by returning to my own home base in Split, the old Hotel Park, left intact from the 1920s, with its Viennese terrace under linden trees. Set in the pine woods of the shore-line park, where people enjoying the sea dominate a "low-key" setting, it offers a sharp contrast to the historical urbanity of Roman, Venetian, and Austrian influences on Croatian Split.

The New Skyline

A thoroughly jolting experience awaits us in the remaining two-thirds of Split, its "New Skyline," a product of the 1970s and 1980s. We shift from the pedestrian experience to the "speed reading" of the car. We drive to the green heights of Mount Marjan, that peninsular limestone bastion fragrant with pine that preserves a sample of the indigenous landscape of millennia past and restrains the growth of the inner city. From the heights we can easily make out the Roman and Venetian parts of the core, blending geometric outlines and campaniles, as well as the surrounding belt, with its low red roofs set among an assortment of newer, low-rise building types. Beyond, over the rolling hills, is the surrealistic New Skyline. Spire after spire emerges, like menacing outsiders from an alien planet. This jagged skyline, stretching out for miles, might be taken for an architectural mountain range rather than a human living place housing a hundred thousand people on a site that was almost vacant in the 1960s.

Down at street level once more, we take the tunnel built through Mount Marjan in the late 1970s to this new, mushrooming Split. The tunnel through the base of the mountain from the sea to the bay represents a positive use of modernization. It relieves the old center of traffic in front of the Palace and transforms the *riva* into a pedestrian haven where the pleasures of socializing in public places, so much a part of Mediterranean cultures, can be enjoyed. When we emerge from the tunnel, we have direct access to the major sports complex, built along the shore of Poljud Harbor, on the Bay of Kaštela. Built for the 1979 Mediterranean Games, the complex represents a positive initiative of the sports-minded people of Split toward reinforcing their European identity. Ironically, these inventively designed structures, with their undulating rooflines and grand scale, stand just across from a restored ancient monastery that houses the regional agency for the preservation of traditional architectural monuments. From the six-lane highway—quite out of proportion to the low level of car ownership—we see one of the earliest of the offending housing developments: two white low-rise buildings so boldly cut across the urban scene that they earned the nickname "The Chinese Wall." The controversy over their impact during the 1960s provoked the sudden shift to vertical construction, with buildings fifteen to twenty stories high being built on vacant land to the south.

As we continue along the several miles of the arc of the highway, we see the industrial land along the bay shore below. To the right, the white wall of the New Skyline stretches on relentlessly, revealing no apparent organization of land use or circulation routes intended to encourage community formation. The asphalt routes and walls of concrete go on without end, unbroken by small-scale neighborhood centers, which would have provided the kind of social life, diversity, and flexibility of urban patterns and forms that so enrich Split's old core. In spite of the high concentrations of housing, this environment, so recently built, stands stark and lifeless, as if no one lived there.

Traveling back toward old Split, we come across a bold attempt at a modern commercial center on the scale of an American shopping mall. Far from displaying the functional style of its model, this complex has the static quality of an architectural monument. Steps of enormous width lead up from broad traffic ways to large terraces and several levels of shops; the effect is one that Diocletian's architects might have wished to achieve. The scale, range, and quality of the shops seem too advanced, too modern, and a far cry from the traditional shopping customs of Split's population.

Finally, we drive to that portion of the New Skyline originally named

The high density of
Split's New Skyline.
Photo by author

Split III, where this high-rise building boom began. Here the architectural and urban design and character are more refined. Towers rise in identifiable clusters, each with its own character, on landscaped "superblocks," accessed by pedestrian ways. Apartments of fifteen stories look out to the sea and down on a mix of low-rises, some rowhouses, and a few single-family houses previously established.

This project was ambitiously planned for fifty thousand people, ten thousand more than Split's entire population at the time of my 1937 visit. By 1990 the district as a whole held a hundred thousand people. It originated in the late 1960s as a move to offset the monolithic character of early high-rise housing characteristic of the Le Courbusier movement, as well as "The Chinese Wall," and in response to the pressure of the new population moving in from the nearby hinterland. A nationwide competition was held in 1970, and the winning design included provisions for a system of pedestrian ways, a diversity of shops, and a sense of self-containment. Participation of the residents at various phases of the construction became a part of the design process, thus generating community identity. It is an impressive achievement. Its scale and the spirit of the architecture recall Manhattan, but without that city's public transit, diversity, and lively street life. The extensive system of landscaped walkways and steps connect the high building clusters more to parking areas and bus stops than to nodes with shops and cafes where people might gather and collectively gain a sense of identity. Therefore, the apartment dwellers of Split III continue to return—by bus—to the traditional center for social life, shopping, and employment. They seek that intangible collective experience of urbanity and environmental support for the human spirit that we witnessed in Split I, Diocletian's Palace, and II, the surrounding belt of recent years. Now, as in the case of Zadar, the image we have gained of Split and its unique combination of time and space must be darkened by the shadow of shells falling there in late 1991. The euphoria of 1990 has turned into deep anxiety.

DUBROVNIK: AN AUTONOMOUS CITY-STATE FOR 650 YEARS

Internal Urbanity versus External Forces of History

Autonomy, self-containment, and self-management distinguish the independent city-state of Dubrovnik from Zadar and Split. Dubrovnik achieved these qualities not through conflict but by responding intelligently to its isolated, rugged location and to the same external forces that shaped Zadar and Split. Whereas these pressures inhibited the full flowering of the native potential for uniqueness in the latter two cities, Dubrovnik's ability to deal strategically with these turning points ultimately led to an environmental identity all its own, outstanding in all of Europe. The massive rocks the city was built upon, the towering mountain ranges rising above, the unprotected exposure to the open sea, and the almost dead-end coastal position in relation to Dalmatia as a whole were hardly attractive or adaptable to urban development. Yet these challenges generated a self-sustainable city of original and durable site-building relationships, as well as exemplarily resourceful municipal institutions, second only to those of Venice itself.

Dubrovnik's natural site was a rugged peninsula rising well above the Adriatic Sea and separated from the mainland by a swampy channel that was filled in by the twelfth century. Together they formed a jut of land, completely walled in, tilting upward at a dramatic pitch to cliffs in the west that drop straight down to the sea. On the east side, the land rises to the steep base of Mount Srdj, an elongated ridge that cuts all buildable sites in the vicinity off from access to the sea. This variety of landforms in close proximity was the main source of the lively quality of the built environment, fitted to a small site and producing a broad spectrum of human experiences and environmental responses.

The immediate subregion has its own constraints. In the north are the two peninsulas of Lapad, both edged with precipices that drop to the open sea and virtually prohibit access. Nature, however, makes up for the challenges of the site with a deep indentation of the sea to form the long and narrow harbor of Gruž, from which the people of Dubrovnik became destined to sail throughout the entire Mediterranean. Up the coast, access to the site by land was impeded by the Dubrovnik River and the Ombla, a deep, fjordlike waterway. Down the coast, the site's

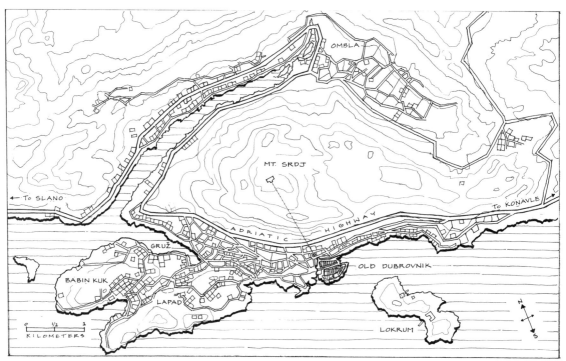

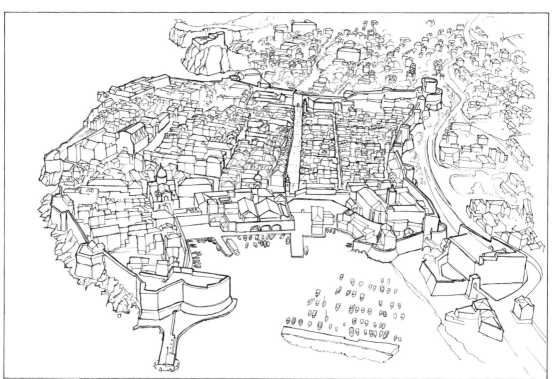

jut of land, together with the island of Lokrum, shields from the sea a tiny cove that became the Stara Luka (Old Harbor). The narrow band of usable land high up above the sea continues for some five miles, and then the landscape opens up graciously to the rolling uplands of Konavle.

From Early Settlement to City-State

The rugged coast restricted the various Illyrian tribes who peopled the Dubrovnik area to the nearby fertile soils of Konavle. This region served as a focal point of unified Illyrian culture at its peak, during the last half-millennium B.C. Well suited to settlement, from early times this territory was home to a network of rural villages, and it became the breadbasket for Dubrovnik. The Greeks, who traded with the Illyrian tribes, left no certain traces of permanence in the area other than at Epidaurus, today's Cavtat, a coastal site serving as a center for the Konavle area.

Vinko Foretić, Dubrovnik's prime historian, described the sequence of events and human innovation that shaped this remarkable union of urban planning, design, and management from the seventh century A.D. to 1808. Drawing on material in Dubrovnik's municipal archives, which cover all the centuries of independence, one of the most complete records for a medieval European city, he wrote in detail about the city's cultural, political, and economic history.[16]

One version of Dubrovnik's origin is that Epidaurus, once a Roman outpost, was destroyed either by an earthquake of major proportions or by invading tribes. The refugees made their new home some fifteen miles to the north, on the highest part of a steep and rocky islet separated from the mainland by a narrow channel. Ragusa, the name used during the early years of Venetian influence, remained in use up to the twentieth century even though the Croatian name Dubrovnik emerged in the twelfth century. The name Dubrovnik comes from a particular oak that grew on the site; *dubrava* is also Croatian for "a grove of trees." New archaeological evidence of the 1980s confirms the general belief that even in earlier centuries, a small Christian settlement occupied the uppermost corner of the promontory, known today as Pustijerna. By the mid-ninth century the Ragusans had become skilled at fortifying the settlement by building stone walls straight up from the great rocks, and in 866 they withstood a fifteen-month siege by the Arabs. In the tenth century, Constantine Porphyrogenetus, the Byzantine emperor and historian, sailed the Adriatic coast and his writings document the self-sustained nature of the settlement established there on the rock.[17]

Beginning in about the year 1000 the Venetians targeted Ragusa, along with Zadar and Split, in their attempts to dominate the entire Dalmatian coast and thus weaken the hold of Byzantium. Throughout the eleventh and twelfth centuries the Byzantine presence lessened,

Dubrovnik's regional growth is restrained by its rugged medieval site (opposite top)

General view of Dubrovnik (opposite bottom)

favoring Dubrovnik's goal to govern itself. By 1022 this community had obtained an archbishopric from the religious authority governing all of Dalmatia, located in Split. By the twelfth century Ragusa had become a city-state in its own right, with religious authority and considerable maritime trade supported by commercial navigational skills and facilities. Serbia and Bosnia became important inland markets that established Ragusa as a strong regional center because of its accessibility both by land and by sea. This drive for local self-identity made Dubrovnik a symbol of a rising Croatian urban culture based on a blending of Eastern and Western values. In 1204 Dubrovnik had achieved its own identity as an autonomous city-state, free of Byzantium. Until 1358 the growing power negotiated coexistence with Venice, whereas Zadar and Split had to settle for the status of protectorates.

Dubrovnik's history is characterized by a smooth sequence of progressive steps forward rather than uncertain turning points and crises. As Bariša Krekić a native son and leading scholar of the city, points out, through this skillfully maintained autonomy the city gained the social and economic stability necessary to create a physical form expressing a clearly integrated urban identity. The patrician class formalized its role in the city government as early as 1332. The members of this class were not the usual landed nobility, since little cultivatable land existed close by. Rather, their skillful role as cultured and cosmopolitan merchants and shippers gave them their status. They shared a collaborative interest with the common citizen, who developed skills necessary to shipping and the transportation of goods.[18] In contrast, in distant places such as Pelješac feudal conditions generated an antagonism that eventually led to popular support for the Napoleonic reforms of the early nineteenth century. In 1333 Dubrovnik made Pelješac, which it considered a valuable agricultural and strategic asset, part of the city-state. The peninsula became a focal point for planned settlements unique to the Dalmatian coast.

Francis Carter describes Dubrovnik as an impressive, internationally oriented city-state. Small in area, it maintained a network of trade relationships by land with the Balkans as a whole, reaching as far as Istanbul through its system of consulates at key points along the way. When extensive mining began in Serbia in the nineteenth century and in Bosnia in the fourteenth, Dubrovnik played a major role in advancing the industry as the Balkan middleman with an ever-expanding territorial sphere of influence on land and on sea. Foreign merchants depended on its cosmopolitan people for their knowledge of foreign languages. By 1426 the republic extended some forty-five miles along the coast, including Pelješac, Primorje, Konavle, and the island of Mljet. The population numbered about twenty-five thousand, of whom five thousand lived in the immediate urban area.[19]

Dubrovnik's urban planning traditions may have been learned from

Rome's past and Venice's present; however, the remarkably detailed municipal records for the five centuries up to Napoleon's takeover for the most part demonstrate an ability to effectively manage its own affairs. For example, Milan Prelog points out that as early as 1272, the city's statute first incorporated some forty-five sections for regulating and guiding urban development. This consistency through the centuries produced the coherent general plan of the city and its system of walls that we know today. Clearly, the people of Dubrovnik consciously planned according to a model that was superior to those of most cities of that time, especially if one takes into account the relentless site of stone to be built upon.[20] These municipal archives and the city itself show that Dubrovnik's leaders took seriously the need for continuity and an orderly basis for making decisions over generations as a means for defending their identity as a place.

Dubrovnik had traded its status as a commune for that of a republic by the early fifteenth century. From the Balkans, Dubrovnik imported not only minerals but also livestock, wax, wool, hides, leather, spices, salt, and foods. In order to pay for these, Dubrovnik exported products, such as textiles, for which there was a demand among the Italians and the French. These opportunities for economic growth also advanced skills in shipbuilding, masonry, and carpentry, thus increasing Slavic immigration to Dubrovnik from the republic's inland regions. The city's role and share in the Mediterranean maritime trade should not be underestimated. From the twelfth century on, Ragusa had the privilege of free commerce in all parts of Byzantium and was a leader in shipping activities, along with Venice, Pisa, and Genoa.

With the coming of the Renaissance, trade increased and Dubrovnik became known throughout Europe as a leading example of stable and productive self-government. This was particularly true in England and France, where various objective studies of cities were carried out and published. Among them, John Streater's study of model governments carried this title: *Government Describes: What Monarchie, Aristocracie, Oligarchie and Democracie Is Together with a Brief Model Government of the Commonwealth or Free State of Ragouse, London, 1659.* Streater concluded his elaboration of these four components that constitute a model self-governed state by applying them to Dubrovnik: "This Commonwealth or Free-State maintaineth itself by its Just Impartial Policy, in perfect Freedom and Strength."[21]

In 1453, as the Ottoman Empire took over Constantinople, it assumed a dominant presence and became Dubrovnik's immediate neighbor to the East. Rather than acting belligerently toward the Turks, Dubrovnik negotiated the paying of tribute in exchange for free trade. This was to their mutual benefit, since the Turks needed products from the West, which could pass safely through Dubrovnik, either via the Neretva River

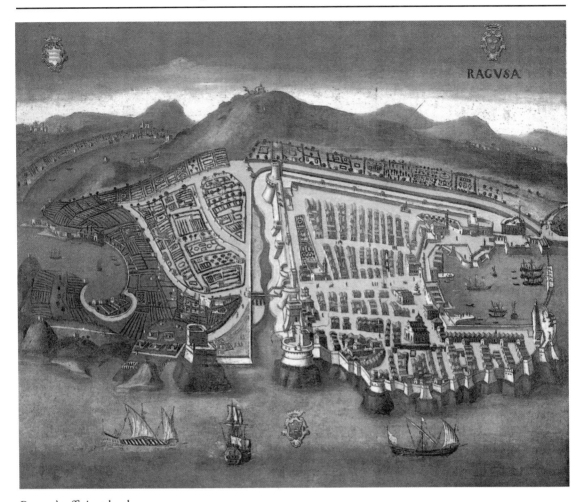

Ragusa's efficient land use in the sixteenth century: Urban gardens, harbor, and walls for defense. *Croatian State Archive*

valley to Bosnia or further east by an old medieval road through Montenegro.

Planning and maintaining the physical city as a reliable framework for a stable international economic and political life became a policy for Dubrovnik that could serve as an example to other places. In 1667 a severe earthquake all but destroyed the city yet reinforced planning through rebuilding. According to Krekić, the earthquake had "enormously tragic consequences and it was really a miracle that [the city] managed to survive."[22] The historic disaster damaged Dubrovnik's financial and administrative ability to carry on foreign trade, and without the skills of the stonemasons and shipbuilders in rebuilding the city, economic recovery would have been prolonged. Turning inward, Dubrovnik increased production of goods and improved building regulations, health codes, census-taking, and civic administration in general. This comprehensive approach to urban management is apparent in the spatial order and elegance of the Stradun, today's main street, which unifies the city and invigorates its social life, and in the regularity of housing up the slopes to the walls. This now famous street was created from the wetlands that were filled during the tenth and eleventh centuries, increasing the urban area within the upper walls by about one-third.

Bruno Šišić, Dubrovnik's landscape historian, has shown how this experience in urban planning and architecture during the early centuries laid the basis for designing gardens from the fifteenth century well into the Renaissance. It was because of the limited space within the walls that the fine examples of landscape architecture developed in the summer villas that spread up the relatively gentle slopes above the Pile Gate and in the monasteries as part of daily life.[23] (Again in the mid-1990s Dubrovnik is called upon to rebuild, this time after a human-made disaster, the 1991 shelling — the most traumatic destruction since that of the 1667 earthquake — rather than a natural disaster.)

Under a peace treaty concluded at Srijemski Karlovci, on the Danube, in 1699, Dubrovnik successfully negotiated with the Turks to revive the republic's land trade, though on a diminished scale. Dubrovnik continued to play its unique role as a point of contact for the Turks with the Mediterranean world up to Napoleon's arrival there in 1806.

From Napoleon to Austria and Croatian Independence

Dubrovnik took the brunt of the decisive blow given to Dalmatia by the French, but the city's strategic relations with the East and its uniqueness as a medieval city-state made it something of a special prize. Long independent, the city was caught between the French troops and the Russians, who had staged a violent siege with the help of the Montenegrins from their mountains high above the fjordlike bay of Kotor. On 26 May 1806 Dubrovnik decided to let the French in, and Marshal Marmont, with some four thousand troops, took over the city and in 1808 abolished Dubrovnik's independence. Marshal Marmont became the duke of Dubrovnik and in time developed a particular affection for this staunch, self-created city.

In the new order of *égalité*, against the will of the nobility, the lower classes were given a more representative governing council that included a new class of bourgeoisie in addition to the traditional nobles and plebeians. Major urban improvements were carried out during the brief occupation of some eight years as the French, already experienced in previous occupations and equipped with ample manpower, put their military engineering skills to work. These projects included access roads and bridges connecting settlements, public buildings, and a fortress atop Mount Srdj. The British laid siege to the city in 1813, the French having been weakened by the fall of Napoleon. Austria, already strategically well established in Split and Zadar to the north, took command from the British in 1814. For the next one hundred years — up to the close of World War I—this self-managed city-state was dominated by a Germanic culture of Europe's hinterland, distant from the Mediterranean context of Dubrovnik and its use of the Croatian languages. (I well recall this basic cultural antagonism during my youth in San Francisco, when I heard newcomers from Dalmatia express their unwillingness to serve in Franz Josef's army.)

Our expressive nineteenth-century traveler Paton provides some telling images of Dubrovnik in 1847. Arriving by ship at Gruž, he is impressed by the "dark blue waters and clear air" as he walks the hilly road to the Pile Gate. The "musical quality of the Illyrian language" appeals to him, but not the replacement of the centuries-old name Ragusa with the Croatian name Dubrovnik, a part of the Croatian Renaissance movement of that time. (Even in my youth in San Francisco both forms were used.) He finds that "the older people of the city, long a seat of culture, carry an aristocratic bearing," while "the peasants are full of natural ease and politeness."

Paton recalls the city, "the Slavic Athens of the seventeenth century," but the "taste, learning, wealth, commerce, science and politics have melted away. Only the outward city remains to nourish the patriotism of the Ragusans." He attends a ball given by titled nobility, echoing past opulence, and speaks of the "beauty of the women," the Hungarian Gypsy music, the Viennese waltzes, and "social customs becoming liberalized with the French and Austrian occupation." Afterwards, "in the moonlight," he climbs up the hill above the Pile Gate to see "the summer villas of the past, destroyed by the Montenegrins in 1806 and still in ruins."[24]

Dubrovnik's era of sailing vessels gave way to steamships when in the second half of the century Austria took over the shipping field by providing direct accessibility to the Adriatic ports from Trieste and Rijeka and by railroad from Vienna. At the turn of the century the modern harbor at Gruž, where I arrived in 1937, and the city's first modern hotel were built. At about the same time the streetcar line, the *tramvaj,* from the new harbor was built over the scenic route to the Pile Gate, together with the Hotel Imperial, shelled in 1991. A railroad line from Gruž to Sarajevo and road systems connecting Dubrovnik with the interior marked the beginnings of Dubrovnik's shift from the Middle Ages to the twentieth century. These facilities launched the "jewel of the Adriatic," with its attractions of history, climate, and theatrical setting, as a tourist destination, only to be blocked by the two world wars. Only after the revolution of the 1940s did tourism fully assert itself, especially after the building of the Adriatic Highway in the 1950s and the airport at Čilipi in the 1960s. During my first visit, in 1937, I saw Dubrovnik in a sorry state of physical and social disarray; on my return in 1968 and subsequently I found enormous progress in sensitive restoration of the city's rich array of urban treasures. As in Split and Zadar, this was largely due to the initiatives of locally devoted specialists. In contrast, in May 1990 I found Dubrovnik suffering under what Donald Appleyard has called "the stifling embrace of tourism."[25] Big-time hotel development, large-scale tourist promotion, and consequent new growth on the limited coastline beyond its walls has exceeded a reasonable balance with environmental history and identity. The walls have held the population to some five thousand for centuries, while the population of the narrow coastal region increased from thirty thousand in 1970 to more than fifty thousand in 1990.

The peninsulas of Babin Kuk and Lapad have become neighborhoods of hotels. On the slopes above the harbor at Gružž and along the heights above the Dubrovnik River multistory apartment buildings have been added. Boxlike, concrete single-family homes have been built, mostly without permit. Plans were drawn up in 1990 for a bridge high above this major inlet from the sea, long an obstacle to direct access; this would bring more congestion to the area and expedite urbanization up the coast. The latest environmental outrage, severely intruding on Dubrovnik's unparalleled scenic setting, opposite the island of Lokrum, is the Hotel Belvedere, a massive complex of connected units some twenty stories high that descends the face of a landmark promontory to sea level. Should permanent and tourist population increase, Dalmatia's finest urban heritage and source of regional identity could be lost to the artificiality of modernity. Ironically, the Belvedere was heavily damaged in the bombardments of 1991.

The Konavle region, the traditional rural area to the south, survived tourism but not the flagrant attacks of 1991. Its clusters of villages and small fertile valleys within easy access to the city made it an integral part of Dubrovnik. Inland from Cavtat and about ten miles down the coast—a few hours from Dubrovnik by *mazga* or by foot—the plateau land of Konavle, whose recorded history reaches back more than six thousand years—including Illyrians, Greeks, Romans, and Slavs—has provided a microcosm of the history of the Dubrovnik region. In early times the hinterland for Epidaurus (Cavtat), this beautiful site became a part of the Dubrovnik city-state in the 1420s.

The combination of valleys, hills, fields, and bays enclosed by mountain ridges fostered a network of interdependent villages, not unlike that of the Kuna region. The inhabitants kept their rural traditions alive, and their farming products maintained the breadbasket for the urban folk of Dubrovnik. That mother city's European shipping linkages nurtured cosmopolitan living and kept Konavle's cultural development in tune with the times. Indeed, only decades ago the airport located there brought the world to its doorstep.

Stijepo Mijović Kočan described this unusually varied landscape as "both gentle and dramatic, mountainous and coastal, flat and rocky, evergreen and ever changing . . . [lying] between the rustic agricultural life of its ancestors and its modern urbanized inhabitants, between yesterday's mule caravans on narrow paths and today's machines on asphalt strips, between two confronting worlds, whose quarrels and wars raged through Konavle, leaving deep and indelible wounds."[26] These varied qualities and ties to Dubrovnik became part of the deeply rooted identity of its emigrant natives in places like Watsonville, a coastal valley of similar character south of San Francisco. Little could they or Mijović have known that more "deep and indelible wounds" would be left when the village of Konavle became the site of the worst destruction by the Yugoslav army's wanton attacks on the Dalmatian coast in the fall of 1991.

An Urban Reading of Dubrovnik

Having "read" both Zadar and Split, we might expect Dubrovnik to be easier to read. We note in our reconnaissance how the city slopes down from its north and south walls to the Stradun, spread out before us like an open book. The simple up-and-down topography of the geometric street system displays itself so directly that the first-time visitor can readily perceive the city as a functional and visual whole. Deeper meaning can be gained, however, by experiencing block by block the predetermined continuity and coherence of the centuries of guidance expressed in this visual order. In Split, for example, Diocletian's Palace seems to have taken shape in slow motion, reflecting the external turning points of centuries. In contrast, Dubrovnik is almost like a snapshot of a single time period transported intact to the twentieth century. Thus, we must make a thorough reading at ground level and forgo the drama of reading the walls and towers alone.

Two choices suggest themselves as starting points: the Stradun and the walls. Having learned the importance of the experience of arrival, we choose the walls for their solid, enclosing character and all-encompassing elevation, beginning with the area well beyond the Pile Gate to heighten the sense of entry as an event in itself. Walking the walls high above the tilting rooftops, we can experience a 360-degree overview of the two halves of the city straddling the Stradun. We can then come down to eye level and walk this central spine, sampling the cross streets as they slope from the Stradun up to the walls and down again. We shall rule out entirely any thought of reading Dubrovnik from the cable car to Mount Srdj, rising inharmoniously above the city, aloof in its modern technology from the rich opportunity to experience the medieval pedestrian world at its best.

Now, to put ourselves in a frame of mind remote from the present century, let us contemplate the words of one who read Dubrovnik some

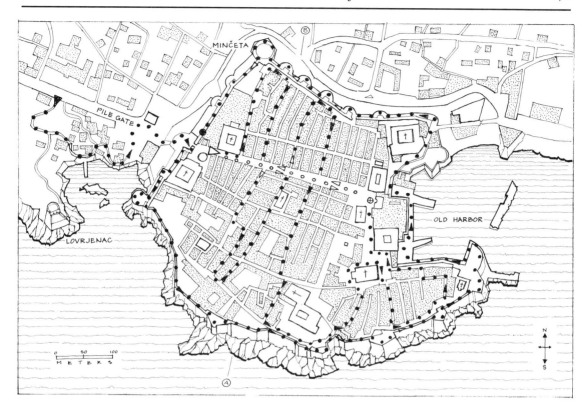

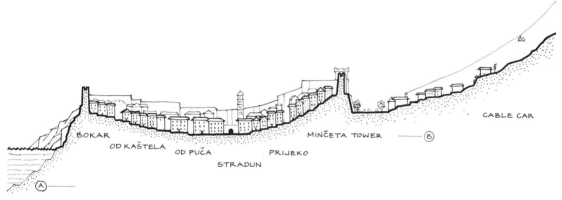

Dubrovnik: Urban structure and reading paths

five hundred years ago. Sir Richarde Guylford, a member of the court of Henry the Eighth, wrote in 1511 of his experiences in Dubrovnik on his way to Jerusalem: "It is ryche & fayre in suptuous buyldygne with marveylous strengths and beauty togyther with many fayre churches and glorious houses of relygyon. . . . there be also many Relyques, as the hed and arme of seynt Blase. [It is] . . . 'the strongest towne of walles, towres, bulwerke, watches, and wardes that euer I sawe in all my life."[27]

The Pile Gate Environs

Putting our thoughts in order for this Dubrovnik reading, we sit on a bench within a walled enclosure high above the sea, on the rocky promontory of Dance, less than ten minutes from the Pile Gate. Behind a fence of grillwork nuns in dark habits, from the convent of Sveta Mihaila, Gospa od Dance (Our Lady of Dance), work with spirited diligence at cultivating their terraced vegetable gardens. Close by and overhead on a tall ladder, we see a nun pruning grapevines, oblivious to our presence. The ancient cemetery behind us, where I found the family name of my music teacher during my childhood in San Francisco, lends an appropriately somber note to this medieval scene.

The terraces look down on almost vertical cliffs to a rocky shore and then across Dance Bay to the incongruous modernity of the Hotel Libertas, built in the 1970s, on the steep cliffs opposite. In spite of this visual invasion of a contemplative daily life, the nuns display a sense of belonging, connectedness, and commitment to their venerable environmental complex, a relic of Dubrovnik from centuries ago. Their blend of energy, concentration, laughter, and fulfillment in their work shows them to be enviably at one with the world. I found this place on the rocky path that begins across the street from my living quarters in a private home and leads down to a stone-lined cove with crystal-clear water, ideal for swimming at the end of the day.

Our thoughts in order, we ascend the path to the hilltop pine woods of Gradac Park to begin our reading with a purely nineteenth-century view to Dubrovnik's Pile Gate, towers, and walls. We stroll down through the spicy pine woods to the old city's main approach route, the Maršala Tita, which after the shelling of 1991 became Ulica Ante Starčevića, honoring the nineteenth-century Croatian reformist. During my visit in 1937 the hilly street overlooking the sea, with the simple, nonpolitical name Put od Pila (Pila Gate Road), was home to the streetcar line, the *tramvaj*. This product of the Austrian period did for Dubrovnik what the cable cars do for San Francisco. Villas from the past line the road, and an imposing building that was once a school for teachers now functions as the Dubrovnik Inter-University Center. We note that this neighborhood, Boninovo, displays many vestiges of Austria's turn-of-the-century tourism, including, on our left, the Hotel Imperial, restored in the 1960s to its original character. Like the Inter-University Center, it was bombed and burned in 1991 and restored by 1994.

These large, wooded properties contrast with the fine-grained, irregular layout of the ancient Pile neighborhoods on steep slopes overlooking the sea below the main avenue. We decide to explore them. There, winding byways surely would have been the homes of common sailors and fishermen centuries ago. Our unstructured wanderings reveal a geo-

Fort Lovrjenac and the winding byways of the Pile neighborhood.
Photo by author

graphical order that documents their very early settlement: two deep, protected coves with tiny pebbled beaches, the lofty Fort Lovrjenac, which we saw from heights of Danče, rising skyward between them. The second cove is called U Pilama. It is not the source of the name of the Pile Gate, as I mistakenly assumed; rather, the word *Pile,* which means "gate" in Greek, gave the name to the area. The promontory of U Pilama sweeps around to the foot of Fort Bokar, marking the southwest corner of the city walls we shall mount. The cove's bounding street, Tabakarije, became well known to me in 1937 as the home of family relatives in San Francisco. Their dwelling looked down on a dramatic waterfront setting whose tumultuous geologic formations evoked eternity, while the gentle rocking of small boats in the clear waters of the cove, ready to seek food from the sea, evoked human security. This image changed when the family was forced to seek refuge in Italy after the 1991 shelling destroyed their roof.

We mount the old ramped walkway up to the Poljana Brsalje, just outside the Pile Gate itself. This paved public open space, now a major terminus for thundering buses, taxis, cars carrying international tourists, is filled with townspeople, tourists, and travel agencies. The contemporary character of the scene belies any sense of connectedness to the imposing walls that serve as a backdrop. These monuments to human achievement were celebrated more by the peoples of the past, who built the broad viewing terrace above the sea. We pause against the rail to reflect on this striking conflict between past and present. There, high above the tossing sea, it became a tradition—one I myself enjoyed in 1937—for people to gather as an orchestra played each evening adjacent to the Dubravka

Restaurant. This turn-of-the-century flavor survived Communism only to be swept away by the motorized tourist world of today.

Our strategy has worked. After sampling the environs outside the walls, entering through the Pile Gate becomes truly a theatrical experience. Crossing the drawbridge over the former inlet to the sea, then passing through the outer wooden gates and finally the inner gate, we make a grand entrance as if onto a medieval stage set. This puts us all at once in a frame of mind to identify with the city as an integral whole. The stage in this case is the town square, the Poljana Paskoja Miličevića, with the Franciscan Monastery on the left and the convent of Sveta Klara (Saint Claire) on the right. The stone of the domed reservoir called Onofrio's Fountain matches that of both the paving and the building, unifying the whole scene. Here we mingle with townspeople and tourists, who seem to bask in relief from the stressful reminders of war and modern city life. The walls function effectively to block out the external buildings, people, cars, and asphalt. The push of that world gives way to the pull of Dubrovnik's medieval purity. Bursting into the sunshine on the square's ivory stones, polished by centuries of use, heightens our sense of joining a performance of another era.

The Stradun stretches out before us in perfect Renaissance perspective, lined with shops neatly shaded with blue awnings and thronged with people. This inviting vista terminates with the Clock Tower at the Luža Trg (Loggia Square). (Should we return at night, when the lights of the centuries-old buildings are reflected in the paving stones, our identity with the early citizens of Dubrovnik who created this environment becomes even stronger.) As we feel the dominance of stone in the Stradun's massive form and even its worn texture of stone, we absorb something of the the collective subconsciousness of its builders. Indeed, as we sit on the ledge of Onofrio's Fountain and contemplate the meaning of Dubrovnik, we are moved to a direct and deeper dialogue with the place than was the case at Zadar or Split, where overlapping turning points have blurred the qualities of contemporary urban life and those of the past.

The Walls

From the intimate scale of stones and Stradun you and I now seek the long stairway to the top of the walls to perceive the vast scale of the city in its entirety. Bear in mind that as we walk this mile-long aerial circuit you are also traversing a time-distance of the seven centuries it took to build this "skywalk," parts of which stand seven stories high. Urban form and time have joined to create a walking experience not to be equaled. Yet we walk here under very different circumstances from those of the people who built the walls and walked them for their own survival. You can to some degree share their sense of identity with Dubrovnik by gain-

The Stradun: From
above the Pile Gate to
the Old Harbor

ing your own sense of the massive walls as truly formidable and impenetrable. Inland they come tumbling down to us from their corner towers on the foothills of Mount Srdj, and seaward they rise up from the crags above the Adriatic, then drop to a low point of entry through the Pile Gate. There we enter the stairway through a tiny arched opening cut into masonry some twenty feet thick. Above, a niche holds a statue of Sveti Vlaho (Saint Blaise). The walls seem a proud reflection of what man's intelligence can accomplish in the face of nature at its most rugged.

A climb up the enclosed steps of ancient stone masonry brings us to a kaleidoscope of views. The Stradun stretches eastward below us like a straight edge dividing the city into two sloping segments. To its left, note the regularly spaced cross streets and coordinated facades of the seventeenth-century addition; then look at the broken skyline to the right, evidence of an evolution over time along the southern ridge, where the city had its beginnings. Red roofs cascade downward at varying angles, and small, domed churches speak upward for the reassuring role of religion in such an awesome place. The whole scene seems choreographed. The land and the pitched red roofs together toss in imitation of the sea at the foot of the walls. Down below, the stepped streets,

Sveta Klara's convent
and the Franciscan Bell
Tower. *Photo by author*

anchored in place by the relentless geometry and power of the Stradun, gather into one place the collective life of the city. In the foreground the courtyard of Sveta Klara's convent and the Franciscan Campanile speak for the protection offered to old Ragusa over centuries.

Joining this perceived movement of stone and tile, our vision swings landward, where, to our left, the fortified tower Tvrdjava Minčeta rises to old Dubrovnik's highest point at the northwest corner of the city. Behind, Mount Srdj proclaims the city's traditional defense against the interior. Now tracing our vision seaward toward the southern stretch of wall, we see the freestanding Tvrdjava Lovrjenac, which we saw from Gradac, rising more than one hundred feet on its own great rock in defiance of the sea and potential interlopers since its construction began in 1050. And to the west between these two fortified towers lies the residential area that extends out from the Pile Gate to Gruž and beyond to the tourist hotels of Babin Kuk and Lapad, safely away from the visual integrity of the Old City. Below us, under the bridge of the Pile Gate, a green, sunken park provides a quiet amenity recalling the leisurely times of the turn of the twentieth century in what once was the moat that repelled enemies. The open sea captures our gaze as we walk the crest of

The wall atop the cliffs, the sea far below.
Photo by author

the walls to the small bastion of Bokar, attached on a promontory as if holding this southwest corner of the city securely in place. Surely this must have been the point where the families of seafarers could last see their departing ships and where they could first catch sight of those returning from the long Mediterranean voyages of Dubrovnik's seamen.

As we leave Bokar, the geographical tension of this human-made citadel of security pulls us onto the narrow and most precipitous of the south-facing walls. On the right, the sea lies far below; on the left, the rooftops appear to toss in every direction. Although the footing seems precarious, the walls speak out across ages of use with such confidence in their hazardous position between sea and mountain that we feel assured of our safety.

We follow a series of curving segments of wall marked by six smaller towers, each carrying a saint's name, that served as chapels where prayers for safe return could be said within hearing of the magical sea. On a cold and windy day, as we look down on the red-roofed dwellings huddled together in seeming disarray, our sense of the walls' protectiveness overcomes our sense of their austerity. If, at intervals in our walk, you look more closely at the dwellings below, their basic order reappears in the straight and narrow streets that plunge downhill toward the Stradun. Beyond, the rooftops roll upward to the mountain backdrop. The entire panorama appears to shift continually upward and downward even as we stand on the solid, immobile top of the wall as if on the rim of a bowl. From this vantage point, so loaded with the forceful weight of the past, we look down on the human life of today. This quiet vitality can be sensed in images of people returning from market up the steep streets,

Inside the Gradska
Kavana, boat haven of
old. *Photo by author*

in clotheslines strung with wash in precise order by clothing types, and
even in the topless sunbathing on precarious ledges high above the sea
accessible through tiny openings in the wall.

We have now reached the southeast corner, where Tvrdjava Svetog
Ivana (St. John's Fort) looks boldly down on the Stara Luka, where in
the old days the man-made harbor protected sailing vessels from the tur-
bulent sea. Today, tourists embark from there to visit the pine-wooded
island of Lokrum, where Maximilian of Austria and his wife, Carlota,
lived in the restored Benedictine abbey and its gardens before the tragic
Mexican episode in their lives. As our minds retreat some thousand
years, we reflect on this rugged promontory of Pustijerna as the origin
of Dubrovnik. Its oval shape fitted to the land, and the ten or so streets
running north to south speak for the intelligent use of geometry by the
people of Dubrovnik in even the earliest times.

The walls turn us abruptly west on a long straight segment past the
harbor to the steps that come down in the vicinity of the historical center of the city, Luža Square. There we enter the ancient Arsenal, transformed in the twentieth century into the Gradska Kavana (City Cafe).
We have coffee and contemplate all that an unparalleled walk can tell us
in terms of the human values drawn on in the creation of Dubrovnik's
walls. For here are expressed attributes essential to society's well-being:
courage, persistence, security, protection, and continuity from generation to generation. Yet, in our times these ingredients seem no longer
necessary because of the powerful technologies and their personalized
strategies for building urban environments. Has our sense of identity
with a given place become endangered?

After our rest at the Gradska Kavana, our energy restored, we mount the steps to the northern section, starting our walk at the Clock Tower. On this eastern portion of the route we encounter the freestanding Fort Revelin, set apart from the walls by a bridge at the eastern Ploče Gate. Paralleling the Tvrdjava Minčeta projecting from the northwest corner, Fort Revelin, together with the massive complex of the Dominican church and monastery, has guarded the strategic northeast corner since the fourteenth century.

Walking the heights of the north wall between these two bold embattlements brings to mind the irony in their role as visible symbols of security in an insecure world. Inclined upward toward Minčeta, the highest point on the walls, the route choreographs us up ramps and steps, around the five remaining semicircular bastions. We feel like the smallest of creatures atop a gigantic abstract sculpture. In contrast to the south wall, with the threat of constantly moving open sea far below and the echo from the past of attack on the north wall, the steep slopes of Mount Srdj rise protectively, offering security. At this point one sees beyond the walls, not sea, but something of the modern world of cars, streets, parking lots, modern homes, and the cable car traveling on its steep suspension lines, seemingly thin as a spider's thread, to the overlook high above the city. All these elements intrude and force us to look back with awe and wonder to the combined daring and civic consciousness reflected in these works of the past. We question whether the technological wonders of today offer us a comparable degree of human advancement.

We look down once more on the interior of the city, where the regularity of the fourteen geometric streets that drop down like slots in a monolithic stone assembly of buildings suggests the city's quality of sheer intelligence, as if it had a directed and organic life. These slots reveal deliberate planning in the interest of efficiency, security, and sanitation by people of Dubrovnik, a legacy left to us. But from this vantage point we also have intimate views that offset long-range planning; ubiquitous clotheslines stretching across the narrow space of buildings, flower pots on window sills, and neighbors in friendly conversation from window to window. On a summer day the bright sun lights up a desirable and densely urban way of living. On cold winter days the scene clearly loses its appeal, and urban life draws inward.

As you look from street to street, your vision reaches clear to the Stradun at the bottom of the gradient and is then quickly pulled across the Stradun. There you follow on up a similar street that first runs on flat ground and then charges boldly up to the south walls along the sea, where we have already been. An image of Dubrovnik as the hull of a ship comes to mind, with the Stradun as its keel, and the north-south streets running up the slope as the ribs. The wall we have been following then becomes the remaining structural elements of the city-ship as a whole,

Looking down from
the walls to the
Stradun

holding the ribs in place and topped by an encompassing deck. This
contemplative experiencing of the walls has revealed a potential source
for identity with Dubrovnik, more mind-expanding than the readings
of Zadar and Split brought forth.

Shifting to a diagonal angle of vision, as we reach Tvrdjava Minčeta, a
peak on an "architectural" mountain ridge, we lose the image we had of
an orderly arrangement of straight sloping streets. Now the rooftops toss
in a delightful disorder, as if chasing each other for the sheer pleasure
of movement from one side of the walled city to the other. Besides the
Stradun, this landscape of rooftops is probably the strongest identifying
image of this medieval wonder. Of the millions of tiles, each of a different weathered red, how many have been shattered in the cruel shelling
of 1991? From Minčeta, gravity takes us down the sloping west wall high
above the Pile Gate; we again see the Stradun as it knifes through the
stone blocks.

The Stradun

We now expand on our image of the ship's hull by reading its "keel," the
Stradun, and walking the "ribs," narrow streets running north and south
up from the Stradun. Our attention focuses on the details of this "slot"

The strong converging facades of the Stradun

of stone. Note how the absence of curbs creates a oneness of facades and ground surface; the paving glistens in the sun, even reflecting the bright colors worn by pedestrians, and tells us of the social function the stones have served to achieve their sheen. The facades appear to be parallel but are not. Greater width at the Clock Tower adds variety to the Stradun as it is perceived from each direction. The street, more like a hallway, makes a clean cut, and with awesome determination this piece of human-made geology brings both sides of the city into mutual coherence.

Moving along the north side of the Stradun, we can count the fourteen identical blocks, implying both a sense of age-old structural solidity and the strong hand of responsible authority. The blocks are equal in width, yet variations in the design of the shop openings express the individuality of commercial uses. Entrances to the residences on the upper floors are on the side streets, a desirable separation. The several doorbells on each building tell us that more than one family lives in each dwelling.

On the south facade block sizes differ in length and door and window openings are irregular, even though both sides were built after the earthquake of 6 April 1667. For example, some have no openings onto the Stradun; others have a series of arched doorways on the street, usually with half-doors built into the arches to provide for selling products from the interior directly to the street, a practice left over from the Middle Ages. There are two stories above the ground floor, and some facades have shields or crests, indications of the residents' wealth or authority. The second block, twice as long, has eight openings, all typical store-front openings. The seventh and eighth blocks have ten openings, and

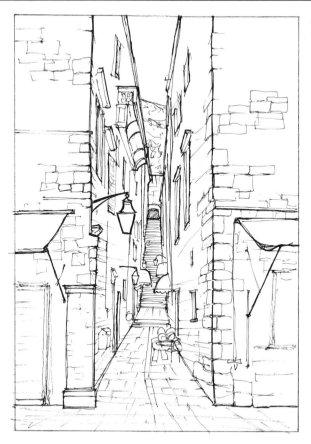

Stepped street up from the Stradun

the lack of cross streets breaks the continuous view down to and up the other side of the Stradun. Yet, this lack of continuity provides a welcome sense of enclosure, diverting one's attention to the strong line of the Stradun, and a relief from the regularity of the north side, with its "slots" of stepped streets.

Our dialogue with urban designers of centuries ago terminates at the Luža Trg with the Clock Tower, where the Stradun ends. At the square, with its variety of freestanding buildings lacking in regularity, we have a range of choices. The surface on which we walk now flows freely around the eighteenth-century church of Sveti Vlaho on our right, its baroque facade standing free, as if to break with the rigid discipline of Dubrovnik's earlier centuries. To the left is the Sponza Palace, whose arcade and lacy stone windows reflect the heights of prosperity of the patrician class and the intimacy of cultural contact with Venice's masters in architecture.[28]

Our sense of space continues to lead our vision from the stone face of the Clock Tower outward to the blue waters. This reminds us of the diminutive physical space occupied by this once powerful and internationally influential city-state—in commerce, diplomacy, and culture. To the right of the Clock Tower, the open flowing space moves us along the facade of the Arsenal, now home to the Gradska Kavana, which once housed the munitions for defending the harbor. Its facade joins the Rector's Palace, whose arcade of seven columns repeats the pattern of the arcade of the Sponza Palace. In the loosely defined space beyond are churches, dwellings, and the Gundulić marketplace, which together form a gracious focal point of Dubrovnik's unique governmental and social life.

As this level space terminates, the facades and rooftops of the city begin to climb the slopes to the south segment of the wall at Pustijerna. On the slow contemplative walk back along the Stradun to the Pile Gate, we consider again the role this grand street plays in Dubrovnik, binding its several neighborhoods into a powerful whole.

The North Neighborhood

To sample the neighborhoods, we begin on the north side of the Stradun, at the first street past the great Franciscan Monastery. There I am pleased to discover the name of Celestin Medović, the well-known nineteenth-century painter from Kuna, my father's village, carved into the street marker of native stone. This illustrates the reverence given the cultural heritage and reaffirms my own in-born sense of belonging here. We rise on a gentle gradient. An assortment of small shops probably associated with the families of the dwellings—a pinball-machine arcade, a beauty parlor, a film shop—give way as the stone steps become steeper. Our climb to the north wall and the base of Tvrdjava Minčeta breaks only at Prijeko, the north district's Stradun, which runs parallel to it and level for the same length. *Prijeko* means "over" and probably refers to the time when the Stradun was a channel and the unbuilt shore was "over there." The Italian version of the name, Priecho, was mentioned as early as the 1280s. We are transported to the purely pedestrian, nonmechanized world of centuries past, in which the human body has full responsibility for moving itself onward and upward. Beyond Prijeko the steps become even steeper.

As we count the 130 steps, our hearts pound with the unaccustomed exertion. A handrail appears, an invention of long ago and just as needed today. How many hearts—young or aging—have pounded their way up the steps over the years, eager to reach their homes at the top in response to the human instinct for protection from enemies, cold, and wind! And how far have we kept our bodies from experiencing the direct relationship between such elementary elements of life, the forces of gravity and heartbeats! I count my own—one hundred per minute—and try to divorce myself from this twentieth-century world of cars and technology and identify with a person of my age mounting these steps centuries ago.

But the reward is great, for at the top, with the Minčeta tower above, we feel reconnected to the environment. We let our eyes sweep down, just as they might have done in the sixteenth century, to the bright morning sun of May on the pavement of the Stradun, where people walk on level ground with an early-morning sense of purpose. The down slope of the north neighborhood leads our vision across the Stradun and upward to the ever-rising rooftops and the creamy stone facades of the south neighborhood and the wall overlooking the sea. The clean, legible geometry that defines the routes of human movement we experienced on the ground is replaced here with the same bold up-and-down and side-to-side movement we discovered atop the walls. The variety of perceptions we are capable of, according to our individual sensibilities to the visual experience, have overcome the rigid pattern of this north neighborhood dictated by the urban planners of the Middle Ages.

Standing here, at the Minčeta corner, we become enmeshed in the fifteen hundred years of urban history at our feet. A dominating theme develops out of this reading of the walls, the Stradun, and this sample of residential urban form. A theme of rhythmic movement pervades our body and moves the mind upward, from the "keel" via the "ribs" of the "hull" to the "deck" atop the wall, then down to the "keel" and back up to our viewpoint on Minčeta, in a sense high on the "mast." Giving way to this rhythm, upon leaving the level Stradun we must go straight up, knowing that we must again come straight down, following the parallel lines of the regularly placed stepped streets. The rhythm of the urban choreography inspired by the city's structural system is echoed by the voices of past planners. The contemplative experience evokes the rocking sensation of a ship, the very symbol of Dubrovnik's reason for existence and the source of the character of its people and its autonomous government.

But we can learn from other perceptions at a closer range of experience how early men and women fashioned this "city-ship" to fit their needs. In places we can see where austere buildings are attached to solid rock, the buildings almost seeming to be stone outcroppings themselves. By chance, a visiting geologist points out to us this wedding of nature's stone and man's stone as an example of geology and architecture working hand in hand with the help of human beings. We determined that prior to the devastating earthquakes of the sixteenth and seventeenth centuries people cut the wood of long-gone forests, but nature's fires and earthquakes turned the citizens of Dubrovnik to the permanence of stone. A grapevine stands rooted in a tiny opening in the stone paving, and a huge, ancient trunk protected from vandals by a modern casing clings to a house. Three stories up, at the roof, the grapevine rests on a trellis that crosses over the narrow street to offer shade for neighbors above and pedestrians below. Nearby, a fresh generation of boys play basketball in a court built between the walls at the Minčeta Tower, where certainly the sixteenth-century guards of the tower would have played the games of their times. Now we have come to perceive time measured in centuries as an integral element of the nature of a place.

To go down to the Stradun we choose the next street over, the steep slot of Ulica Plovani Skalini (Clergymen's Steps). The first flight breaks from the rigid geometry to curve around a rocky outcropping and form a small piazza. Eighty-two steps straight down, our anticipated view below to the Stradun is blocked by a jog to the left, providing relief from the dizzying descent. Twenty steps further down, the street straightens out to the pure geometry of the slotlike view to the Stradun. The facades of the dwellings here are narrow and without embellishment, more likely belonging to members of the middle rather than the patrician class, a reflection of the more humble status of this section of Dubrovnik. With each step downward, rather than a pounding heart, we feel a need for caution against falling, a sense of the rhythm of feet impacting the ancient stones against the weight of the body. Again, over the centuries, how many feet have taken the weight of how many bodies through this downward cadence of the built-in choreography of Dubrovnik? Leaving the silence of the residential areas, we hear the sound of tourist-oriented bars and restaurants on Prijeko, replacing the native social life of a decade ago. Stone firmly blocks sound, whereas these slots of streets convey it, and the wind as well. Like the Stradun, the level Prijeko serves as a counterforce to the up-and-down motions of body and eye that dominate our experience. Even at eleven o'clock in the morning music blasts forth from the bars on Prijeko and pink and yellow parasols are in place for this other world that has invaded Dubrovnik. Fifteen more steps down, and we reach the Stradun.

The South Neighborhood

Crossing over the Stradun, we note that the opening of the Ulica we have just descended is diverted by a block of buildings. This requires us to jog over in order to cut a cross section down the north "rib" and up the south. We do not ascend immediately, for Široka, a wide street, runs flat for a long block to Od Puča, the Stradun of old Ragusa before the north section was added. It becomes clear that the only level area within the walls for several centuries was precious and reserved for commerce and larger buildings. Today, narrower than Prijeko, Od Puča serves as a busy shopping street for Dubrovnik residents; the area has larger and more sedate restaurants than Prijeko, a number of them on side streets, occupying old palaces and using terraces for outdoor dining. A wide range of shops—for children's clothing, toys, stationery, books—is evidence that the street is used by residents, not tourists. A second, shorter and wider cross street, Za Gradom, is more oriented to household needs. Široka jogs a bit to the left and suddenly becomes Od Domina, with fifty steps straight up to Od Rupa. This meandering street, with its ups and downs and passageways under arches, and the Muzej Rupe surely were part of the original, incrementally planned city.

Turning right, we are attracted by a tiny, parallel alley named Ulica Puzljiva, which features a hand-drawn sign reading "Panorama Guest House." On closer look, we see the name Violić, an invitation for me to enter. A small restaurant on a garden terrace that is walled on one side looks down upon a verdant open space to a full-scale "panorama" of Dubrovnik's interior. Rooftops drop down to the Stradun and roll up to Tvrdjava Minčeta. After I introduce myself and my mission, there ensues an experience that only a person with the name Violic could have: wine, dinner, and life stories of origins and Bridges to other bearers of the name in Chile, Australia, and California. I learn of the other locality of persons bearing this name, my new friends' family: Osojnik, badly shelled in 1991, just north and inland from Dubrovnik. Again my dual identity with place is reinforced.

We return to Od Domina, and finally, after sixty steps more, with hearts pounding, we arrive at the final east-west cross street, Od Kaštela, the site of a castle in Dubrovnik's earliest years. The vertical distance we climb appears more than the horizontal blocks walked and creates a false sense of distance from the Stradun. The knowledge that this challenging ridge is where Dubrovnik began also creates a sense of distance in time from the modern life and its tourist orientation on the Stradun below. The distant past is everywhere, in the irregularity of the streets and the placement of dwellings, as well as in the smaller, more primitive structures.

With the wall rising well above us on the right, the route along Od Kaštela offers rich evidence of human adaptation to today's daily needs, which are very different from the survival needs of ancient times, symbolized by the ever-present wall. We take in image-forming bits of daily life: the sound of Beethoven's "Moonlight Sonata" floats out of a large school building as a young pianist practices; Siamese cats stretch out at ease on a stone doorstep in the warm morning sun. The walk widens between the dwellings that form a residential wall opposite the massive fortified wall. A slender iron bridge spanning the gap provides access to a space borrowed for vegetables and flowers. A grapevine rooted at its base has climbed two stories to the bridge. An elderly woman pulls in through her window white clothes precisely spaced out on the line according to type, as is the Dalmatian custom, from the support anchored in the ancient wall.

Further down, the wall facilitates recreation activities. An opening leads out to cliffs high above the sea where tanned bodies, with less than minimum coverage, casually sunbathe on precipitous ledges. Nearby,

Od Kaštela widens to form a rectangular space, backed and shaded by the wall, surely needed in the 1490s for movement of men and arms out of sight and out of the line of fire. In the 1990s the space has become a basketball and soccer court for the school where Beethoven played for us. Oblivious to our contemplation of the meaning of this mixed environment, the boys sense that we are intruders until their coach halts the shouting and shoving to allow us to pass in benign safety.

Dropping down a little, we reach the stately Poljana Rudjera Boškovića, named for Dubrovnik's Renaissance Newton and Galileo, known among European scientists of his time for his accomplishments as an astronomer, mathematician, and philosopher. At the top of its grand nineteenth-century baroque stairway, note the abrupt transition we have made from a compact residential neighborhood to the monumental splendor of Renaissance Europe. The towering mass of the Jesuit church behind us forms two sides of the square. Below, at the end of the Stradun, we see again the linked open spaces that irregularly weave the Rector's Palace, the Clock Tower, the Archive, and the church of Sveti Vlaho into a glorious complex. These contrast sharply with the byways, stairways, and archways of Dubrovnik's "ribs" and "hull." An image of majesty in a built environment comes to mind. Just dwell on these great buildings backed by the boats in the *luka,* people on the squares, and the east walls sloping up to Fort Revelin. Reflect on the massive Dominican monastery that anchors the northeast corner. Together, these symbolize both the civic achievements of this self-sustained city and its underlying human yearnings for meaning in the places we create. At this moment of reflection, the bells of the Sveti Vlaho ring forth and continue to ring with a glorious sound celebrating that very day the first meeting of the democratic majority recently elected into the Hrvatski Sabor (Croatian Parliament) in Zagreb.

We move along the route close to the great wall to trace on foot the long, pointed promontory,—itself shaped not unlike a ship—where settlers from Epidaurus established themselves. From this oval site old Ragusa evolved into the Dubrovnik we have just experienced. Pustijerna is cut by nine narrow streets that step steeply down from the crest above the sea to the level of the Palace and the Cathedral. We zigzag up and down these streets, under archways, all the while protected by the wall, rising above us four stories high in places. We try to imagine what life was like when Pustijerna was all that people had built on this rugged peninsula.

We complete our reading of Dubrovnik by returning to the Stradun that evening, Wednesday, 30 May 1990. Jubilance is in the air, and thousands of people have gathered—here as in all other Croatian cities—to celebrate the new Parliament. Dubrovnik's historical setting heightens the mood of optimism for the future. Restaurants and cafes serve free *rakija* from flower-decked tables in the streets.

Only one year later this event would be reversed. Around the world, the media would expound stories of tragedy, sorrow, and fear, showing pictures of these very walls pock-marked from shelling, rooftops crushed, and windows spouting flames. The human life of this proudest of cities would be brought to a standstill. The euphoria I experienced in this reading of Dubrovnik in 1990 made it impossible for me to imagine the Stradun, the walls, the historical monuments, the churches of Dubrovnik undergoing the bombardment of 1991. This is a prime example of how a built environment can be made a target for the sake of destroying a place as a symbol of identity.

CHAPTER FOUR

Three Seaside Villages of the Islands

It is just that the people who have been living on the island for a long time and have been bearing its burdens and civic works should also benefit from the resulting good, rather than the newcomers who have not shared the burden. — Statute of Brač, 1305, amended 1432

VILLAGES AND TOWNS COMPARED

At first glance, shoreline settlements of Dalmatia look very much alike: creamy-white stone houses with slate or red tile roofs stand shoulder to shoulder; dark green pines rise against a brilliant blue sky above a matching blue sea dotted with ships; shoreline walks give shape to the town harbor, the focal point of community life. However, a closer examination reveals important differences. On the island of Brač, across from Split, the gentle slopes, upland plateaus, and accessible shores facilitate contact between settlements. These geographical advantages foster a sense of collective belonging to the island as a whole among the residents of Brač's twenty-two villages and towns. The urban forms of Sutivan, Pučišća, and Bol, however, generate differing local identities. In contrast, on Hvar a sharp main ridge stretches the entire length of the island, thus concentrating settlement in the northwest, where rare level land and natural harbors meet. Thus, the town of Hvar dominates in the island's affairs. To the south, the topography of the island of Korčula is somewhat similar to Hvar's, but with numerous, less divisive ridges, permitting scattered settlements. Furthermore, ample harbors fortunately occur at both ends of the island, resulting in two large towns—Korčula to the south and Vela Luka (Big Harbor) at the north—widely separated, each with a strong sense of regional leadership.

A determinant of distinctiveness among places is the particular way water and land meet in those places. In Sutivan, Pučišća and Hvar the land "wraps around" the harbors, though in different ways. Sutivan creates a "theater" with five "stages"; Pučišća tightly encloses its harbor and village activity, and Hvar splays its arms open to the sea. In Korčula land and town boldly thrust out into the sea. Finally, in lineal Bol water and land stretch out side by side.

These variations in urban scale and form have anthropological and design origins, both essential to understanding the nature of the indi-

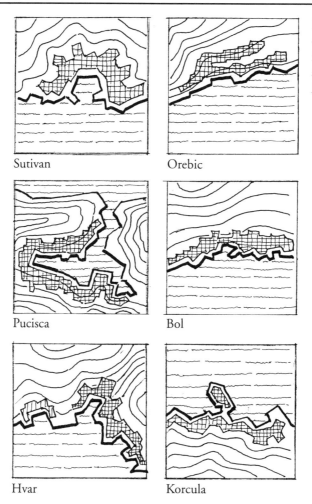

Reading reveals contrasting urban forms in four villages and two towns. *Diagram by author*

Sutivan   Orebic

Pucisca   Bol

Hvar   Korcula

viduality and the local identity of a place. Generally, in Dalmatia a town *(grad)* is larger than a village *(selo),* which in turn is larger than a nearby hamlet *(zaselak),* the latter housing closely related families in a *zadruga* who are the current descendants of forebears going back centuries. The differences may not be readily defined; for example, Sutivan, Pučišća, and Bol are equals in an islandwide system of compact, urbanlike villages, small in size and population and localized in function. On the other hand, the towns of Hvar and Korčula, to be read in chapter 5, are considerably larger and play central roles on the islands where they are located.

Following this hierarchy up to mainland cities, Zadar, Split, and Dubrovnik serve as focal points for the subregions occupied by towns and villages. These cities have maintained close relationships with these smaller places, since much of their growth has come from the islands. Their identity with them has continued because of close family and cultural linkages.

From an anthropological perspective, a certain sense of intimacy and continuity of relationships between people distinguishes the villages from larger places. As the anthropologist Lawrence Wylie has put it, "A village is a place where everybody comes to know everybody else."[1] His unique study of a village in Provence parallels my own study of Dalmatian villages for the way that the intimacy of daily experience distinguished "insiders" from "outsiders," a key to true identity with place.

As to uniqueness of design, it was Thomas Sharp's drawings and words in his *Anatomy of the Village* that inspired me to study further the Dalmatian villages I had first experienced in 1937. Cautioning us to protect the pedestrian scale, he made clear how individuality of urban form can grow out of a given site's physical characteristics, local cultural history, and community consensus: "Since there is an infinite variety in sites and a wide variety of functions . . . and since nearly all villages have grown naturally, there is no set pattern to which village plans conform. No village is quite like any other. And this, of course, is the glory of it, that every village is an individual place."[2]

SUTIVAN: THE FIVE-STAGE THEATER

Brač: The Regional Setting for Sutivan, Pučišća, and Bol

Sutivan, like many small towns and villages on the Adriatic coast, lies between two threatening forces. On the one hand, the younger population has shifted to the modernized mainland, attracted by opportunities for material advancement and urban cultural life. The jagged skyscraper skyline of Split stands less than an hour's ferry ride away. On the other hand, Sutivan's small-scale maritime urbanity and perfection of form could well be overrun with the cult of tourism in the quest for sun, sea, and relief from the monotony of increasingly large-scale apartment complexes of inland cities.

The plateaus, gentler slopes, and shores of most of Brač facilitated a common identity among the island's settlements. Its single peak, Vidova Gora (Mount Vitus), rises high above the south shore. Although Brač had been occupied for centuries by prehistoric peoples—the Illyrians and Romans—these communal qualities evolved after the seventh century, when the first Croatian tribes settled there permanently.

More has been written by Bračani scholars about the historical sources of identity with their island than natives from other geographical subregions of Dalmatia have written about theirs.[3] Even before World War II and the revolution, a group of intellectuals in Split representing various fields and localities on Brač banded together to form the Croatian literary group Hrvatski Skup and carried on serious research on Brač as a place, its past, and its people. Led by Andre Jutronić, they revived the

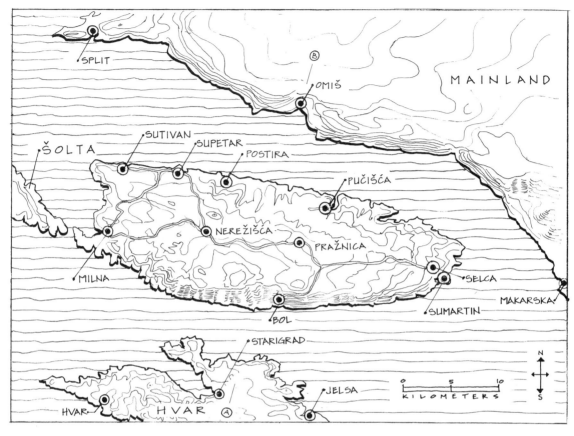

The island of Brač

earliest known writings on Brač, those by Bishop Dujam Hranković, re-
tained in manuscript form since 1405. To read his descriptions of Brač
now is to bring the past to life and reveal a level of awareness of an environment and identity with it parallel to our own today. Hranković portrays the life of the six thousand inhabitants on the hills, plateaus, and
wooded mountains that supported many kinds of animals. He records
the history of the preceding centuries and names the governmental and
civic leaders of his time. In so doing, this Bračanin of old creates an
image of an orderly way of life as real as ours is to us now.[4]

Hranković's writings were republished in 1802 by Andrija Cicarelli,
a pastor and native of Pučišća. Contained in Cicarelli's own *Observations on the Island of Brač,* Hranković's work is the result of studies of
vital documents and records kept in churches and monasteries. Hranko-

vić expressed his wish to leave behind an image of the homeland for later generations.[5] Cicarelli recognized in those writings a strongly felt identity with Brač resulting from love of place, echoing down through generations to our century and to myself as I write about his words.

Following this theme of identity with place, Cicarelli wrote in 1821 *A Historical Review of Pučišća,* which was translated from Italian to Croatian in 1918 to commemorate the founding in Pučišća of a branch of Split's Hrvatski Skup.[6] My own copy, given to me in Pučišća in 1937, was inscribed then by a descendant of the Cicarelli branch of my grandmother's family. Our shared Bračani great-grandparent was a contemporary of this revered pioneer historian.

Taking up this thread of connectedness between these two historians, in the 1930s Andre Jutronić collected population information from various parish archives covering some thirty-one villages and hamlets, going back as far as 1566 for Pučišća. Arranged by family, this information covered their origins, localities, churches, and cultural monuments that determined their way of living. Interrupted by World War II, the work came out in 1950 as the first work in the series of intermittent works Brački Zbornik.[7] In 1954 issue number 2 came out under his editorship, bringing together articles by seventeen members of the Brač group written "in the spirit of the times," that is, just after the war, when suffering and destruction had to be put to rest and identity had to be geared to a changed future. Topics ranged from Hranković's history of five centuries before to heroism in the war and revolution, to schools, electrification, health, roads, and tourism.[8]

In 1960 Zbornik number 4 focused on history told through cultural monuments and updated Jutronic's population studies from 1579 to 1958, village by village. (For example, in 1579 Pučišća had a population of 250; the number rose to 450 in 1708, to 1,200 in 1857, and to 2,256 in 1900 but dropped to 1,616 in 1958.)[9] In 1968 Dasen Vrsalović put Brač's entire history into Zbornik number 6, reaching back to the Stone Age and following the Illyrians, the Greeks, the Romans, and the early Croatian settlers, who entered from the mouth of the Neretva River. The work covers the Venetian, French, and Austrian periods and concludes with Brač's status after World War I as part of the new Kingdom of South Slavs.[10] Vrsalović takes us up to the emigrations from the late nineteenth century through the 1920s of large numbers of Bračani to California, Chile, New Zealand, and Australia. A full account of this distribution worldwide, demonstrating the continuing connectedness to the homeland, is presented in Zbornik number 13, published in 1982.[11] Outstanding among writers from Brač is Vladimir Nazor, a native of Postira, near Supetar. He is known in Dalmatia for his poetry inspired by the special qualities of the island, its villages, and people.

Today, a strong sense of belonging is expressed by each of the twenty-

two of the island's villages and hamlets. They are fairly evenly distributed over the island, which measures about twenty-five miles long and
three to seven miles wide and has a population of some twelve thousand
persons. About two-thirds of the villages have fewer than five hundred
persons; the remainder average around twice that number. This demographic pattern has contributed to a sense of commitment to and identity with a single small urban place on the part of most of the people.

The wild, harsh landscape of Brač, studded with *gomile,* gives the
island a potential for a more personal type of tourism, related to the
natural environment and village life itself. Paved roads and a bus system,
completed only in 1979, greatly advanced the social quality of life for
island residents, improving their access to schools and health facilities
and reinforcing the existing collective sense of identity with the island
as a whole. In the nineteenth century Brač's geography permitted comparable economic growth in the interdependent mountain and seaside
villages, such as Pučišća and Pražnica. Wine, olive oil, fish, dairy products, and wool were more or less standard products in all the subregions.
A number of places had fine stone to be quarried. Local access to these
limited resources allowed the residents of Brač to be more self-sufficient
than those of other islands.

The consistent smallness of Brač's urban places and their proximity to
Split have been advantageous in fostering a self-reliant social, cultural,
and economic life. The introduction in 1980 of car ferries year round
made possible part-time commuting, continued interaction within extended families, and use of old family homes for weekend and vacation
purposes. Each summer, international airlines bring Bračani emigrants
and their progeny from overseas.

I have referred to Brač's tradition of institutional self-sufficiency with
the formulation of its own governmental statute at the Benedictine monastery at Povlja in 1184, an event that lessened the impact of Venice's
control. The self-governing charter included forceful, detailed environmental regulations concerning community grazing and forest lands and
individually owned farmland. This same spirit of self-management continued under the former Yugoslavia's early policy of decentralized socialism, which favored smallness and uniqueness. Each village has expressed
its attitude toward tourist development through its *mjesna zajednica*
(local improvement committee). For example, Sutivan, the subject of
our first reading, took a strong stand against new hotel development,
which it saw as a threat to its traditional identity.

Reading Sutivan As a Whole

Essentially, Sutivan's system of streets and land uses is boldly structured
around a central port area, the *luka,* whose loosely radial pedestrian
system gently brings the town populace to the maritime terminal and

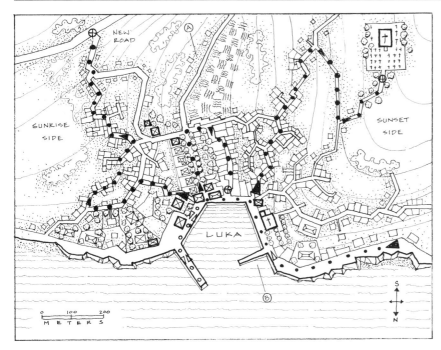

Sutivan: Urban structure and reading path

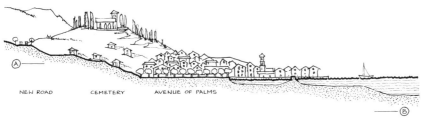

various community services, and the Central Greenbelt and Avenue of Palms, lying in the lowlands dividing the village into two sloping sectors. I call these two sectors the "sunrise side" and the "sunset side" of Sutivan, for the dominant image of nature's role in structuring the village. The clarity of this urban form greatly facilitated my first steps toward developing this "urban reading" method.

First let us examine the configuration of the shoreline site and compare it with the nineteenth-century map of Sutivan on the wall of the *turističko društvo* (tourist society) office. This map showed a small inlet from the sea that provided a bare foothold for a port in early times. Its continuation inland as a low-lying drainage route divided the town site from the beginning into two gently sloping and quite separate settlements. The inlet became the *luka,* around which the leading families built their villas. This made possible the export of olive oil, wine, and other products from the sixteenth century on. The drainage route facilitated access to rural areas inland, where settlements were concentrated on the safer highlands. With the advent of the steamship in the late nineteenth century the inlet was reshaped by the construction of a five-sided *obala,* with sea walls, docks, and a promenade. This basic system served a peak population of eighteen hundred persons in the late nineteenth century, even more than today's summer population and three times the present year-round population.

The Luka *and the Five Stages*

A brief reconnaissance of Sutivan tells us to begin our reading at the *luka,* the magnetic heart of Sutivan. As we approach, note how the jutting curves of the rocky shorelines to the east and west emphasize the all-embracing, compact, and intimate quality of the *luka.* From either direction one experiences an element of discovery and surprise, even comfort and refuge, in comparison with the rugged character of the adjacent shore. The wide-open exposure to the sea, the meandering stone beaches, and loosely strung-out dwellings contrast with the hidden, protected quality of the geometric *luka,* clearly a bold work of man.

Our eyes define a strong image of maritime and urban social life, brought together by the five-sided harbor that dominates the *luka.* This rigid form becomes a water-surface piazza furnished with fishing and pleasure boats, a highly receptive point of interest that all the residents of Sutivan can easily reach on foot. The multisided form of the harbor provides several stages where the people can act out their individual roles in daily life: sober housewives en route to market at dawn, tranquil old men resting on benches, schoolchildren headed for the bus, handsome young people eager to see and be seen, and visitors such as ourselves looking around. And it allows them a chance to play their parts as members of the community: at the early morning departure of the ferry for Split and its arrival in the afternoon; at the fish dock after that; strolling on warm summer evenings; and at the start of a funeral procession to the hilltop church and graveyard. Brian Bennett, an anthropologist I was able to lure into a one-year study of Sutivan, warmly describes the drama and cultural significance of this human scene: "The villagers' lives are acted out in front of the community. The quay is as a stage and the villagers are actors upon it. They make their entrances throughout the day, according to the routine of their individual lives. The daily round of life is open. It is this open quality of daily activity that reveals the village's persistence in maintaining many cultural patterns against the onslaught of tourism."[12]

We sit on a bench to observe these activities taking place around the *luka*'s *obala*—the sea wall and walkway of stone—as if we were in a theater with five stages. The community serves as performers according to the ever-changing scenario of daily life, or are they the audience for us as intruding performers? On three of the stages the facades of buildings provide a truly theatrical backdrop. Two architectural masses with towers to the east and west call attention to the opposing hillsides of the town

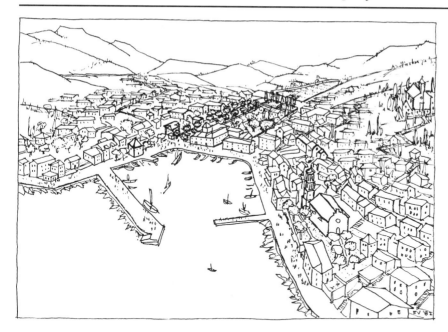

Sutivan's community life is shaped by its five-sided harbor. *Drawing by author*

and give a sense of balance, integrity, and definition to the encircling five stages. We feel firmly and comfortably placed—both physically and socially—in the heart of Sutivan. We feel a sense of identity and connectedness, even as outsiders. Then, too, the distance across the water from stage to stage is small enough to stir human contact: individuals can be observed, recognized, and even hailed. Indeed, Sutivan's *luka* appears to have been sensitively designed, not by a specialist in urban design but by the collective subconscious of the community residents over the centuries to meet their own psychic and functional needs.[13]

The robust arm of stone that is both breakwater and pier juts out from the shore to keep out the open sea. It serves as the main stage for the lively scene when the early morning ferry from Split docks or when it returns from Supetar, Splitska, and Pučišća: the anticipation of arrival, the frantic rush of passengers to get off, others pressing to board, the emotions of greetings and farewells, the unloading of food and goods from Split. Whether "insiders" or "outsiders," we feel securely integrated into the mellow yet rugged life of Sutivan.

The second stage begins where the *obala* leaves the sea and has as its backdrop the dominating Marjanović *kaštel,* built of ancient stone with a sundial high on its walls, a reminder of the Middle Ages, when fortification was essential and the sun—rather than the clock towers—told the time. A solid row of stone buildings and a gracious villa with large trees, entry terrace, and extensive gardens complete the setting. The villa belonged to the eighteenth-century poet Kavanjin, who turned to his native Brač for inspiration and identity. All these manifestations of urban development in harmony with people together demonstrate the

The ferry's daily arrival from Split. *Photo by author*

power of place in shaping the cultural level of Sutivan. Here the towns-people are en route to boat, market, or beach, and the old men of the village spend summer mornings on the benches in the shade of the date palms on the grounds of the poet's villa, now the Tourist Society's office.

The young men frequent the *kavana* (cafe bar) located where the *obala* intersects with the Avenue of the Palms to form the Times Square of Sutivan. Constant movement from 5:30 in the morning in summer to early afternoon and again in the evening makes this an active intersection. Here we find—readily visible and accessible, yet discreetly spaced—the facilities one finds in all Dalmatian towns and villages: the town news kiosk, the pocket-sized Dalma supermarket, a restaurant, the public telephone, a *kavana,* and a small steamship office. Their agree-able, intimate placement in Sutivan represents the prime source of the village's uniqueness of identity and social interaction.

The third stage is delineated by the facades of two mansions that were home to the town's leading families in the nineteenth century. One is the Ilić Palace, on whose walls we can read the dates 1878 and 1885. The stone shingle roof of one wing, however, speaks for this family's origins in even earlier and less affluent times. With only a few elderly members of the family remaining, we ponder over the future of this mansion and its extensive walled-in grounds rich and cool in verdant trees, rare on the stony island. Together with the Avenue of the Palms, it forms the Central Greenbelt. Questions come to mind about the future: Restore the villa to serve as a small hotel in summer and a place for community activities in winter? Open the park to the public and integrate it into the

Avenue of the Palms.
Drawing by author

town's structural system? Find ways to attract year-round population as a means of justifying its restoration?

At the bend in the road where the fourth stage begins is a minor node, formed by the open fish stand and the bread supply shelter. People come here to enjoy the afternoon shade on a long stone bench under the Tamarisk trees in a tiny triangle of park whose ornate pattern recalls past elegance. A path leads to the west neighborhood above, which we shall read later on. Its sheer architectural quality and cultural symbolism make the backdrop for the fourth stage the most theatrical. Here a row of very different residential facades, representing many periods, even the contemporary period, is anchored by the creamy stone mass of the church of Sveti Ivan (Saint John, the source of the name Sutivan) and its neo-baroque Campanile. Both rise majestically against the blue sky from a tree-lined terrace high above a retaining wall softened by stone planters billowing with oleanders. Stone benches invite us to contemplate Sutivan's daily drama. We observe the last stage, a small but important one: the pier and access ramp where smaller private boats and the fishermen dock. In summer this is an important center of activity throughout the day. The pier offers the best position for viewing the other four stages, which become a single, ever-changing panorama of human activities against a changeless urban landscape of stone.

The Central Greenbelt and the Avenue of Palms

A fresh and very different character challenges our perception as we take to our feet again to read the second major component in Sutivan's overall system: the Central Greenbelt and the Avenue of Palms. Strong in imagery, this component tells of the changes in social mores that came with the new large-scale urban design of Austria's Hapsburgs. The elite urban fabric here represents the deliberate neo-baroque planning for the rising bourgeoisie and contrasts sharply with the everyday needs of the community, expressed in the central *luka* and the sunrise and sunset neighborhoods on the hillsides.

The Avenue of the Palms is a bold green axis some two hundred meters long and forty in width bordered by date palms—a nineteenth-century import from North Africa to virtually all Dalmatian towns—placed with absolute symmetry. On the east side a row of large villas, the last of which is occupied by the post and telegraph offices, continues the grand urban manner and contrasts arrogantly with the irregular pattern of the more humble east neighborhood—the sunrise side—rising above. The formal front gardens add to the sense of spaciousness. Along the entire west side runs the high wall of the private park of the Ilić Palace, furnished with benches from times gone by. These elements make up the Central Greenbelt, which is, after the five stages, a second source of Sutivan's uniqueness, a green and lush basis for urban identity.

The garden's forbidding high wall contrasts with the openness of the sea to the five stages. Palms and pines of extraordinary height mingle in leafy chaos, suggesting the horticultural enthusiasm and curiosity of the creators of this garden as they challenged the naturally harsh and stony landscape of Brač. Seagoing families that they were, their source for plant materials may have been the entire Mediterranean, and their inspiration may have sprung from the botanical gardens of London or Liverpool. This hidden landscape would provide a rich cultural reading in itself were one allowed to enter.

This nineteenth-century formality continues beyond the Ilić Park to a tangle of trees and shrubs that invites exploration of its history and future. The regular spacing of neglected oleanders and other plants reveals a true, though forgotten, formal square divided by crosswalks into four quadrants, each with a date palm and the remains of fragrant laurel hedges. Stone benches and a central monument set in a grand design, now lost in a jungle of growth, suggest values of a way of life that came and went, leaving its message in this green heart of an otherwise stone Sutivan.

Finally, an open-air vegetable market and ornamental water-supply facility are evidence of linkages between the town and the open farmland of the narrow valley beyond during the earlier periods when Sutivan relied on products from its own stony fields. The date 1888 on one large house overlooking the square further coincides with the times during which Sutivan rose from an ordinary village status to become an urban gem.

These discoveries make the dynamics of the city's past clear to us and bring the essence of its spirit of place within reach. For here are embodied the regional results of nineteenth-century prosperity: a grander urban quality and a more sophisticated lifestyle. This became the setting for a heightened intellectual and professional life as the sons of leading families returned by steamship and rail from study in Vienna and Prague. Surely, the unprecedented urban development of Vienna accomplished by Franz Josef reverberated in Sutivan. Sutivan's leaders too aspired to rise above the lifestyle of the simple fishing village of earlier times.

We can now see how this low-lying bottom land with its tidal inlet was reclaimed when the sea wall was built in the 1870s. Thus, the soil was cleansed of its salt content, and the conditions were created for a formally landscaped park and promenade as a unifying feature of the village. The new identity with display and public life of this tiny Ringstrasse contrasted with the rural ways of the farm-oriented families of the two neighborhoods on the flanking slopes we are about to explore. This contrast in environments today dramatizes for us the cultural history of Sutivan and contributes to its uniqueness. We wonder how unaware visitors might be and how much more this Central Greenbelt would be enjoyed if its story could be told, its environment made more "legible," and its potential for contributing to the cultural enrichment of Sutivan's residents developed.

The Sunrise and Sunset Sides

In this perspective of contrasts, reading the humble east and west neighborhoods heightens our visual and intuitive experience of Sutivan as a whole. We note that here the sun's position, its rising and setting, where there is shade and where there is not, have been dominant ecological factors in determining Sutivan's distribution of people according to their social position.

We begin on the sunrise side, at the intersection of the second and third stages. How the east neighborhood evolved is plainly written in the miniature main street—some three meters across, wide enough for two *mazge* to pass—which leads from the *luka* up the slopes to the Brač plateau. The tightly spaced stone houses, some with roofs of the flat, shingled stones of earlier centuries and small yards for farm animals, recall a time before the building of the Avenue of Palms. Down this stone-paved route still comes an occasional *mazga* loaded with greens for the few livestock housed in the neighborhood. Today it is only a short walk to the Dalma supermarket for the milk, cheese, and eggs that traditionally were produced at home.

This kind of free and subjective reading of the environment opens our mental door to the identity with place of villagers of centuries ago. We can understand the social and economic life of the past through details of architecture: stone openings rounded at the sides to admit huge wine or olive barrels, dormer windows for the traditional top-floor kitchens that leave the living floors free of odors, the side wall remains of the former peaks of cottage roofs where second and third stories were added as families grew and life shifted its emphasis from rural to urban life. We reach the top of the hill and the outer edge of Sutivan. Below, the two neighborhoods nestle, separated by the green heart of Sutivan. The Campanile rises out of the red roofs, backed by the blue sea. Beyond, the airy violet mountains of the Dinaric range seem to rise lightly over the mere hint of the ever-lengthening skyline of metropolitan Split.

Here the illusion of identity with the past fades as the newly asphalted main access road to Sutivan unceremoniously intrudes. As a sign of change to come, new "modern" vacation houses begin a suburban pattern that blurs the clean edge of the traditional village, each introducing an individualistic image that contrasts with the mellow, collective structure and the compact urban texture we have experienced so far.

Descending from the hill by tiny side streets, evolved through decisions by humble people in response to social realities, we approach the glistening blue sea. Note the large, pretentious houses on the preferred sites close to the waterfront. These villas of the turn of the century with their private parks and great trees attest to the geographical segregation of social classes from hills to shore.

Finally we experience the sunset side of Sutivan. It has an appealing unplanned quality similar to that of the sunrise side, but here the houses are more consistently strung along a single, main road up the hill to the fields, where steeper terrain than on the sunrise side allowed fewer by-ways. The larger number of older houses, many in a state of abandonment or ruin, suggests that this district dates back several hundred years. Furthermore, unlike on the sunrise side, a number of dwellings along the narrow main route up the hill have combined window and door openings on the front facade at street level. These served as shops in early times; in fact, this neighborhood very likely played an active role in the life of Sutivan as early as the seventeenth century.

Again, as we approach the edge of the village on this side, we see the Central Greenbelt covering the narrow floor of the little valley. The clarity and unity of Sutivan's overall structure, the very core of its sense of place, is reinforced. Nearby, the contrast between the monumental stone gateway to an ancient vineyard of landed gentry and the simple dwellings on the hill suggests that the latter are homes of the farmworking class who in formative centuries peopled this west neighborhood.

We complete our reading of Sutivan by following a secondary neighborhood route to the seventeenth-century church of Sveti Roko (Saint

The sunset side's rural origins. *Drawing by author*

Rocque) and the hilltop Cemetery. The stone of the stepped ramp glistens from centuries of human wear and patiently awaits repair or, worse, burial under asphalt. Ascending to the church, we sense the force of religious functions in shaping the sunset side. The unchanged mood of the landscape evokes an identity with the past inhabitants—a sharing of their sorrows—as we follow this route of centuries of religious and funeral processions. At the crest of the hilltop the past lies about us in the vast array of tombstones set amid the fragrance cypresses, pines, and rockroses. They represent the very families and individuals who over the centuries created the distinctive environmental qualities we saw in Sutivan, lying below, a mixture of red roofs, cream-colored stone, and blue waters. Rising above us, a tower built by the British at the time of the French occupation breaks the serenity and sense of "oneness" of place gained from visually experiencing Sutivan.[14]

What We Have Learned from Sutivan

The overall pattern of Sutivan and the sequence of its basic components reveal four definable sources of uniqueness and collective identity, all of which are determined by the original configuration of the natural site: the well-articulated juxtaposition of its major and minor circulation routes, their close relationship to the centers of human activity, the contrasts between the neighborhoods, and the way the neighborhoods are framed by landscape and seafront. Sutivan's smallness enhances the potentials for perceiving these qualities. The result is a high level of human appeal in terms of sociability, diversity, community containment, and authenticity. A sense of wholeness in both the site and the design of Sutivan provides strong mental images on which to establish our identity with the place. In short, we conclude from our first village

Up the steps to Sveti Roko and the hilltop Cemetery. *Drawing by author*

reading that the method can serve to open our minds; that identity, uniqueness, and a sense of place can be found and defined; and that criteria for use in today's urban design can be sifted out.

These qualities come together in the visual and social image of Sutivan as a theater with five stages. Its clearly marked system of "nodes," to use Kevin Lynch's term,[15] seems to have been established virtually without conscious planning by a single agent, yet it serves admirably to maximize social interaction. Its lack of self-consciousness, the differing symbolism in the environmental parts, and the overall coherence of the urban form evoke feelings of exhilaration. Discussing this conclusion with a vigorously community-oriented priest on Brač, I was impressed by his remarks: "Left to its own decision-making, a community will unconsciously create the kinds of connections via windows, doorways, paths, and nodes to meet common needs for contact. Indeed, what else is the village, town or city but a vehicle for social communication?"

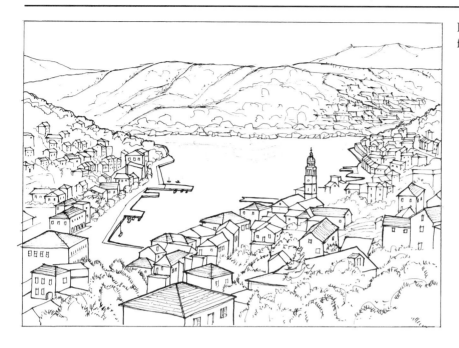

Pučišća's intimacy of
form

PUČIŠĆA: THE ARENA

Uniqueness through Comparison

We now turn to Pučišća, where we learn that comparisons with Sutivan are not only inevitable but also essential for more sharply revealing the essential qualities of each village. Although they share a common history and similar social and economic conditions, the vastly differing land-water relationships and local topography have established major differences in urban patterns, qualities of living, and environmental sources of identity. Nowhere in the system of towns and villages in Dalmatia does terrain play such a compelling role as in Pučišća. The spectacular irregularity of the steep-sided natural harbor has dictated a range of attributes that make "reading" Pučišća an entirely different experience from reading Sutivan. Its form has shaped class distinction, determined the lineal distribution of social and economic activities, and intensified residents' sense of belonging and, in some ways, "outsiders'" sense of alienation.

Bol's horizontal, linear town site has had a role in promoting greater social equality for residents and a sense of openness and privacy for "outsiders." The village stretches out on relatively flat land at the foot of the steep side of Brač, opposite Pučišća. In contrast with Pučišća's readily recognized urban form and the emotional moods it induces, Bol is difficult and frustrating to read.

Pučišća's urban structure stems from the way its natural harbor, shaped like a boot, zigzags in from the sea, flanked by inaccessible and unbuildable slopes facing the deep, watery cleavage in Brač's north coast.

This fjordlike inlet meets three narrow valleys that descend from the hills to join these quiet and beautiful waters. The smaller coves *(udoline)* they form provide connections to the plateau hinterland where settlement began. The most westerly of these, Stipanski, became dominant and established the direct linkages to Pražnica that brought my grandparents together. These small inlets and their surrounding steep slopes break Pučišća into three clearly distinguished neighborhoods that together form an undulating amphitheater where every dwelling has a balcony seat from which to observe—and comment on—the spectacle of daily life below.

Cut off from visibility to the open shoreline of the Adriatic, Pučišća looks inward upon itself. Visible to all from the town's hillside dwellings, the newcomer feels conspicuous in a vast urban arena. The broad *obala* following the undulating sea wall around the harbor exposes us on an endless lineal stage with no wings to run to; the dwellings stacked up on the steep hillside look down on us, and we know that Pučišća's sense of place will grow out of this rare urban experience. Community facilities—shops, cafes, the post office, the school, and several churches—form no strong nodes, and we find them too far apart to work as effective points of sociability. Across the *luka,* we see more separated places as potential destinations, but distance and lack of cohesiveness gradually erode our interest. Not only food, drink, and postage stamps but also casual interpersonal encounters are denied by the disconnected spacing and lack of gathering places. Although Pučišća forms a dramatic oval of populated land around a centralized body of water, it has no clear human center of gravity where people can readily gather. However, in the late 1980s an older building in a central location was remodeled and expanded as a small hotel in keeping with the village scale.

Up to the early 1980s, Pučišća's upper limit was well defined by agricultural land maintained by the rural residents of the upper slopes. Then food supplies became available by boat and truck and reduced residents' dependence on local sources, to the detriment of the already scattered shops. Moreover, a new circulation route now bypasses the *obala* and cuts a gash in the green upper slopes. This has stimulated construction of small dwellings designed for access by car. Thus, the traditional advantages of location are being reversed. Over generations, families such as my grandmother's became members of a privileged class with direct access to their own boats. Today the new families on the upper slopes are favored with vehicular access.

Yet, the village retains an image of self-assurance and self-containment from the past. Its people know that the stone from Pučišća's quarry has been in demand for centuries and continues to give it life. Most of the homes are occupied the year round; relatives who live elsewhere come for summer vacations or for weekend visits. Virtually no accommodations

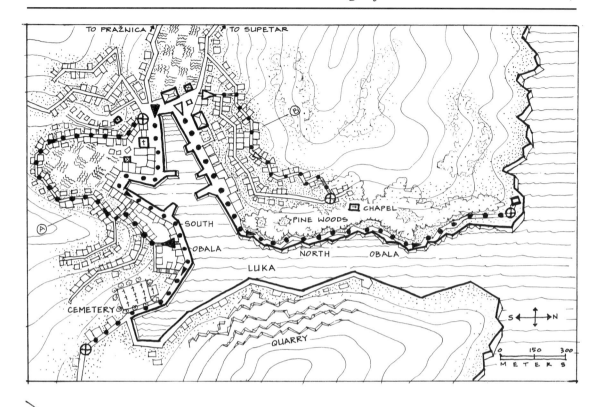

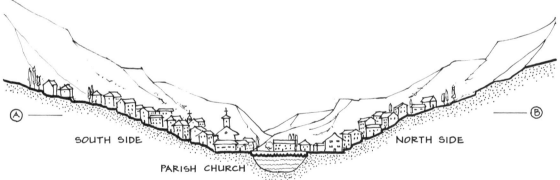

Pučišća: Urban structure and reading path

have existed for visitors, nor has there been any interest in promoting tourism. Indeed, tourism could damage Pučišća's genuineness, privacy, and residents' sense of identity with it. Under these conditions, "outsiders" truly feel conspicuous and, unlike in Sutivan with its five stages, unable to acquire even a minimal temporary identity.

The enclosed configuration of Pučišća may have reinforced its sense of community, which has grown through its response to the leadership of the church on the island as a whole. In 1966 Pučišća celebrated the four hundredth anniversary of the founding of its parish. Its small commemorating publication speaks of the people's pride for "that past built by the inhabitants of Pučišća in their surrounding environment."

Outlining its history in terms of the natives of Pučišća and their accomplishments, the text reviews the hazards run and the successes achieved: the arrival of the first settlers, refugees from Salonae in the seventh century; the reopening of the Quarry in 1455, unused since Roman times; the building of the first fortified tower in 1467 for protection against the Turks; the destruction of World War I; and finally the rebuilding of Pučišća.

The parish demographic records show that from 1566 to 1966 a total of twenty-two *župnika* (parish pastors) were appointed. These respected churchmen were responsible for recording the births of 12,239 natives of Pučišća and the deaths of 7,426. The population increased from a mere 350 in 1566 to 2,953 in 1912 and 1,739 in 1966. This fine little document closes with a sentence that is appropriate for the present: "Let this humble survey of the history of Pučišća be a stimulus to today's generation so that they may be proud of their past and on the basis of their history, build a better future."[16]

Reading the South Obala

Pučišća's meandering *obala* defines both the shape of the blue harbor and the highly extended periphery of the village. Its present geometric form resulted from the construction of a sea wall under Austria's public works programs in the late nineteenth century. Previously, the leading families, many of them descended from the earliest settlers from Bosnia, had their own docks on their own harbor frontage. One section on the south side remains unchanged and shows how the entire shoreline looked prior to the building of a common sea wall to accommodate larger steamships at about the time my grandparents left for America. Ships traveling as far as Trieste then docked in front of the parish church, adjacent to my grandmother's home. Around the turn of the century an extension, known then and now as the new *obala,* was built on the north shore, and docking facilities were shifted to that side. I learned from family members that when their villa was under construction there in 1888, they could dive directly into the sea from the building.

To experience fully the lineal nature of Pučišća's structural system, we start at the old *obala* and walk along the south side from the intersection of the road from Pražnica to the Cemetery and the stone Quarry. We can imagine arriving on foot over the jagged *karst* hills from Pražnica, and I did in 1937. The narrow valley, carpeted with vineyards, gradually descends, and suddenly the natural landscape breaks open to the fully enclosed blue harbor surrounded by the balconied houses of stone rising up the steep slopes. The built environment itself, however, offers little information about how the village is organized. Instead of a clear entry point and gathering place for people, we find an overly large and shapeless urban space that spreads across the area where the north and

The south *obala:* From the starting point to the church

south *obalae* join. Here vehicles move at will among pedestrians who are given no point of focal interest, vegetation, or shade as a relief from the asphalt paving that stretches without breaks, heartlessly in summer heat and winter cold.

Prior to World War II, however, the entire surface of this important gathering place was paved with the same fine cream-colored stone used to build the houses, thus linking land and sea and providing a spacious level setting for the tiered dwellings rising above. As I pointed out in the introduction, in 1937 this grand open space was handsomely crowned with an immense old twisted pine. One of its giant limbs was supported on a shaft of Pučišća stone, as if to symbolize the village's reliance over the centuries on that critical resource mined from the Quarry, which rises out of the harbor. In the shade of its spreading branches, the old men would sit on benches, and the children would feed the birds. A relative of my grandmother's with whom I stayed in 1979 told me that after World War II a violent *bura* brought the tree down, after which this focal point of stone, pine, and people was removed to make way for the coming of the automobile. Then the stone paving that was so much a part of this entry area was paved over with asphalt in the 1960s. My relative pointed out the plaque memorializing this modernization mounted on the wall of the nearby Moro mansion by the *mjesna za-jednica,* ironically demonstrating both its pride in being modern and short-sightedness regarding Pučišća's unique identity with the past.

On our right the building, itself an example of earlier insensitivity to urban design, imposes its pink stucco bulk of Victorian Renaissance style directly in front of the old line of traditional stone dwellings facing the harbor. It belonged to one of the numerous Pučišća emigrants who decided to put their abilities to use in more prosperous endeavors in Chile when the local wine industry failed. A sign indicates the villa's conversion to a vacation home for youths. Nearby, two grocery stores are too small to give life to this stark, open space.

Beyond the Moro villa stand two large homes of ancient origin that were expanded in the nineteenth century. Both were built by families of Bosnians who immigrated in the sixteenth century. The first is the home my grandmother left in 1871, the Lukinović-Bokanić family home where I stayed in 1937; next to it is the Desković mansion, famous for the generations of sculptors who lived there. Both were among the ruins left when the Italians burned the wooden interiors of the houses of leading families during World War II. The semi-ruined state of the Desković home contrasts with the finely chiseled stone plaque mounted on the wall to memorialize the famous sculptor who lived in the house from 1886 to 1937. The Lukinović house, on the other hand, has been completely rebuilt by a native of Pučišća who lives in California.

At this critical point the facade of the sixteenth-century parish church of Sveti Antun (Saint Anthony) faces westward, at a right angle to the fronts of these imposing houses, and breaks the continuity of the *obala*. This paved space formed by the church and houses is occupied only when the bell tolls for Mass on Sundays or when a large bus parks there on weekdays. An imposing military administrative building close to the harbor's edge, a reminder of the heavy hand of Austria, completes the composition.

We make a sharp turn in the *obala* and discover a second deeply indented *udolina* at the base of the next hillside neighborhood. This sudden unexpected openness and contrast in scale give new life to our reading of the *obala*. An inevitable *slastičarnica* (pastry shop), facing the east to receive the morning sun and the afternoon shade, is something of a social center throughout the day. Its brightly colored awning and parasols for outdoor tables and its trim Macedonian neatness provide a rare highlight on the rather aimless and somber experience of walking the *obala*.

At the bottom of this cove the narrow green valley of Solina opens up, again reminding us of Pučišća's three links to the hinterland and its agricultural-maritime origins. As we leave the cove the irregular shoreline without sea wall unfolds, demonstrating how individual docks served waterfront houses a century ago. These docks give way to sloping concrete pads that serve as swimming beaches and the striking view to Pučišća's major landmark, the Quarry, across the large blue cove. Its strange, opalescent greenish color, caused by the infusion of stone dust, establishes a subtle source of identity. As the road swings along the third cove, the shore reverts to its natural condition, houses are left behind, and we have a clearer image of the site of Pučišća at the time of the first settlement by Bosnian immigrants in the sixteenth century.

Finally, we reach the end of the third cove, where the green Stipanski Valley ascends from the bottom of the sea to the ancient Cemetery, a setting of dignity, calm, and suggestions of the past. Here lie the many generations preceding my grandmother. The dark green cypresses brood over the finality of hundreds of tombstones engraved with the family names of the builders of Pučišća and protect the chapel from the winds of the open valley. All of these elements taken together handsomely punctuate the south end of the lineal system around Pučišća's harbor. An old woman trudging along bids me "Dobar dan" [Good day] and asks, "Kako je u Kaliforniji?" [How are things in California?]. She knows where I am from. Watching from her window, she knew on my arrival who I was and why I had come, clear evidence of Pučišća's ready awareness of the daily events in the life of the community, gained from its unique urban form.

Reading the North Obala

We now return to Pučišća's "center" to follow the north side of the *obala* out to the open Adriatic. The massive school, quite out of place at this end of the central cove is our obvious starting point. However, a plaque placed on the wall of the school in 1968 commemorates the founding of the first public school in Pučišća hundred years earlier. The structure creates a small, more active secondary urban space that is fed by the entry road. Here buses from other parts of the island arrive on the new coastal road built in the 1970s, and people come to the news kiosk when the daily papers arrive.

Just beyond, fronting on the harbor, a tiny park and town *kavana*, a minor focal point of social life, welcome us with abundant shade trees and open space, a relief from the stark asphalt. The outdoor portion, used only during the summer months, fills an extensive terrace dotted with tamarisk and pine and well above the *obala* level. Its enclosure with a handsome iron grill gives it a welcome sense of containment. On weekend evenings, rock concerts with dancing in a hearty Dalmatian style turn the terrace into an outdoor theater, with children and older folk crowding at the grilled fence to watch. The strong beats of the young people's bands echo across the water and up to the balcony houses that look down on the changing urban ways. This tiny green social center set in a sea of asphalt and the larger amphitheater of the whole hillside village establish in our minds the uniqueness of Pučišća, so different from the intimate, lush green heart of Sutivan.

As we move northward along the harbor's edge, only a few tamarisk trees shade the side facing south and soften the newer freestanding resi-

The north *obala* out to sea beyond Pučišća's white slate roofs. *Photo by author*

dences of mixed styles, so incompatible with old Pučišća's compactness. We come to a second gathering place where the ferries glide in silently and almost mysteriously, for the sea they come from is not only out of sight but also forgotten in this lake-like setting. Now, at docking time—in 1979, remember, before the coming of buses—it seems that all of Pučišća descends from its terraced homes to greet friends and relatives as goods and supplies from Split are unloaded. My 1937 visit returns to me when nearby we find a restaurant run as a public service by the stone-workers' organization occupying the ground floor of the Kraljević mansion. I visited that branch of my grandmother's family there and remember well its crystal chandeliers and two grand pianos. Partially destroyed during World War II, it was rebuilt as a vacation home for youths.

The spicy fragrance of the pine woods that begin to clothe the slopes tell us that we are nearing the end of the hillside houses. A weathered chapel in the woods above the road marks our last view of Pučišća's urbanized harbor, as the channel to the open sea begins here. The chapel, which has stood guard over the channel for centuries, since long before the *obala* was built, brings to mind images of the simplicity of life during Pučišća's centuries of taking form. The aroma of grilled fish leads us to a small private restaurant tucked away on a terrace among the trees. Beyond the chapel, the pine woods and their wildness take over, and in our minds the urban image of Pučišća is replaced by an image of nature and the elements.

Totally separated now from the built environment, we face the open sea, that contributor to all Dalmatian identity and a reminder of the

rugged, self-reliant life of the people of Pučišća that went into building this place exposed to the raw coast. The contrast with the sheltered,
introverted quality of Pučišća, where the thousand rectangular "eyes" of
the stone houses followed our every move, is sharp, almost threatening.
Alone now in the silence of the past, we move through the pine woods
and out along a forbiddingly steep sea edge where huge rocks mass, pre-
venting even the slightest intimacy with the Adriatic. At the Lighthouse,
the end of the road, the whole sweep of Brač's inhospitable stone shore
unfolds; beyond, there is but one abandoned settlement until the shoreline reaches Postira.

Returning through the woods, we are struck with the contrast between the shaping of this town, built of its own stone through the sustained power of its people, and the formation of its land, sea, and pines
through the force of nature. Pučišća swings into view again, one cove,
one narrow green valley, one hill at a time, until again the sea is gone
and the amphitheater of Pučišća surrounds us. Now, at dusk, we experience the whole winding waterfront, from the wide-open headlands and
sea to the closed-in towns, as if moving through an urban choreography
of grand scale. Fresh perceptions of place follow one after another with
shifting views of the town and changing lights on the tiered dwellings.
With twilight, the windows soften as the sun leaves and the water dark-
ens. Night comes, and the lights in the somber stone houses reflect in
the blackened waters, dappling like swaying lanterns. In the darkness we
are relieved of the sense of being watched from the houses above, and
Pučišća gives way to a different sense of place than by day.

The open sea along the shore thick with pines has revealed to us a
basic source of the identity we seek: the contrast between this timeless
regional setting and Pučišća itself underscores the centuries of struggle to
build the town and reinforces a sense of collective rootedness. We conclude that Pučišća's uniqueness lies in an introverted quality established
and held in place by its environmental form, one that creates a powerful
identity for residents as well as for emigrants and their progeny abroad,
such as myself.

Three Neighborhoods Tiered on Balconies

The dominance of the *obala* gives us a clear and strong pattern of Pučišća as a whole, whereas the three neighborhoods appear to be without
their own systems of paths or nodes. The village's linear *obala* is its *grad*,
home of the urban classes; the hillsides are its *borg*, home of the country
people. We learn this as we walk the dense, random pattern of dwellings
linked to steep entries that take us up steps and ramps of stone polished
by generations of footsteps and laden with local cultural history. From
the few main routes lead many smaller, crudely paved byways that only

Steps up to one of the three hillside neighborhoods

the permanent resident can follow. We feel like intruders; Pučišća is truly a town where people live and work year round, the opposite of Sutivan, with its ready identity for visitors.

The ladderlike steep steps allow us to peer into courtyards, doorways, and windows, learn how people live, and bid "Dobar dan" to residents and children sitting in doorways or passing by. Our reading takes on a human quality that contrasts with the austere, extenuated quality we found on the *obala*. As we near the top of the slopes, houses become smaller and simpler, another indication of the segregation of social classes by terrain.

The only neighborhood focal points we find are the several chapels from earlier times, no longer regularly used. Plaques all link their founding to the Turkish occupation and the subsequent Bosnian emigration to the coast. For example, we find the date 1563 inscribed in Roman numerals on the chapel of Sveta Lucija (Saint Lucia) just above the parish church, while on the north shore the chapel of Sveti Roko's date of 1533 indicates that this neighborhood was established earlier. The chapel's neighborhood is divided by a ravine, terraced centuries ago for truck crops, which joined the two neighborhoods in a common purpose. We

Pučišća at the end of our reading. *Photo by author*

recall Andrija Cicarelli's history of the role of the community in founding these churches.

From the routes we follow at higher levels the roofs of Pučišća's hillside neighborhoods form a stirring and lasting image. Note how their helter-skelter pattern gives the hillsides a sense of motion and vitality, so different from the solid, unyielding quality of the *obala* below. Their frosty-white topping of lime, made from the slabs of local *karst,* used for centuries to seal the stones, have been laid like shingles; this interplay of rooftops unites the neighborhoods into a single urban landscape. However, these qualities are being lost year by year as new roofs are being laid in the more accessible factory-produced red tile.

Experiencing the three free-flowing hillside neighborhoods entirely on foot provides a refreshing sense of freedom from the automobile. And now, as the sun lowers and sets over the Talija Valley leading up to Pražnica, the funnel-like shape of the narrow opening in the hills becomes an enormous floodlight for the entire urban scene. The sides of the houses of all three neighborhoods pick up every last ray of sun, and the walls, roofs, and the windows again mellow into the approaching twilight. The buildings no longer seem to be made of the cream-colored stone, but of light itself, and the darkening water of the harbor takes on a pearly luminous quality that transcends reality. Nightfall unifies the diverse features of Pučišća that have intrigued us by daylight. This effect is hastened as the reflection of each window light stretches across the black, shimmering water. One by one they darken, and the curtain of night imperceptibly falls.

Bol stretching west
from the Dominican
Monastery to Zlatni
Rat with Hvar across
the channel

BOL: THE LADDER PATTERN

Urban Structure

The basic landscape character of Bol has influenced its development and
identity sources in far more subtle ways than was the case in Sutivan or
Pučišća. Yet the experience of deciphering Bol's unclear pattern can bring
rewarding returns to the visitor with an exploratory frame of mind. For
that reason Bol's high level of tourism and promotion—begun as early
as 1932—greatly exceeds even that of Sutivan. The town was home to
1,200 to 1,500 people from the seventeenth to nineteenth century, and
its year-round population today is about 1,000. On the southern side of
Brač, Bol stretches for some three miles along the seafront and parallel to
the low hills at the foot of some of the most rugged limestone terrain on
Brač. Across the open channel lies the sharp ridge of the island of Hvar.
One senses tranquillity and repose in a setting of such vast marine space.
The absence of any natural harbor and the ready accessibility of Bol's
shoreline gave rise to a lineal pattern in which all dwellings have direct
access to the sea and its economic benefits. The main circulation route
reaches from the historic Dominican Monastery *(samostan)* to the east
and to the vineyards far to the west where the monks established viticulture in the fifteenth century. The road terminates at the phenomenal
slim promontory of golden sand known as Zlatni Rat. Bol's lineal town
site evolved along this basic route.

A second, parallel route was built on filled land in front of the sea-

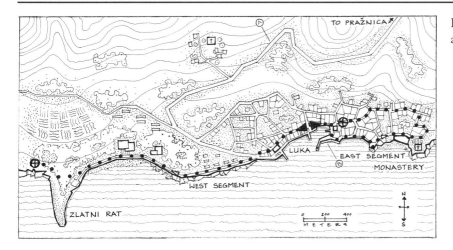

Bol: Urban structure and reading path

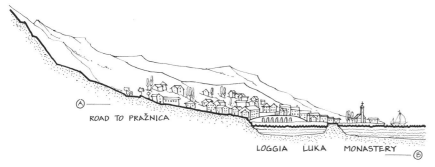

front dwellings around the turn of the century. The wine boom and better access to the vineyards at Zlatni Rat gave a central port facility priority over individual piers for seafront property. As development was consolidated by filling in gaps along the spine, links were established between the two parallel routes. In this way, Bol's irregular "ladder" pattern emerged. The "rungs" of the ladder have become meeting places for social groups—old men seeking company, housewives en route to market, mothers with small children, and teenagers in constant chatter—each with its own distinctive open space and trees that break the waterfront facade and corridorlike inner road; these breaks provide a sense of rhythm to walking the entire system.

We find the central rung to be the logical place to begin our reading of Bol's ladder system. We take first the east segment, from the Church to the Monastery, then from the parish church to Zlatni Rat. It becomes difficult to capture a "mental map" of Bol's visually elusive lineal plan in its entirety. Rather, we learn to focus on the myriad unrelated yet superb details of urban composition that have evolved on their own. In the three miles from the Monastery to Zlatni Rat a veritable parade of environmental delights, both human-made and natural, present them-

Villas from the past on the sweeping curve of the Loggia. *Photo by author*

selves: seafront streets and inland routes, shops, parks, piazzas, fountains, churches, pine woods, beaches, all together revealing Bol's own uniqueness among Dalmatian urban places.

The East Segment: From Church to Monastery

At the central rung of Bol's ladder system, we look around the loosely formed street piazza that serves as the main meeting place and the site where the few shops Bol supports are concentrated. The news kiosk—always a sign of centrality in Dalmatian towns—offers us benches nearby where we can sit with the old men of Bol and contemplate the sense of place at this image-forming node. We set off from the kiosk to begin reading along the *obala* in the direction of the rising sun and the Dominican Monastery, which in itself marks the dawn of enlightenment in Bol's early years. Bol's skyline, lacking a campanile or church tower, appears rather ordinary; yet it emphasizes an intriguing lineal quality.

Ahead of us, where the *obala* meets the fishing pier, the medieval *kaštel* with Venetian windows rises. Near its base the fish market, where deliveries are made directly from the fishing boats, attracts buyers, and the meat market stands nearby. The grand curve of the Loggia, with its six elegant arches, draws our eyes up the ramp from the seaward side to its handsome rooftop terrace. There Sveti Antun's tiny chapel rests as if on a giant stage framed with an enormous cypress on each side. The ramp descends to join the building facade at the street. On my first reading, on Saint Anthony's saint's day in June 1969, a procession led by priests and fishermen came from the parish church up this ramp to the chapel, then down to the waterfront to bless the fishing boats. This grand design, more alive and penetrating than that of Sutivan or Pučišća, makes up for the monotony of Bol's lineal pattern.

We mount the sweeping curve of the ramped street along an unbroken row of superb seventeenth-century residences softened by billowing trees and the shade, which give way to the tiny chapel of Sveti Antun we saw from below. It was originally dedicated to Saint Bernadine, but a popular painting of Sveti Antun placed there brought about a shift in its identity. Looking out from the quiet landscaped terrace, we see Bol backed by the harbor with its ships and the island of Hvar across the channel.

Experienced from above, the crescent-shaped Loggia, built against the retaining wall of the terrace, becomes Bol's most distinctive piece of urban imagery. Although a plaque dates it as 1584, in fact it was rebuilt after the severe bombing of World War II, quite faithfully, judging from photos taken in 1910. The view of the *kaštel* in the foreground further expands our image of Bol. Restored for use by the *turist biro,* this old landmark has become a point of social attraction and houses offices for Bol's hotel management and tourist activities. From the *kaštel* the road swings out to the far side of the U-shaped harbor to the principal boat dock, where the larger ships arrive carrying tourists in the summer months.

Eager to see Bol as a whole from a distance, we descend toward the *kaštel* and head for the end of the dock, watching the attenuated skyline unfold. At the end of the wharf an expansive view of Bol's elusive urban structure in its entirety awaits us from the precipitous promontory on the east to the Šume Borova (Pine Woods) marking the urban terminus on the west. This reading has already taught us that only by experiencing each component of the entire length of the system can we perceive its overall uniqueness.

We now return to the beginning of the dock, at the base of the *kaštel.* Here the shore road takes a sharp turn to the east, where the contrast with the spacious nineteenth-century promenade we have experienced so far is vivid. The environs of this narrow route between the sea and the ancient buildings of this very old district takes us back several centuries. The quiet view of a shallow cove with no sea wall, the helter-skelter pattern of the semirural houses like those one would find in the countryside, accessible by stone-paved ramps and steps, could represent the image of Bol at, say, the time of Napoleon's conquest of Dalmatia. At the end of the cove the shoreline road passes through a more rural area, turning inland to join the upper inland road. Dwellings are spaced well apart, with fruit trees and gardens recalling the rural life of the past. But Bol's shape for the future is forecast by new and wider roads that lead off to developments up the gentle slopes, where houses are being situated for automobile access.

Our shoreline route now provides an impressive approach to a second major architectural treasure, the Dominican Monastery and its former

The east segment from
the Loggia and the
luka to the Monastery

high school, the only one on Brač. Founded by the Dominicans in early
times, after World War II under Communism the high school became
the Hotel Bijela Kuća (White House). As we pass this pine-studded
complex overlooking a glorious white beach and a shallow cove, the
handsome mass of Bol's principal monument to its past looms large on
its rocky, wooded promontory at the eastern end of Bol. The Monastery, one of the choicest examples of restored architecture from Brač's
past, played a major role in Bol's development. During five hundred
uninterrupted years the Dominicans offered social, cultural, and even
economic guidance in education, literature, religion, and agriculture.
Now housing one of Brač's finest museums, the Monastery gives depth
to our reading of Bol. The complex of buildings, gardens, and cemetery
jutting out on the rugged promontory of Glavica is an ideal blend of the
natural and the built environment. We respond with a deep sense of the
human element that went into its evolution, driven by the social and
spiritual purpose of centuries ago.

The documents in the museum attest to the Dominicans' arrival in
1462 and the permission given by the Venetian-appointed duke of Brač
to build the Monastery on this striking site. The island's oldest written parchment document, issued in 1184, describes how Bol played an
important role in the early cultural development of Brač by calling a
meeting of the ruling clergy of Brač in Bol. The original copy of this
document was presented to the town of Pučišća by native son Marko
Lukinović at the time of his retirement from his position as *opat* (abbot),

The west segment to the *obala* and Zlatni Rat

just below the level of bishop, which he held from 1776 to 1809. He was a forebear of my grandmother and a contemporary of Andrija Cicarelli.

At vespers we enter the chapel and quietly observe the white-robed Dominican monks in the state of deep contemplation that comes from daily prayer. The bells ring for a full ten minutes as the brothers murmur, their prayer books in hand. The harmony and rhythm of the bells and the whispered prayers, unchanged over centuries, carry a message: identity with place can have a life of its own in the context of a permanent environmental setting and a human use that goes beyond the material ingredients that dominate our urban places today. As we leave the Monastery to watch the sunset from the ancient cemetery, we realize that our sense of identity with Bol has multiplied tenfold in this rich environment.

The West Segment: From the Kiosk to Zlatni Rat

We return to the news kiosk and the central rung of Bol's ladder system in the late afternoon, when the harbor area is filled with shoppers. The activity gradually slows down and turns into the customary evening promenade, characterized by the leisure that visitors to Bol seek, mainly in the west segment, where most tourist hotels are located.

The next day, our reading takes us along the old inner road, the outer "rail" of the ladder. Fitted tightly between this old inland road and the waterfront stands a smaller *kaštel,* now handsomely restored as the Hotel Kaštel by joining it to several adjacent buildings, as I learned when I

1890s town houses on
the inner road

stayed there. Across the road, on the land side, a series of nineteenth-century buildings—one dated 1893—offer evidence of the period when Bol's previously separated neighborhoods grew together and became more urban. A major break occurs in the compact facade, the first of those we identified from the dock. Here a small park offers shade and a view of the placid sea, steps, and access to the waterfront road below with its *slastičarnica.* This open space is regularly occupied solely by women, who have come from both sides of the linear system.

Just beyond, Bol's first gas station, built in the early 1980s, violently disrupts our rich experience as it juts out into the sea on fresh fill, jarring us from our absorption in piecing together the elusive essence of Bol's urban form and its hidden messages. Designed self-consciously as a very contemporary pavilion that fits neither the scale nor the character of Bol, the building and the cars it attracts are anathemas to the architectural integrity of the town and its maritime origins.

Further on, the buildings stand shoulder to shoulder as if to protect one another from the sea. They break again at a space of monumental urban design quality. A slender terrace extends from a gentle curve in the road, and an immense, multilevel piazza, some two hundred feet across at the *obala,* opens dramatically below the road. The piazza is backed by a high retaining wall softened by trees and far enough back from the *obala* to define a popular gathering place for tourist activities, all with a view of the sea. The terrace, dominated by a larger than life-size bronze statue, terminates at the side wall of the huge and architecturally handsome Vinarska Zadruga winery, from the wine boom of the nineteenth century. This wall marks off a broad level open space for teenagers' soccer and basketball games.

A large monument to the revolution stands out of place in the midst of the paved space and breaks the continuity of our reading. Massive steps on one side and a ramp on the other lead up to the interior street. The open space marks the western end of the evening promenade and of Bol's strung-out tourist attractions. According to local historians, this grand piazza gives no hint of its revered and utilitarian beginnings. For here in centuries gone by natural springs served Bol and the countryside. The horses and *mazge* that carried water up the hills required the extensive space, the wall, the trees, and the ramps. When modern water lines installed in the late nineteenth century replaced the horses and *mazge,* the elaborate design we see today was carried out under a major public works program of the renowned civic leader Ante Radić, who became *načelnik* (mayor) of Bol in 1883. He is known for establishing one of Brač's first *čitaonici*. In 1904 he developed Bol's first urban plan, encouraged the formation of cultural societies, and instilled pride in the people of Bol for their public buildings. A stone plaque commemorating Radić reads in Croatian:

Djedovi nam	To honor our forefathers
od iskona iskopali	who from earliest times dug here,
općina ogradila	the *općina* built
ovu zlatnu vodu	this golden well
ljeta gospodnjega	in the year of our Lord
MDCCCLXXXV	1885

Moving further west along the *obala,* our sources of Bol's identity run out. Finally, the inner and outer roads join at the point where Bol's elongated urban development ends abruptly, as though by regulation. There the Pine Road along the untouched rocky shore and open sea wall leads us toward the modern resort hotels of the 1960s and Zlatni Rat. Below us the blue-green Adriatic Sea breaks into white on small beaches of cream-colored gravel, quite a difference from urban Bol. Benches invite us to sit and contemplate this contrast in environments. Above the road a few modest cottages from before World War II occupy extensive acreage of level land where building sites are indicated, as if a subdivision of summer villas once begun had later failed.

This drive of about a mile and a half in length in a parklike setting obviously was intended to connect the town to the tourist and

Serene Zlatni Rat, Bol's sandy "golden point," marks the west end of our reading. *Photo by author*

recreational development near Borik Beach and Zlatni Rat. The original road, however, connected the Monastery, the village, and the vineyards planted by the Dominicans. Highly productive in the nineteenth century, the vineyards never fully recovered from the phylloxera plague of 1894. These lands were taken in the 1960s for building the very large hotels just above Borik Beach. This abrupt change in economic activity demonstrates the dubious exchange of year-round use of level, fertile land—so rare in Dalmatia—for a style of tourism that is unrelated to Bol's cultural heritage and operates for a limited part of the year.

We conclude our reading at the most remarkable beach of the entire Mediterranean, Bol's Zlatni Rat. This spit of white sand framed by pine woods and blue, clear water projects out from the shore like no other in Dalmatia. Virtually the only sandy beach in Dalmatia, it has been known throughout history. So attractive to European vacationers, today the beach, rather than the village itself, is now Bol's main source of identity. For example, I inquired of a British tourist who had been through a large-scale promotional program what his reactions were to Bol a few days after arrival. He described the comforts and conveniences of his new hotel and the pleasures of the white sand and clear waters of Zlatni Rat. When I asked him about Bol itself, he replied in a tone of depreciation that he had neither been to nor intended to visit what he called "the village."

Bol's Unique Sense of Place

Our reading has led us to the source of Bol's uniqueness: its diverse, lineal structural system revealing a wide variety of urban forms. These forms produce experiences that have the potential to appeal to various kinds of users, whether visitors or residents. After a few days in Bol, during which I discovered all of its features on my own, the experience became a continual back-and-forth movement stopped at one end by the Dominican Monastery and at the other by Zlatni Rat This experience, which generated a continual sense of involvement with the changing forms, was just the opposite of my experience in internally structured Pučišća and Sutivan.

Unlike Pučišća and Sutivan, Bol seems to have consciously opened its doors to visitors and to be satisfied with the light impact so far. This may be due to early initiatives in this direction. I discovered a small exhibit on the history of tourism in Bol, held, ironically, in one of the ultra-modern hotels near Zlatni Rat, documenting an extensive international promotional program for tourism as early as 1932. The accompanying advertisement, published in both German and Croatian, preceded any similar local initiatives by other Dalmatian *općine*.

Bol's particular lineal and dispersed pattern presents a challenge to the visitor trying to learn his or her way around its rich environment. Yet, one who has curiosity about the culture and the time to stay in the village itself can become a part of the daily life and be rewarded with the continual discovery of subtle relationships within Bol's environmental fabric. In my experience, synthesizing an overall image from many fragments, each in itself a pleasure to grasp, became the main goal. The individual can develop a personal sense of belonging and thus achieve in a brief visit a more lasting "escape" from his or her own world.

For permanent residents, the ladder pattern discourages regular contact with large numbers of people, since there is no single "stage" for casual social interaction. Meetings must be deliberate and arranged, as I found in attempting to communicate with local historians. At the same time, the number of open spaces at the "rungs" allow residents to group themselves according to their own interests. On the other hand, residents at the far ends of the attenuated system can move into a portion of the center by different routes and return without making contact with one another. Thus, Bol's structure offers a high level of privacy and choice not found in Sutivan, where close encounters are unavoidable, and in Pučišća, where surveillance of people's movements becomes a daily way of life.

CHAPTER FIVE

Two Towns That Dominate Their Islands

In our little towns of the nineteenth century there lies no greatness in any single element; rather, they reveal a uniqueness and simplicity of life so organized in three-dimensional form that its coherence can be "read" in their "faces." —Grgo Gamulin, 1967

HVAR: ARMS OPEN TO THE SEA

We now cross the channel from Brač to Hvar and then to Korčula on our route to Pelješac and Dubrovnik. Although these islands are close in terms of travel time, because of their very different forms they are far apart in terms of identity. Indeed, while reading Bol I inquired about sources to consult with in Hvar. To my astonishment, several people I spoke with had never been to Jelsa, just across the channel. Such is the force of identity with a single homeplace on the Dalmatian coast.

Unlike the three villages on Brač, which are part of a system of small settlements, the towns of Hvar and Korčula play a central role on the islands that bear the same name. However, the two represent opposites in terms of urban form as a source of identity. At the town of Hvar, water indents the land and the town becomes a symbol of receptivity, built tightly around a protective harbor made by "arms" of coastline that beckon out toward the open sea as if to welcome the world. At Korčula, the historic Old Town pushes prominently out into the water. Entering this model of medieval town planning is like boarding an urban "ship," surrounded by water on three sides and thrusting aggressively out to sea.

Hvar— The Island

The phrase "arms open to the sea" not only describes Hvar's physical shape but symbolizes its historical role in the international maritime life of Dalmatia. Shipping facilities and development patterns alike grew out of the basic form of both the town site and the island as a whole. An undulating shoreline offered seven shallow coves, the deepest and most protected of which lay at the center, like a pair of outstretched arms, inviting urban settlement. At its base lay the natural land route to Starigrad, the "ancient town" created by the Romans that predated Hvar by more than a millennium, and Jelsa, both port towns on the opposite side of the island, facing Brač.

Towering directly above this central harbor, Mount Sveti Nikola offered visibility in all directions from this westernmost tip of the long, narrow island. Lying offshore, the Pakleni Otoci (Infernal Islands) chain provided a protected entranceway to a deep harbor and a natural site for settlement by the Illyrians, the Greeks as early as 385 B.C., the Romans in the third century A.D., and ultimately the Croats from the mainland. These advantages made Hvar useful to the Venetians as a safe base on their trade routes to the Near East and, in the late nineteenth century, to the steamships of the Austrian Empire. Finally, up until the attacks of 1991 Hvar brought the international tourist seeking Hvar's sun, blue sea, and reminders of a colorful past.

Hvar is Dalmatia's longest island and the third largest as well. Its rugged flanks stretch roughly parallel to the mainland for some forty-five miles. Grga Novak, a revered scholar and native of Hvar, wrote warmly of Hvar's history in his masterwork, *Hvar kroz stoljeća* (Hvar through the centuries).[1] With particular sensitivity for Hvar's source of identity, he tells us not only the sequence of significant events over centuries but also how these events shaped its economy, culture, governmental institutions, and the city as a physical place. Novak writes of Hvar's western end breaking into relatively dispersed ridges and thus created the bays and valleys of Hvar, Starigrad, Jelsa, and Vrboska, the island's main urban places and sites of the greatest concentrations of population. These lower ridges ascend toward the center of the island to form knifelike ridges with two peaks, Sveti Nikola and Hum, both more than eighteen hundred feet high.

Leaving Jelsa at dawn by bus in 1979, I traveled this raw highland from Jelsa to Sućuraj (Saint George), at the far eastern tip, and found the heights almost without settlement. Topography clearly played a major role in determining the distribution of urban places in those centuries before the asphalted roads, buses with music, and other comforts that I enjoyed. From the main ridge I could see north to the mainland and south to Korčula as if in a low-flying plane, until this landscape in the sky dropped down to Sućuraj, then made more accessible by ferry to several points on the opposite mainland. In 1981 the island had a population of about twelve thousand. Hvar is the largest town on the island, with some three thousand people, and in the summer its population doubles.

Sources differ on the origin of the name Hvar: either the colonists derived it from the name of their own Greek island Pharos or they took the name used by the earlier settlers. Under the traditional Greek institutional system, Hvar became a "free polis" with its own laws and coins. According to Novak, in the middle of the second century B.C. the Illyrian state, well established on the mainland, entered into an agreement to pay tribute to the Greek colony in exchange for continued autonomy (36). This move brought Rome to Hvar first as a protector, and after

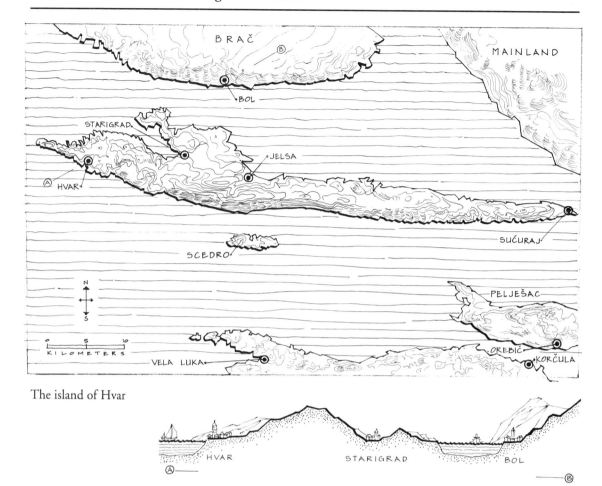

The island of Hvar

Caesar's reign began, in 59 B.C., Hvar came fully under Roman rule. It was then that the *centuriae* system of meting out land in parcels of about 124 acres for agriculture was applied on the rich flatlands near Starigrad. On my first visit there, in 1979, and later on maps I could identify ancient patterns of roads, ownership boundaries, and rock walls, which have recently been a source of controversy since construction of an airport has been proposed for this land.

Not much is known about Hvar after the Romans. During the seventh and eighth centuries the recently arrived Slavic tribes from the Neretva River area took over the island, as others had done at Brač. They erased all traces of Greek and Roman development and engaged in piracy (43). With the formation of the Croatian state in the ninth century, Hvar achieved a dominant Croatian identity and a certain level of self-sustainability. With the Venetians' rise to power in 1276, however, Hvar nobles gained from them a level of autonomy matching that of nobles in Split, Trogir, and Dubrovnik. Hvar's own institutional system gradually evolved its own statutes similar to those of Brač. These favor-

Hvar in 1572, a striking blend of urban form and landscape.
Croatian State Archive

able conditions made it possible for Hvar to hold its own even after Venice's domination, from 1420 to the end of the eighteenth century.

Novak writes how, in the process of accepting Venetian rule, "the people of Hvar promised to build a 'real city,'" to be achieved fully by the erection of a "Communal Palace, Town Campanile and Communal Loggia": "Three public buildings were the pride of every commune. When Venice chose Hvar for the center of its navy on the Adriatic, it persisted in reorganizing the old ways of governing the commune, which for a thousand years had been without a town center, and in providing one, as was the custom in the rest of Dalmatia and in Italy" (162). His detailed description of the sequence of constructing these buildings and others, beginning with the Communal Palace in 1282, only four years after Hvar joined forces with Venice, brings vitality to our experience of the remarkable integrity of Hvar.

In order to maintain the upper hand, the Venetians preserved the centuries-old statute but used the promise of potential reforms to play the people of Hvar against their nobility, thus sustaining a substantial gap between classes. In the mid-sixteenth century there were thirty-eight noble families out of the several thousand inhabitants of the town.

During the following two centuries only six new families gained noble status. The tension between the Hvar and Venetian nobilities fostered in the common people a strong sense of identity and belonging. Ultimately this contributed to the high level of architecture and urban design that we shall experience as we read the Hvar of today.

During the eight years of French rule, when equal citizenship was established, the fixed class distinctions of the past centuries were erased. Throughout the nineteenth century Hvar was virtually a colony of Austria. Novak reminds us, however, that "the consciousness of its inhabitants was significantly enhanced" when the Croatian language became the official language of all Dalmatia in 1909 (202). As we walk through the city, we seek a sense of relationship between the city as a physical entity and the spirit contributed by its people.

The Town Structure

To read history into the streets, squares, *obalae,* and buildings, I borrow from Grga Novak a description of Hvar by a visitor, Vinko Pribojević, telling what he saw in the year 1525, shortly after the beginning of Venice's second ruling period:

The port of Hvar is conveniently sheltered from storms and, as such, is an attractive spot for all ships sailing the Adriatic. On the middle hill stands a well-built, impenetrable fortress surrounded by high walls and rocks. The port can accept many ships . . . on the West there are three castles. . . . Under the fortress is the town of Hvar itself. It stretches over the whole hill and has a significant part that is flatland, as well as a suburb of almost seven hundred houses luxuriously built from square-shaped stones.

The rest of the flatland comprises two big squares. The beautiful cathedral, together with the bishop's residence, make up the surrounding buildings on the higher square, with its several wells and green gardens. The lower one stretches all the way to the monastery of Sveti Marko [Saint Mark] the Evangelist; that is, until the seashore at the base of the western hill. In this monastery Dominicans reside. . . . What should I say about the beautiful palace of the duke with four magnificent towers, which are located on the western side of town near the church of Sveti Marko? (104–5)

In our times, even though the town has grown, that image remains. Hvar's natural inlet has become a U-shaped *luka* wrapped by continuous building facades facing the broad stone *obala,* called the *riva.* The result is one vast "piazza of water," furnished with boats rather than with people. At its innermost end the harbor turns at right angles to become a "piazza of stone," flanked on both sides by more building facades from the past and terminated by the sixteenth-century cathedral of Sveti Stijepan (Saint Stephen) and its dominating seventeenth-century Campanile. To the left, the walled Old Town rises toward the Tvrdjava Španjul (Spanish Fort) high above. To the right, the Arsenal and the

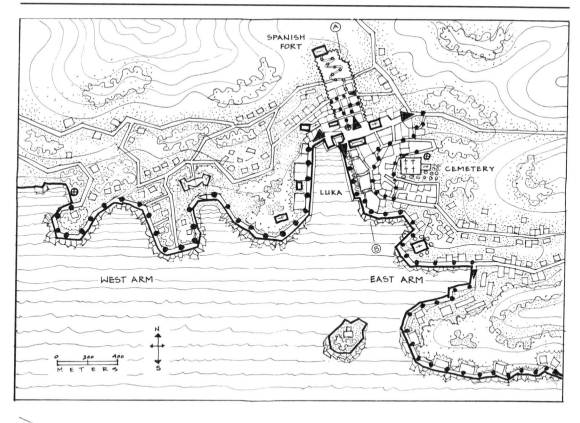

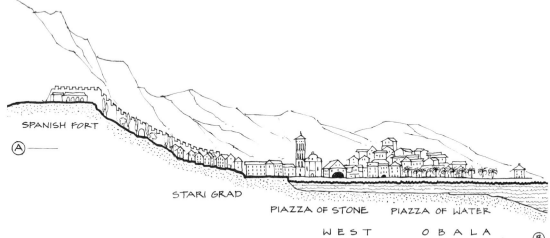

Hvar: Urban structure and reading path

Theater above have as a backdrop the *borg* and the hilltop Cemetery. Directly beyond this superb architectural ensemble in old times lay the traditional, winding land route of carriages and *mazge* leading to Starigrad and Jelsa, today an improved road for buses and cars.

Hvar's pattern of roads and pedestrian ways grows out of its grand geographical configuration. A single broad, waterfront pedestrian route of some seven kilometers hugs the undulating seven coves that fan out from the "piazza of water." Walking this pedestrian route, we can learn Hvar's structural system. For us, as "outsiders," the U-shaped core, where we begin our reading, immediately sparks our visual attention, and for the "insiders" of Hvar it provides a highly effective vehicle for daily social contact. It is like Zadar's Kalelarga and *obala* put together, or Dubrovnik's Stradun at the water's edge. No other Dalmatian seaside town has such an appealing combination of natural and built environments, and this is reflected in Hvar's immense popularity among vacationing young people, who congregate along the *riva* on summer evenings.

Seen from the heights of the Tvrdjava, the outstretched "arms" of Hvar present themselves as an open book, with one set of "pages" laid out to the west and the other to the east. The "binding" runs on the axis through the north end of the *luka,* to the Old Town, and up the walls to the hilltop. Hvar's urbanized openness derives from the undulating shore, which provides constant contact with the sea on one side and the urban patterns around the central cove that is at once the heart and brain of Hvar. To walk Hvar's entire shoreline is to give oneself up to an extraordinary choreography. Our movements are prescribed in grand terms by the way the geography of the seashore serves as a stage for the drama of daily life, and as we move along, Hvar's receding and ascending hills convey to us a strikingly strong sense of place. Hvar's clear urban pattern is made to order for a reading, for here a reading can be readily experienced and recorded.

AN URBAN READING OF THE TOWN OF HVAR

The West Arm of the *Obala*

Sitting quietly on a bench in the tiny park that defines the small harbor for small craft at the far end of the *luka,* we may observe the flow of people converging at this juncture of the *obala*'s east and west arms. The clarity of Hvar's urban form tells us directly that observing the view is the logical way to begin our reading. The park's planting and the lively foreground of little boats rocking on a breezy spring day offer us a sense of containment and order on an intimate scale, in contrast to the grandeur of the Cathedral and its Campanile, which rise as a backdrop for the ancient Piazza.

Look down on the worn, cream-colored stone paving and you will see how it serves superbly as a unifying element extending the U of the *obala*. Visitors arriving on foot from the car and bus terminus first see Hvar from a narrow, almost hidden opening between the walls of the Old Town and the Cathedral. The great Piazza, created long ago, unfolds before them perfectly intact, as it was for visitor Pribojević arriving in the sixteenth century.

Piazza of stone and small boat harbor

Now we rise and begin to trace with our footsteps the patterns of this wondrous blending of land and water along the west side of the *luka,* our first segment. It will become clear that although the Piazza we leave behind is the most commanding paved open space, its role is no longer to join the Old Town on one side with the *borg* rising above on the other, as in early times. Today, it serves largely as a spacious circulation route, its airy, stagelike presence, without the bustling retail shops of the *riva,* seeming to draw people like a magnet to linger in its nostalgic space. Young people collect on the balustrade of the tiny harbor, on the broad steps of the Piazza, or at the Loggia. Children feed pigeons, which outnumber the people, and passersby hail their friends. Only before and after Sunday's Masses does the Cathedral become the focal point of social gatherings. In awe, we are reminded of an urban life pattern we no longer know how to perform in.

By contemplating this interplay of people, place, and water over a period of several days, you will gradually regain that lost sense of the pulse and rhythm of collective life that is generated by Hvar's unique physical structure. Hvar as a place maximizes a richness of daily life for all categories of users. By day all of the town, the sea, and rounded landscape backdrop are there for us to see; by night the integrated whole closes down to a tight urban area, particularly beautiful to experience.

Let us look now at the grand array of buildings along the north end of the *luka,* which represent Hvar's determination over the centuries to grow and survive the impact of foreign domination. They mark the peaks of Hvar's cultural evolution: the Clock Tower, the Loggia, reno-

The *obala:* West arm

vated Palace Hotel, a vestige of Hvar's persistent local government, and the Venetian details on the windows of what remains of the monumental Palača Hektorović.

At the northwest corner of the *luka,* note the increased number of shops: a barber shop, the Jadrolinija office, the popular private Plus Caffe (almost always jammed with the young), the institutional OMC Caffe (which is not as busy), the Atlas Travel office, shops selling books, stationery, and clothing, and a few others. Here stone steps crowd between buildings to provide access to the parallel, uphill route for vehicles and to the historic church of Sveti Marko, whose towering Campanile dominates this corner of the *luka.* History tells us that this church (the former cathedral) and its campanile formed the west end of an axis running along the then open shore to the present cathedral. The two major churches and their campaniles facing each other from opposite ends of the *luka* would have been an imposing sight for us, as well as a bold symbol of the important role of religion. Sveti Marko, however, no longer serves as a church; its importance was drastically cut when, under Napoleon's health code, the main roof was removed so that the cemetery, formerly in the floor of the church, would be in the open air.

As we walk now along the west arm of the U, we note that the stores cease and several smaller hotels take over. From seasonal open-air restaurants comes the appetizing smell of grilled fish and *ćevapčići* (barbecued ground meat). As evidence of the shipping activities that filled this side during the Middle Ages, the offices of the port captaincy are still located here, along with the Weather Bureau and the *općina* headquar-

The *obala:* East arm

ters. Above, a mansion of pink stucco speaks for a wealthy family of the nineteenth century, who perhaps prospered in Chile or California.

As we walk along the west side of the *obala,* note how the construction of the sea wall differs from that on the east side. Here the great stone slabs are laid in steps below the water line rather than in the vertical form of poured concrete. This construction dates back to the sixteenth century and was the first of its kind in Dalmatia, preceding a similar sea wall in Dubrovnik by seventy years. Looking closely, we see that they are notched ingeniously to fit together tightly. I learned from a sailor of ships that this had to be done to offset the force of heavier seas when the east side was filled to form the piazza of stone. That shallow branch of the harbor had provided a place where the power of stormy seas could be released through nature's own balanced system.

Across this watery "piazza" dotted with boats the rounded landrises; to the left we see the chaotic pattern of red roofs of the *borg,* and to the right, the more regular collection of residences built at the turn of the century. The whole is beautifully unified by the low, elongated, dome-shaped hill and the cream-colored stone and red tile roofs, then crowned by the dark green of the plumelike trees of the nineteenth-century cemetery. The union of unplanned geographical form and man's urban works over the centuries is splendid. Its quality is boldly underlined by a solid row of Dalmatia's ubiquitous date palms, the nineteenth-century stone promenade and sea wall, and the blue water of the deeply set *luka,* which on this calm day offer us a most skillfully articulated urban composition.

We leave this rich complex of the past where the *luka* abruptly gives

Down from the fort
through Old Hvar to
the *luka. Drawing by
author*

way to the open sea. The original rocky shore takes over, and the vast
pine woods on the peninsula of Veneranda begins. A tiny chapel located
just above the sea suggests the last place for sailors to pray before leaving
Hvar and the first place for giving thanks after a safe return. Likewise,
its inscription, "Zvijezda Mora" [Star of the sea] speaks for the church's
role in protecting sailors and the guidance of the starlit heavens when
they are at sea.

Beyond the chapel the woods and the uncontrolled sea take the place
of Hvar's urban environment. Here we sense that we have left behind the
security and protection of the town—even as sailors once did—and the
sense of exposure to the elements, illusionary as it may be today. These
senses are heightened by our discovery of a great semicircular terrace that
may have been a nineteenth-century amenity, a terminus for the evening
promenade like one built by the British that we shall see in Korčula. But
we find that its handsome, curved stone bench and high wall were actually part of a gun emplacement built by the French to secure Napoleon's
foothold in Dalmatia. It was built in 1811 on the foundations of the
old monastery of Sveti Veneranda, which was abandoned in 1807. Now
the rambling remains demonstrate the strategic importance of Hvar in
international maritime conquest during one period and, much earlier,
the religious and cultural influence of the Venetians. How different from
the quiet environmental histories of Sutivan, Pučišća, and Bol!

From the hilltop we look back to yet another such symbol, the
Tvrdjava Španjul, rising high above the walls of Hvar's Old Town. When
we attempt to imagine Hvar in the prime of its urban renaissance, free
of twentieth-century intrusions, the rocks, pines, and open Adriatic sea
suddenly engulf us as the excitement of town and people are replaced by
the calming, meditative quality of this wild, coastal place with its remnants of struggle between power and spirit for one's daily life.

Out of the elevated woods and down to the sea, our continuous
shoreline walk unifies and reinforces by comparison the true urbanity of
Hvar. Our walk describes a great, broad curve, bringing visual experiences as we head toward the first of three western coves. But there our
meditative illusions generated by the newly discovered world of nature
are shattered by the monstrous impact of the Amphora Hotel, an alien
form that cruelly interrupts the sensitive experiences we have enjoyed
thus far: it is too big, too white, too geometric, too harshly suggestive of
an artificial urban life style. Yet its location there has one saving grace:
the ill-fitting structure can not be seen from the town itself, and thus it
will not be part of our lasting image of Hvar.

At the bottom of the cove an incongruous pedestrian bridge further
breaks our continuity, but shortly the hotel is behind us and we focus
again on the pines and the open sea below. The walk offers a close look
at native shoreline plants that can be found all over Dalmatia—the fra-

grant Aleppo pine, rockrose, maqui, and rosemary—suggesting how the natural environment appeared to the native Illyrians themselves.

We swing out to the point of the promontory, and a second cove greets us, forcing us to shift our focus away from the open sea and toward the island's rugged hills. Only a few scattered villas left from the years between the two world wars interrupt this fusion of sea and land. Around this shallow cove the landscape becomes wilder, and we experience the purity of pines and sea alone. At the bottom of the third cove this illusion comes to an end. Here the spell is broken by the invasion of the automobile and the newer, road-oriented houses, accessible from the upper hillside.

On the return walk to Hvar's center, all these experiences are reversed, and this in itself is a memorable experience, especially when in the lowering light of the sun setting behind us. The coming of darkness and technology's artificial daylight reminds us how well Hvar's social life and its structural form fit each other. With our backs once more to the Hotel Amphora, we see the tiny island of Galesnik, with its single house backed by scattered villas on the wooded mainland beyond, a preview of the the central urban area ahead.

Then, house by house and roof by roof, the entire panorama unfolds. From the fortress high above, the walls and dwellings of the Old Town cascade down to the Piazza with its Campanile. From the arch of the Arsenal's we can see the chaos of the *borg* surging up the low hillside neighborhood and the bold line of date palms marching above the level line of the *obala* out to the firmly placed anchor of the Franciscan Monastery.

The East Arm of the *Obala*

With an image of the west segment now deeply imprinted in our minds, we return to our bench at the tiny waterfront park facing the piazza of stone and the Cathedral, this time to start our reading of Hvar's east arm. At the corner where the Piazza joins the *obala* the old Arsenal, with its wide, arched opening, is a compelling image of Hvar's maritime strength and cultural level. On the floor above the arch the oldest theater on the Adriatic coast was built by the town itself, for all to use, in 1612. To see this example of architectural diversity as a mere object without reading into it the nature of the society that produced it is to miss the major point of experiencing Hvar.

Along the *obala,* venerable stone buildings form a backdrop for today's busy urban edge of the harbor that contrasts with the irregular character of the *borg* above. The medieval setting is broken by three nineteenth-century buildings that jut out in front of the original line of buildings, with their balconies and eighteenth-century architectural details. Clearly, this resulted from a major public works project to widen the *obala* and especially to establish a dock for the new, larger steam-

ship lines. The wide Esplanade, the planting of date palms, and the scale
in general reflect the influence of new, more formal urban design ideas
from the period of urban development in Vienna under Franz Josef.

Further along, an array of rural nineteenth-century influences appears
that are not found on the west side of the *luka:* the use of reinforced
concrete in the sea wall, the evenly squared-off stone fitted without the
interlocking notches of the primitive east *obala*. As we continue, the
Esplanade widens greatly and separates the pedestrian at the water's edge
from the several popular enterprises. Breaking the continuous facade of
buildings on our left are steps that go straight up the hill to link the *obala*
with the old interior pedestrian street above, parallel to the *obala*. The
date 1593 inscribed on the wall of a tiny chapel tells us that it was there
long before the new buildings were added during the nineteenth century
and thus must have been visible from the harbor and a comfort to sailors.

At the foot of these steps, nodes rich in social activity and urban design quality have established themselves: a sun-oriented cafe with open
terrace and covered arcade, a long bench for the old men of Hvar, the
town news kiosk, the post and telegraph office. Here too is the starting
point for the row of stately date palms. Planted at twelve-foot intervals and stretching some three hundred feet, these palms, imported
from North Africa since the early nineteenth century, contribute a commanding presence to Hvar's identity. They offer a welcomed shaded
background for the social life on the Esplanade in full summer. Lifeless
in the off-season, the *obala* is alive with the coming and going of giant
steamers, which glide silently up to the stone-faced sea wall. Passengers
disembark, others board, and human interaction takes place suddenly
among both surging crowds and joyous individuals. Small motorboats
marked "FKK" (for "Frei Korpus Kultur," or Free Body Culture) take
those who enjoy sun and sea unencumbered by swimwear off to the privacy of the tiny island of Jerolim. The cafes fill with people, and the
vendors selling rugs and handicrafts from the interior of the now former
Yugoslavia set up their stands.

When we reach the last of the giant date palms, the Esplanade becomes a mere passageway, narrowing from some fifty feet to four. The
tight building facade gives way to the long, high retaining wall of two
large nineteenth-century villas and the Hotel Dalmacija. The social contact we have been enjoying ends as we make the sharp turn south, where
we are greeted by an abrupt change of view. We take in the first of the
coves along the east shoreline, ringed with residences and a seaside walk
around to the Franciscan Monastery, handsomely sited on its promontory by the brothers in 1561. The wind and sea spray that assail us when
we round the bend are further signs that we have now left the protective
luka behind, just as the sailing vessels of Hvar did in its formative years.
Inside the monastery, as we contemplate the collective struggle that was

necessary to create the Hvar of today, the setting sun directs a beam of light through a circular window and spotlights a figure above the altar with arms outstretched toward us, the very metaphor of Hvar itself.

As we round the bend of the promontory, the wind carries the sea spray across the pavement. Our view is diverted away from the nearby curvilinear Sveti Klement Island, four miles long with twenty coves and, in the deeper, second cove, the Križna *luka*. At its innermost point the growing New Hvar, under construction, contrasts vividly with the Monastery and our images of Hvar. The severely geometric forms of two hotels further threaten Hvar's genuine identity. We turn our backs on this disruption of nature's landscape to reach the point of Križni Rat, where rocks, pine, and sea once more win out over the scattered, smaller villas. On the hilltop above, called Baterija, stands the French-built fortress, a counterpart to Veneranda, which we saw on the west arm.

Our experience builds as we round point after point and cove after cove, with each one removing ourselves further from the human-made town and harbor. We reach a climax when suddenly we have a clear view across the sea where no protective islands lie between Hvar and Italy. Were it a clear day, we could see the west end of Korčula to the south. The rush of wind, the untouched stony hillsides, and the formidable meeting of sea and land evoke images of the peoples who in a common environmental home joined together over centuries to create Hvar and its identity out of the resources of this island. At our third, and last, cove—Pokonji Dol—we step down onto the beach of snow-white limestone pebbles. Here, we contemplate the urban evolution we have experienced, from the glaring twentieth-century dominance of man, money, and technology over nature, to sixteenth-century cultural refinement, to the glory of untouched nature.

Beyond the small white beach, we reach the time of the Illyrians and Hvar's neolithic beginnings, for the paths are no more than stony trails through nature on its own. Far beyond, a chain of hamlets at Zavala, halfway down the length of the island, are accessible by a tunnel from Pitve, near Jelsa. The slopes, terraced with vineyards, are still virtually free of cars. In 1990, however, I witnessed from the ferry to Korčula harsh cuts into the stone slopes in preparation for a new roadway from the mouth of the tunnel to the new shoreline developments yet to come.

Stari Grad—The Old Town

Hvar's Old Town provides a vivid contrast to the mix of sea, land, and town we have experienced thus far. We turn our backs on the present—the boat-studded "piazza of water," the assemblies of people, their sounds, the flying pigeons, and the openness. Entering the stone gateway, we find ourselves in the fully enclosed urban area of Hvar's fifteen-century Stari Grad, a solid mass of tightly packed houses form-

Urban canyon of stone in Old Hvar. *Photo by author*

ing architectural cliffs sliced through by a grid of narrow, straight streets all held together by the old city walls. Each dwelling presses against its neighbor, and all face one another from a distance of only meters across the narrow walkway. See how they stand three and four stories tall, too high to be taken in at a glance, and require us to move our eyes slowly upward, thus intensifying our experience of urban reading.

A fixed grid of streets and blocks adapted to a hillside location dictates our choice of routes. Three main east-west streets that are level connect the north wall to the south, and some five cross-streets go uphill either by ramps or steps. The steps, built of the same stone as the buildings, give us an uneasy sense of having entered a megastructure masterminded by a single individual whose influence has governed people's movements for centuries. Steps and ramps intersecting level cross streets create spaces so much alike that we lose our sense of direction. We note how this discipline contrasts with the freedom we experienced among the open and varied landmarks of Hvar's shoreline, clearly the work of many. We feel relief when we reach two piazzas at the Sveti Duh (Holy Spirit) church and the Benedictine Monastery, which, although tiny, appear spacious to us.

This purely pedestrian world, where compact cars would only barely fit, draws intimate responses from us. The feel of stone paving under our feet, still intact, polished and sculpted by centuries of wear by man and animal, is a delight. The surface of the bottoms of these tiny canyons gives off a lightness and reflection that suggests the pebble-polished stones in the bed of a cascading stream. We have access to the outside world by only four gates; there is simply no other way to leave the Stari Grad in the heart of Hvar. At the tiny Sveti Duh Piazza, at the cross-street on the second level, we turn right to reach the East Gate. There the wall forms a sharp urban edge, and an opening cut in it large enough to admit small cars prompts an opportunity for escape—if we wish—from this enclosure. Here too cars are parked for the residents who must carry in, on foot, food and other household goods and even equipment, sand, cement, and building materials for the considerable amount of re-building that keeps the old, seemingly dead place quite alive.

We continue up the steps of the north-south street as if in a miniature Dubrovnik and emerge at the North Gate, where we are greeted by the asphalt road, the cars, and the contemporary world. From here the up-hill path takes us directly to the hilltop Tvrdjava Španjul, an experience in itself, up through the pine and stone landscape that incrementally re-veals to us Hvar's essential source of identity: the now familiar arms of Hvar's shoreline flung open to the sea. From this perspective the Old Town now seems all the more enclosed. Its walls built for security in past eras seem to serve in our times to fend off the threatening complexity and aggressions of today's urban life that are changing Hvar.

As we move toward the setting sun to locate the remaining West Gate, our focus is captured by a small empty piazza where a secluded convent continues its medieval life; nuns committed to isolation sell their hand-made lace through tiny openings in their monumental doorway. The nuns' inward life becomes a metaphor for the introspection and self-confrontation that the Old Town begins to produce after several hours of deliberate submission to its tight enclosure. In contrast, the openness of Hvar becomes a metaphor for the vast changes that have taken place in its societal outlook since Hvar's urban evolution began. In this sense, a reading of the Old Town will make us aware of how this precious ves-tige of an earlier urban environment tells us whence we have come.

We sit on a doorstep awaiting a welling up of thoughts on the mean-ing of the Old Town. This introspection in the confined space and dwelling places, unchanging for generation upon generation, causes us to respond to otherwise uneventful experiences. A sole, aged woman in a black kerchief passing by a few feet away suggests the life of centuries. A plant growing in a crack in the stone masonry high above a third-story window flowers toward the sun. Details bear down on us relentlessly— how the place smells and feels, sudden variations from warm to cold or damp—as we move from street to street, immersed in the stone canyon. Hvar's sense of place permeates our responses as we turn inward to respond in new and more sensitive ways to what at first seemed a stark, foreboding enclave.

Within these walls virtually no vegetation offsets the solidity of stone construction; little soil or even sun can be found to sustain trees or shrubs except for the audacious and gymnastic weeds and plants that grow and even flower in the most unlikely ledges in the walls. The ubiquitous grape arbor and bougainvillea appear only rarely here, on a terrace or balcony or roof garden. Indeed, a sundial placed unusually high on the wall of a house overlooking the small piazza tells us how precious sun is here. Nor do we find dominating environmental details in this viewless enclosure; in the end, the pervading image we leave with is of the consistent urban grid and the repetitive architectural forms that typify this very small area. Unlike in the other urban environments we have experienced, no single tower or building dominates the Old Town.

We read something of the social character of the past into these forms: the residents were of an urban culture, held urban occupations, were well-to-do, and represented the upper class, all positions supported by the Old Town's proximity to civic and maritime facilities. From the current uses, maintenance, and appearance we can also learn something of today's character: the core of the Old Town's population are older families who continue to be tied to the past by a strong sense of identity with this place. Some of the homes of these older families appear to be used as weekend or summer homes. A number are clearly vacant, or abandoned.[2]

We have deliberately allowed ourselves time to experience deeply the sensations of enclosure, isolation, introversion, and intimacy of self in relation to place in order to evoke insightful responses and meaning from Old Town. Now, as we leave by the South Gate, through which we entered, we find one vivid image being replaced by another; we move from the closed geometric grid of Old Hvar, where our movements were rigidly prescribed by urban planners of centuries ago, to Hvar's open Piazza and then to Hvar as a whole.

Our hearts leap when we experience the joyous feeling of release in passing through the gate and into the great piazza of stone. The unrestricted movement of people through seemingly limitless space and the freedom to choose our direction gives us a tremendous sense of opening out to the world. We come to life when we hear the sounds of people conversing; we gain fresh meaning from watching children feed clusters of pigeons as they wheel about, defining the three-dimensional space with their flight, their fluttering of wings, soaring to rooftops, and then returning to the children, who are fascinated at this wonder.

Leaving the Old Town and all that it had served to "gather" (in Heidegger's sense) from the past evoked in me a fantasy of movement through a succession of environments of ever larger scale.[3] First, we move from the confining space of rooms, houses, streets, and blocks that is the Old Town into the larger realm of the piazza; there we turn and realize that these environments are but part of a third, yet larger one, the town of Hvar itself. Then the openness of the *luka* tells us of yet another environment, the island that houses Hvar; and the island becomes but a part of Dalmatia's maritime region, and these linkages continue reverberating out to the world at large. This concentration on a hierarchy of subenvironments one at a time taught me that fully achieving identity with a place, such as the Old Town, is a function of one's ability to perceive spaces of varying scales—environments within environments—and to sense the interrelationship between their identities.

Route up from the Esplanade to the *borg* and the hilltop Cemetery. *Drawing by author*

The *Borg*

We cross the great Piazza to enter the *borg*, the district that is home to lower-class people of rural origin. As our first lesson in its incremental structure, we find no clear entrance. In contrast to the Old Town's four gates located in absolute symmetry, there are three or four significant entry points on the south side of the Piazza and well over a dozen to be found during this reading. Few readily appear to be more important, nor are they marked by any architectural features. With no guidelines regarding structure or direction, we suddenly feel spatially illiterate, unable to read the street pattern.

Following a whim, we begin to enjoy the greater freedom to move at will and to anticipate the unknown. No two intersections are alike. We follow walkways that turn and twist in innumerable ways, some following the contours of the ever-curving hill, some steeply ramped, some formed in steps. When streets widen enough to become linear urban places in themselves, we discover views to the harbor and the open areas of Hvar; others are extremely narrow, and even a few pass beneath buildings seemingly without outlet. We soon find that the system is not a labyrinth, and its rational pattern begins to appear. A variety of building forms and placements become landmarks for mentally mapping the area. The introspection induced by the Old Town dissolves away, and the experience of reading the *borg* combines the oppressive compactness of the Old Town with the unexpected glimpses out to the openness of Hvar as a whole.

Clearly, the route to follow is the longitudinal street above and parallel to the Esplanade, which in old times provided the main circulation route and defined the edge of the *borg*. To experience this area in the chronology of history, we reenter the *borg* from the flights of steps up from the Esplanade, where we saw the church of Sveti Duh, which once faced on the harbor. As we walk south, changing architectural elements indicate that we are moving through the Middle Ages and into the Renaissance. More vegetation brightens terraces and balconies, and further up the hill many houses are still as they were in earlier times, that is, on parcels large enough to allow space for maintaining a vegetable garden and keeping livestock close to the house. Toward the end of this main route the style and size of the dwellings and the width of the route increasingly express the nineteenth century. As we round the bend of the hill, the view opens to the first cove and the Franciscan Monastery on its promontory. We have now reached the twentieth century with some dwellings from the period between the two wars. One large mansion of pink stucco—a fashionable symbol of economic success a century ago—stands near the end of the route, just behind the Hotel Dalmacija. Finally we arrive at a point just above the first cove where the route becomes a true vehicular route of asphalt with curbs and gutters and surrounded by recently built, smaller apartment houses.

The route continues to the south and west, where it becomes a thoroughfare providing access to the extensive New Hvar, on the hills above the first and second coves. In this new development the parcels of land are wide, houses are separated, and the architecture expressive of each owner's taste, a far cry from the consistency and continuity of the Old Town and the *borg,* which so well symbolize the collective spirit of the past. We have experienced the full cycle of change in the image and meaning of Hvar's urban environment over five centuries through a thought-provoking and observant short walk.

The Hilltop Cemetery

The logical place to end our reading of Hvar awaits us at the hilltop above the *borg.* We follow the pleasantly disjointed routes of the *borg* and wind our way up the hill to the cemetery, a lush green crown of vegetation covering the crest. Since the cemetery was built early in the nineteenth century, when the French ruled against burials within church buildings, there are no particular historical features for us to read. We do find, however, the recent grave of Grga Novak, Hvar's distinguished historian, who understood the sources of his homeland's sense of place. In his words, "Hvar was really an international town, more closely tied to the Mediterranean trade and to Venice than others in Dalmatia yet far enough away from Split to have a cultural and economic life of its own."[4]

The essence of the cemetery environment lies in its monopoly of the past and its silent reminder of how temporal the roles of generations of inhabitants are compared with the works they leave behind. But this thought also inspires reveries of eventual outcomes of the continuing process of urban evolution. The abrupt change in environmental quality in New Hvar threatens the images of Hvar of the past five centuries that we have come to identify with. By objectively applying this awareness to planning the sequential growth of the future, however, we can promote

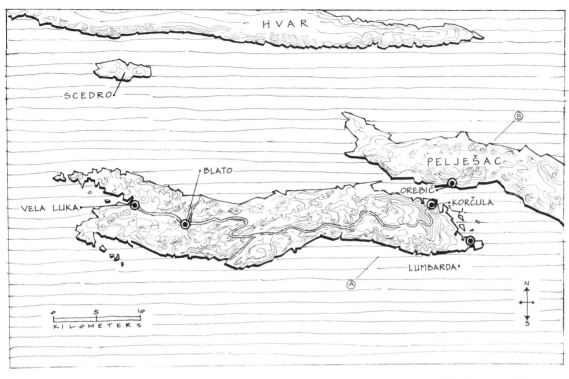

The island of Korčula

the qualities of place and people produced in the past without slavishly reproducing design patterns of the past. Our qualitative responses to Hvar's "arms open to the sea" have shown the town to be broad and out-reaching. With insightful guidance in urban planning Hvar should have the opportunity to maximize the directions to take in the future based on the identity with place evolved over the centuries.

KORČULA: THE URBAN SHIP

The Island Setting

While Brač's geography has resulted in a lineal settlement system in which the population is distributed evenly over the island as a whole, on the island of Hvar the population is concentrated at one end. Differing from both of these, Korčula has two urban centers, one at each end of the island, with a number of villages in between. Korčula and Lumbarda at the east end and Blato and Vela Luka at the west end together were

CRVZOLA

The Town of Korčula in 1598: A geography made for Renaissance maritime urban life. *Croatian State Archive*

home to about 75 percent of the 18,400 inhabitants of the island in 1981. Korčula, was the dominant center during the late Middle Ages. However, it was surpassed by Blato and Vela Luka in both size and economic growth during the prosperity of the nineteenth century. Certainly, all three can be considered towns since each serves as a center for some of the island's thirteen villages.

Blato, settled even in Roman times in a small, rich valley, became self-sustaining in a central position inland with access to ports on both the north and the south shore. Its situation at the natural meeting place of six radial routes gave Blato a distinctive circular pattern and many of the advantages that Korčula enjoyed because of its size and sea access, though without responsibility for defense. During the nineteenth century Vela Luka grew enormously because of its protected harbor and because it served Blato in wine export and trade. However, Korčula's superior site and design have made its image the dominant one.

Going back to the ninth century, Croats established in Herzegovina were the first Slav settlers on Korčula, as they were on Hvar and Brač. Self-government has been the rule throughout Korčula's history in spite of the proximity of Dubrovnik to the south and the Venetians at Hvar to the north. A heritage of Roman times, the first statute for the island, adopted in 1214, is one of the oldest in Dalmatia, along with that of Brač. It helped Korčula to maintain its autonomy under the protectorship of Venice from 1420 to 1797. In this period the town survived the siege by the Turks in 1571 and enjoyed three centuries of stability as a strong urban center. In 1813 the British kept Napoleon's troops from invading, and in 1815 the island came under Austrian rule.

As Vinko Foretić points out, by 1848 liberal ideas that were revitalizing local cultures in Europe had their effect on Korčula. The Croatian National Renaissance movement brought its leader, Ljudevit Gaj, to Korčula and Orebić to advance potential democratic reforms. In 1871 local elections were held under the limited autonomy granted by Austria, and the People's Party won. This success was the source of my grandfather's being named mayor of Kuna district on nearby Pelješac.

Also by 1871 the Narodna Čitaonica (People's Reading Room) was founded, a timely demonstration of the progressive nature and self-identity of the people of Korčula. Throughout Dalmatia societies were formed to establish reading rooms. These became the crucibles of Croatian cultural identity, which was tied to an "environmental homeland" well defined by the societies. In the reading rooms no Italian could be spoken, and by 1893 the name was changed to Hrvatska Čitaonica (Croatian Reading Room) to affirm a clear connection between the historical culture and the uniqueness of the island setting.[5]

Unlike Brač, with its prolific series Brački Zbornik, the island of Korčula did not produce a series of its own until the 1970s. The issue published in 1972, however, was far more comprehensive than earlier ones. Forty-four authors wrote in great detail about the island's many aspects. The article by Ivo Kaštropil is particularly relevant to our discussion of environment and place. Kaštropil offers evidence that the landscape of Korčula was both unusually heavily forested and bountiful, far more so than barren Hvar or Pelješac. Individual place names have been derived since the earliest settlements from the vegetation of their locality. For example, the island itself, named after Corfu by the Greeks, became Corkyra Nigra (Black Corfu) under the Romans for its dark masses of wooded slopes; similarly, we find Hrastovice (oak), Borova (pine), Kupinovica (blackberry), Lozica (grapevines), Grahova (beans), and Maslinova (olives). Other places were named for the wildlife found nearby, suc as Lisjak (fox), Učijak (wolf), and Slavinj (nightingales). Kaštropil uses such place names as metaphors for the visual qualities of early times in describing the natural environment as well as the cultural history via family origin. He thus makes permanent the sense of connectedness and identity that evolved and became rooted through settlement.[6] Sir J. Gardner Wilkinson spoke to this point in 1848: "The Isle of Curzola abounds in trees and brushwood . . . offering a striking contrast to the Dalmatian coast. . . . It supplied the Venetian arsenal with timber and the proportion of land covered with wood is still 43,471 acres, out of a total of 57,130."[7]

Thanks to its rugged topography and frequent natural harbors, by the

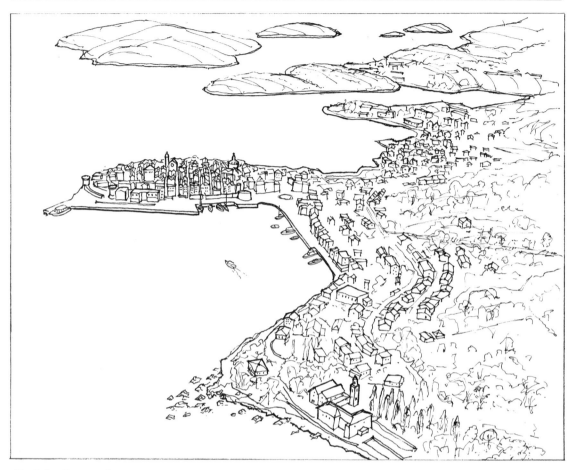

Korčula: Crown of an
undulating shoreline of
land and sea

turn of the nineteenth century the island developed a well-articulated
local maritime system. At the time of Napoleon's conquest, rudimentary
roads were built by the British, who used Korčula as their base against
the French. Only in recent decades, however, has a modern system of
asphalt roads been completed; and a bus service now links the two ends
of the island and the villages in between. By the 1970s the installation
of a regional water supply system following this road system had transformed the urban settlement pattern; the water supply is brought under
the Adriatic Sea from the Neretva River on the mainland to Pelješac and
under the channel to the island.

The concentration of population in towns and large villages has increased the urban character of the citizenry and has given them a greater
stability and rootedness. These qualities are expressed in stronger cultural identity with hometowns and in diverse and competitive attitudes.
This local identity is also due to Korčula's distance from Dalmatia's
centers at Split and Dubrovnik and their spheres of influence. Korčula
remained politically independent of Dubrovnik yet within easy access of

its sophisticated culture. This was especially evident in both the urban design and the community life of the town of Korčula, whose position in relation to Dubrovnik to the south and Venice to the north placed it, like Hvar, directly on the international trade route. However, lacking Hvar's ample arms reaching out to the sea, Korčula, thrusting itself out to sea, became more independent, particularly of Venice. Korčula's economic position rose remarkably during the late eighteenth century and throughout the nineteenth century, fostering the rapid growth of Blato and Vela Luka. With these two centers producing food and Korčula and Lumbarda being home to the leading shipbuilding industry in Dalmatia, the island's economic position reached its peak.

The integrity of the island's development grew out of the unique site and location of the town of Korčula. Thus, a strong urban character developed over the centuries, made manifest in the superb complex of buildings, streets, piazzas, and embattlements. The steep topography rising inland from the town site afforded a sweeping view up and down the Pelješac channel; there the British built a fort dedicated to Sveti Vlaho in 1813 in their effort to suppress Napoleon. Three routes—east and west along the shore and up the mountainous spine—converge precisely at the base of Korčula's promontory and serve as the regional connections for the overall structural system of this most unique of Dalmatia's towns.

Korčula's statute was more sophisticated than Brač's since it had to meet the needs of larger, more complex regional centers. Its development was influenced by contact with Dubrovnik's advanced institutional and legal system. Regulations were oriented to urban requirements and toward a high quality of building design and construction, especially in Korčula's Old Town, where culture and wealth lent support to urbanity at its best. The statute was especially strong in the area of social issues and showed concern for justice and human welfare, although classes were clearly divided. Concern for the environment was expressed in relation to the welfare of the people; laws related to conservation of forests and grazing lands and agriculture in general were highly effective in protecting public resources and stabilizing and protecting the economy. Today the island of Korčula functions as a single commune, with each village or town given limited authority as a *mjesna zajednica*.

The Town Structure

Over the years, the shape of the rocky promontory site of Korčula has evoked in my mind the clear image of a ship docked and ready for departure from a receding shoreline. This image seems especially fitting in light of Korčula's valiant maritime history and its centuries of shipbuilding and trade. The town's overall structural system evolved as a result of its success in these activities as new growth after the Middle Ages occu-

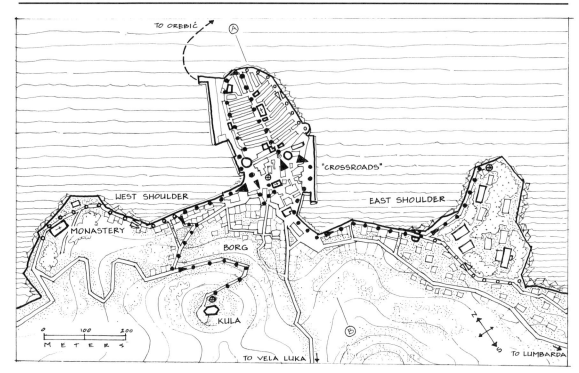

Korčula: Urban
structure and reading
path

pied the flanks of the undulating shore. First came the shipyards in the eastern and western coves of the promontory and then, in our times, lineal extension of development in both directions with the coming of vehicular land routes.

To the east, offshore islands gave further protection as far as Lumbarda, four and a half miles beyond, at the furthest tip of the island. From there, the precipitous south shore of the island prevented shoreline settlement as far as Brna, more than halfway to the west end, which forced development to be concentrated at Korčula and its vicinity. A mile or so from the Old Town, the small promontory on which the Dominican Monastery and the church of Sveti Nikola stand terminates the west "shoulder" of the larger urban area. From there the gently sloping north shoreline allowed a land route as far as Račišće, a distance of nine miles, with its small harbor just opposite the end of the protective channel between Pelješac and Korčula.

In sharp contrast to Hvar's sense of place, Korčula's is revealed not by a focus on the sea but rather by the way the expanse of sea sets forth the skyline of the Old Town, its Stari Grad, and calls attention to its distinctively unified architectural character. The town invites a fresh awareness of the human ability to collaborate with geography and create a place from which to send ships out upon the sea. Because of its symmetrical oval shape, a "mental map" takes form almost at first sight. Two regional routes converge at the point where the oval site connects with the mainland, forming a regional crossroads and an irregular square at the very gate of the Old Town. Here are located markets, shops, services, banks, and civic activities, as well as the dock for coastal ships to one side and the ferry to Orebić on the other.

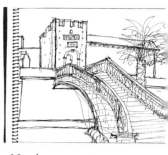

Neo-baroque entrance to the Old Town. *Drawing by author*

Unlike Hvar's Old Town, hidden in the larger landscape, Korčula's is clearly delineated by the remnants of medieval embattlements that dominate the surroundings with memorable imagery. The Old Town rises from the blue waters, a striking work of urban art. Laid out in a fishbone pattern, its symmetrical street system reflects the same kind of order and efficient use of restricted space found aboard ship, and it imposes a similarly but more gracefully disciplined movement. The streets sloping up to the central spine of this system and down to the opposite side suggest the rocking of a ship, and the views out to the surging sea at the end of each tiny passageway, portholes. The town's sharp point, terminating the main axis and free of the mainland, recalls a ship's prow, and the Esplanade, which replaces the walls, become the decks above the sea.

From its rich assembly of buildings comes a sense of the authenticity of the whole urban ensemble, giving Korčula its brilliant sense of place.

An Urban Reading of the Town of Korčula

Stari Grad, the Old Town

We begin our reading of the Old Town by entering from the crossroads via a grand neo-baroque staircase that arches handsomely over a strip of park, once the broad moat filled by the sea. At its wide landing the stone gateway of the Old Town, with the Lion of Venice emblazoned high above, receives us with dignity and establishes a powerful sense of entry.

Only in our reading of Dubrovnik have we experienced such a rush of responses at the very beginning of our walk. Instinctively we turn around now to look up at the fortified square tower of the Land Gate, which we have just passed through, anticipating more specific responses to this provocative point of arrival. We note the incongruity between the baroque stairway and the severe gate tower and the two enormous round embattlements to the east and west. These read clearly as products of the fifteenth and sixteenth centuries, whereas the stairway was added at the turn of the twentieth century. This addition was intended to lend a sense of vitality to the architectural setting after the somber walls of the city were torn in the 1890s. Yet, in spite of its romanticism and artificiality, the arched stairway dramatizes with a certain flair the experience of our entry into the past.

Far from romanticism, the stone bas-relief of the Lion of Venice, high above the gate, stands for centuries of political and cultural reality. The inscription below suggests that those who installed the stone plaque in 1925 wished to remind us of Korčula's historical identity with Croatia, which was established as a state in A.D. 925.

U SPOMEN
KRUNIDBE HRVATSKOGA KRALJA
TOMISLAVA
DCCCCXXV–MCMXXV

[In Memory of
The Crowning of the Croatian King
Tomislav
925–1925]

Where in earliest times the moat below guarded against the constant insecurity, today, ironically, the space is used for car parking, the result of Korčula's present-day ties to this end of the island and its being a focal point for shops and services. To the right of the bridge, a row of six date palms recalls the Victorian era. To the left, high above the former moat, a vast terrace left after the removal of the walls stands bleak and bare in winter and spring but comes alive with vendors and their produce and folk wares in the heat of summer.

Now that we are primed by this period of contemplation of the act of entry, the scale of the outside world reduces to that of a ship's intimate interior. We feel even more like outsiders intruding on the past, ill prepared to gain any genuine sense of identity. We first enter a small piazza; the sense of enclosure of its irregular space and the look of serving some important ancient function invite us to stay. We seek clues to the piazza's original social purpose as a forecourt to the town: the fine open Loggia, to the left, provided a gathering place in wet and hot weather in front of the municipal council building, the symbol of Korčula's traditional self-government, and to the right, across the tiny piazza, is the chapel of Sveti Mijailo (Saint Michael). Beyond, a handsome arch over the narrow street joins the church to other religious buildings. What contrast we find here with today's social purposes! An artist draws silhouettes of tourists, who in turn fill the outdoor tables of the Gradski Podrum, a lackluster restaurant run by Korčula's tourist-hotel management organization. Down the tiny street to the left the typically inconspicuous sign of a private tourist restaurant attracts us.

Straight ahead, and visible only through the narrow opening of this

Old Korčula's inner
heart

solid urban fabric, the superb tower of the Cathedral beckons us. Its imposing mass dominates the image of Korčula we have accumulated thus far and draws us onward. The stepped street reinforces our mounting sense of anticipation, and we will see that this was indeed once the main street of shops, as evidenced by the facades with combined doors and windows. Today these shops sell mainly tourist items, such as *filigran* silver, leathercrafts, and the like. Hemmed in as we feel because of the tiny scale, the openings created by the three cross streets exemplify the diverse experiences that Korčula's unique street pattern holds in store for us. Dwellings run at precise right angles to our main axis, but subtly the cross streets assume varied angles as though to test our increased visual awareness. Those going down to the west side offer glimpses of the sea solidly framed; it is as if we were looking through portholes. We turn to the east and follow that gentle bend in the alignment of the narrow, high-walled street as it curves out of view of the sea.

At the top of a last flight of steps unfolds the modest space with the grand facade of the Cathedral and the Bishop's Palace on the right, flooded with brilliant sunshine. On the left, the Town Museum and the Palača Arneri remind me of the generous guidance given to me by a senior member of that family, Juraj Arneri; his recollections, laden with insights gained from the past, have found their way into these pages. In the late nineteenth century his father, as mayor, brought down the old walls and established Croatian as the official language. When I consider his perspective and my own responses, I think that of all these culture-rich Dalmatian towns, Korčula must surely have the most perfect inner heart. Yet we have traversed, in only a few steps, the noble entry gate, the civil government center, the old core of commerce, the religious center, and the homes of the old elite families. This spirited reading reveals the phenomenon of a built environment compacted into a minute space—like a toy city—yet containing a high level of cultural and economic integrity developed over centuries. This tiny power ultimately fended off the Venetians, the Turks, and the French, and with the help of the English kept their cultural heritage remarkably intact for those, like ourselves, who wish to seek it out.

How different are the cities of today, with their exorbitant consumption of space for living and circulation, for commerce and industry, and how low the level of local cultural identity! In Korčula today we can almost feel its ancient heart pulsating today in spite of having lost its human vitality to the crossroads, that new center outside its entry gates. But so much the better for us, since the revelation prompts us to revive the past in our minds and more fully relish the remarkable tranquility of Korčula as a place in history filled with latent energy.

To penetrate this urban past further, we pass through the open doors of the Cathedral, accompanied by the same stone paving as outside, so that the interior seems a continuation of the piazza. We can mentally reconstruct the sixteenth century, when virtually all of the people of both *grad* and *borg* used the Cathedral, its piazza, and the street system not just as one great work of urban art but for the social purposes of strengthening the spirit, which made them able to cope with survival throughout those arduous centuries.

We continue to the end of the piazza, where it wraps around the north end of the Cathedral to form a piazzeta in front of the chapel of Sveti Petar (Saint Peter). A tiny alley leads to the celebrated—though frequently questioned—home of Marco Polo. Waste paper, a by-product of modern advances in packaging, scatters about in the wind, leaving piles of debris to despoil our memory of the numerous ruined sites. Yet, all suggest that Korčula's identity has survived the conflict between the past and the present. As we follow the main north-south street, steep steps remind us of the rugged shape of the original hill site of the town. We experience a sense of intimacy implied by doorways that open directly to the narrow street.

Steep steps lead us to the prow of the shiplike Old Town, where the unceremonious exit to the sea contrasts with the handsome Land Gate whence we started. Having traced by foot the very line traced on paper by a town planner of the fifteenth century, we are exhilarated by the idea that an urban plan can be the vehicle for communication between individuals centuries apart. Expecting a sense of geographical grandeur on emerging at the very tip of the promontory, we now find our outlook blocked by an enormous, ruined half-dome of stone that was once part of the old masonry walls and the sea gate. Upon its heights, a cafe-bar has been added; its ungainly modernity clashes harshly with the lively past our reading has evoked. When we round the dome, the vast expanse of sea spreads out, calm as a lake, backed by Mount Sveti Ilija and the ridge of Pelješac on the opposite shore. We sense the enormity of the space after experiencing the constriction of the Old Town.

This response wells up within us to a grand finale as we move around the east shoreline to embrace this elongated panorama, which is framed by a protective mass of dark green pines that billow over the edge of the sea wall and toss with gusts of wind. We rest on the sturdy stone parapet that runs the entire length of this rugged overlook and gives us direct yet protected contact with the sea breaking on the rocks below. Our intimacy with nature is broken, however, by the intrusion of small cars parked in front of modernized older homes, quite out of character with both Korčula and the Old Town.

As we return to the point of this urban promontory, the roadway drops down to become a dock, the very one I stepped onto in 1937 on my way to Kuna. Now we find here on the side a busy *obala* where smaller craft come and go, giving a sense of maritime vitality to the Old Town. Scanning this western facade of old Korčula, we sense the ancient scale of the massive Balbi Tower, left in place when the walls came down, which dominates this side of the Old Town and boldly sustains the image of the Old Town. We pass the Milicija (Police Headquarters) and the old Hotel Korčula, which has been handsomely restored in recent years, with its pleasant arbored terrace looking out to the town's west shoulder, the segment we have yet to read. A tile set in the wall of the small lobby tells us the hotel dates from 1912, a time when tourism had begun; its scale and style seem so much more in keeping with Korčula than the newer hotels built further out on the east shoulder of the urban area.

Past the Hotel Korčula we find the town Loggia, a fine reminder of the democratic element in Korčula's early governmental system and essential, along with the Communal Palace and the Campanile, if it was to be considered a real town, as Novak pointed out with reference to Hvar.[8] This was a pavilion open on three sides with built-in benches where the town council could sit and hear from all the citizens of the town. However, at the time of this reading (1981) we find the Loggia sadly abandoned, and we pause on a stone bench in its cool recesses to reflect on the dilapidation we see. The side entries are rarely used; one is blocked by the hotel itself. Inside, there is a plaque on the back wall, chipped as if to destroy purposely its ancient inscription, leaving only four oddly spaced letters on the bottom line: F— D——— I—— B. Ample graffiti recently scrawled on the walls, along with papers, orange peels, and plastic scraps, insult this venerable social facility, still a symbol of Korčula's cultural heritage.[9]

By 1990, however, along with the renovation of the old Hotel Korčula and other "modernization" projects we shall come to, the Loggia's use had been completely transformed, its spirit vastly changed. The architectural gem had become a fully outfitted travel office with all of the usual paraphernalia—telephones, computers, travel posters, promotional advertisements. Now, rather than serving for communication among the people of Korčula as in centuries past, the Loggia reaches out to all of Europe and the world. Nevertheless, in spite of tourism's impact on the environment, Korčula's identity persists.

On one side of the Loggia we mount another neo-baroque stairway, smaller than the one at the main gate and also part of the embellishments made after the removal of the old walls. Two massive date palms on each side add to the image of the nineteenth century and the fashionable break with the image of the Middle Ages. We then go up the tiny street that still handsomely links the port and the Cathedral.

Again in the main piazza, we retrace the steps by which we entered, at the forecourt piazza just inside the Land Gate. Our mood upon entering was one of anticipation; now, we feel a sense of release as we emerge from the geometric confinement of the Old Town, just as we did on leaving the rigid environment of Hvar's Old Town, product of an authoritarian elite. However, whereas Hvar evoked feelings of captivity, Korčula generates emotions of joy, delight, intimacy, and excitement because of the rich variety of forms and spaces within a symmetrical plan that delightfully choreographed our movements.

We now look down from the heights of the baroque steps to Korčula's central urban core, formed by the regional crossroads. As we view this busy scene, our mood is less like one of individual passengers aboard a ship and more like that of free-flowing people with diverse purposes milling about unorganized spaces between buildings seemingly placed at random. We once more become aware of being seen by others and of the interplay between ourselves and those who surround us. As we focus our attention on them, they take the place of the details of the huddled buildings in the Old Town.

The Crossroads Core and the *Borg*

As in Split, the pleasantly scattered, meandering disorder of buildings and movement routes outside the Old Town contrasts with the order and preconceived form within. We find banks occupying restored nineteenth-century stone buildings, a post office in an old palace, travel

The crossroads, Korčula's present-day core. *Photo by author*

agencies and shops in glass-front buildings, a supermarket under an old embattlement, and, in the open air, a farmer's market occupying the circular top of this ancient bastion. In this seemingly disordered setting we find people taking intricate routes that are unknown to outsiders to do their daily shopping or reach their places of work.

Two worlds exist here, each with its own identity. One is the Old Town, a symbol of a disciplined, organized past committed to wresting a livelihood by participating closely with the forces of sea and land. There the clear relationship of people to place dominates, and old Korčula becomes a living thing, standing in immediate juxtaposition to the visual, incremental disorder and nondescript sense of place of the second world, the crossroads, which typifies the nonspatial and meaningless nature of so much of today's urban development worldwide.

We grope for clues to guide a brief reading of the crossroads and the *borg,* where people from the countryside live—their meandering walkways of varying widths, the ill-formed open spaces between buildings of different heights and ages, and, in the Socialist style of the early 1980s, few signs or advertisements to communicate commercial vitality. The open-air market, fitted snugly into a circular terrace with a stone parapet, the base of an old embattlement, gives us a place to begin our reading. Colored parasols beckon us to women selling fruits and vegetables. Set against the sea and the docking area on the east side, this terminal for major vessels becomes a major focal point of social life in the summer, when Korčula is invaded by the outside world. As crowds throng and wait for the ships that so elegantly and effortlessly glide up to the sea wall, we have the feeling that these travelers are but passengers

From the *borg*'s miniature piazza up steep streets to the hills. *Drawing by author*

transferring from one ship to another—one that moves, the other, Korčula, "docked" at the shoreline. The experience of participating in this exchange between ship and land provides another means for identifying Korčula's essential character and ties to the world at large.

After the symmetry of the Old Town, and finding no clear sequence of spaces, we lose our way, then intuitively take to following the flow of people and observing their destinations. This leads to the discovery of a substantial number of retail shops and services we did not find at first. Our response demonstrates the infinite variety to be found in the "sense of place" of urban patterns. A sense of excitement, anticipation, and direct personal involvement develops as we follow individual walkers, learning their destinations and again getting lost. After following several individuals, we gradually come to recognize the functional order underlying the visual disorder. We learn that the pattern comprises six radial routes between buildings and clusters of buildings. These lead to the docking area and the sea to the east and west and to the irregularly shaped main square directly south, which links the east and west regional routes. The crossroads function clearly as the center of daily life in Korčula and the hub of the circulation system for cars and buses for the entire east end of the island.

A glimpse of the *borg,* the small, helter-skelter residential area on the hill just above the crossroads, allows us to complete Korčula's urban history. Here the country folk have lived since the Middle Ages, and today it is an integral part of Korčula's urban fabric. The streets attempt to follow the geometric order of the Old Town, and we are attracted by a rectangular piazza with a tiny church that appears to be an oasis of the spirit. But this gives way soon to a random, incremental pattern similar to that of the commercial center. A single main road cuts through the *borg;* the traditional route to the mountainous interior for centuries, today it is a mere local pedestrian way. Unlike the *borg* of Hvar, Korčula's is lacking in qualities and features that contribute, even in a small way, to the town's identity.

The East Shoulder

The east segment of the shoreline forms the right "shoulder" of the Old Town's promontory and urban expansion in the direction of Lumbarda. We begin our walk at the dock for large ships, where the shoreline road traditionally followed the curve of the original shipbuilders' cove. We note how it now runs straight across the arc of the curve on filled land. We, faithful to the past, take the original curving route in order to look at the dwellings still facing the former cove.

These were the houses of shipbuilders, who had to live close to the shipyards, where often they worked at any time of the day or night in order to deal with the urgencies of maritime responsibilities. The baroque balconies and architectural details attest to the high level of well-being and social status that some shipbuilders achieved, unlike the rigid class structure elsewhere. Individuals could achieve upper-class affluence through their own work and responsible civic leadership in town meetings at the Loggia. The broad crescent of the original waterfront road takes us to a triangular park where the neglected remains of a circular stone band pavilion suggest late nineteenth-century attempts at waterfront modernization and amenity emanating from Vienna.

We return to walk the starkly straight new shoreline road. Asphalt and cars leave virtually no place for pedestrians. Where the road begins to climb the hill to the inland villages, we turn to follow the original road of old times along the shore, now for pedestrians only, to Borak, the beach and resort area.

A small node offers us a *slastičarnica* with an outdoor area where we can sit, alongside a *frizer* (hair stylist and barber), marking the takeoff point for walkers. Across the way, next to the water, a neglected ship-repair area is the last vestige of the shipbuilding trade here. A row of six stone houses blocks our view to the sea from the long outward-curving road. Yet they create in us a sense of anticipation, of passing through a gateway to the open shoreline and enclosing promontory, with its sand, rock, pines, and small hotels. The neo-classic style and mechanically cut stonework of the houses tell us that this area was part of Korčula's nineteenth-century expansion. The date on one of the houses, 1882, puts it at precisely the peak of prosperity and of the ascendancy of the bourgeoisie under Austria's tutelage. When this wall comes to an abrupt end, the shoreline bursts open again, and we sense a harsh intimacy with the wind, the sea, and the rocks below battering the Korčula of long ago.

As we round the gentle inward curve, the town's water-polo pool and bleachers come into view, a major focal point in the early summer evenings. Shouts of players and their skillful movement through the water bring youthful vitality to Korčula and reinforce its identify with the sea. Looking back to the Old Town from just above this vibrant scene is the home of the water-polo coach, where my family and I spent some memorable weeks increasing our identity with Korčula. As we near the end of this east shoulder the neo-baroque walks, balustrades, and steps down to the beach and the once grand villas suggest that this is an area from the nineteenth century. The outermost end became an elite suburb, oriented to a new bourgeois lifestyle unlike the disciplined life of the Old Town.

We come upon a perfect image of these changing styles of the nineteenth century: a grand semicircular terrace and bench overlooking the shoreline park and dappled blue waters well below. Although it stands in a sad state of neglect, the elegance and grace of the half-circle stone bench magnificently offset the abandonment. Its wall wraps around an area of paving some twenty feet in diameter laid out in a precise de-

Korčula's east shoulder before the marina installation

sign; we are reminded of the cannon emplacements in Hvar. We climb the broad steps leading up to the paving to contemplate the meaning of the inscriptions on two bold, stone obelisks some three meters high that flank the terrace:

PETRO LOWEN
HOC CIVIB SOLATUM
VIAQ HAEC CURIB APTA
INCOLIS OMNIBUS
COMODO ET UTILITATI
CONSTRUCTA

LIBERTATE FERVENS
HOC GRATIA ANIMI TESTIMONIUM
COMITAS CURZOLENSIS
POSTERIS TRADENDUM
DESIGNAVIT MDCCCXV (1815)

[To Peter Lowen
Under whose fortunate auspices
this resting place and roadway
have been built to serve for
   vehicles
with comfort and utility
for all residents]

[Deeply stirred by freedom
The Korčula communities
have designated this testimony
to a spirit deserving of gratitude
to be handed down to their descendants.
   (1815)]

Peter Lowen was the officer who commanded the British forces during their two years of rule on the island, from 1813 to 1815. The British, "acting like gentlemen" in taking over the island from the French, let the citizens of Korčula continue their local administration. With increased taxes, the road to Lumbarda and the fortress on the hill of Sveti Vlaho, together with the park, promenade, and semicircular piazzeta, were built. The Korčulani addressed the inscription to Peter Lowen to honor the good relations that they had had with the British following the period 1797–1815, when they had been faced with aggression from the French, the Italians, and the Russians.[10]

Across the road more steps of neo-baroque design descend to a paved esplanade where an ancient wellhead atop a large cistern recalls the purposes of "comfort and utility" referred to in the inscription, echoing the writings of Vitruvius, Rome's founder of architecture and urban design. In speaking with local historians we learn from local historians that several baroque homes, probably sea captains' mansions, had been built here even in the eighteenth century. Even then Korčula was a favorite place of scenic and historical attraction for tourists and vacationers from England.

Sir J. Gardner Wilkinson visited the monument only thirty-three years after it was built. The following quotation from his writings echoes the classic revival in architecture and suggests how, at that time, England was influenced by its archaeological research of the previous century: "These stone seats, in imitation of *hemicyclion* outside of Greek towns, are met with in other parts of Dalmatia."[11]

We sit on the sweeping curve of the elegant stone bench and imagine the social life here. Surely an important element in this complex of terrace and stairways was the romantic leaning of the British toward the scenic landscape, typified here by the splendid panorama stretched out before us, more like that of a lake than one of the sea, with the gray ridge of Pelješac, topped by Mount Sveti Ilija, standing behind. The bench, wall, and terrace represented the new cultural dimensions of Dalmatia. Visual amenity and the passive use of leisure time in a civilized, architectural setting heightened identity with place.[12]

Korčula's growth in areas beyond this nostalgia-inducing terrace has disrupted these qualities of the past. The modest Hotel Park of the 1960s adapted well to the slopes of its pristine wooded promontory. Behind it, where vehicles have access via a new upper road, the large and overly luxurious Hotel Marco Polo was added in the 1970s. These changes appear mild compared with those of the late 1980s. On my return to Korčula in 1990, I discovered that massive recreation and tourist development had abruptly altered the environmental integrity I had recorded earlier and damaged Korčula's genuine identity with place evolved from its historical sources. The energetic arm of internationally financed facilities has taken over, oblivious to the Korčula's continuity of the past. A marina for several hundred boats, together with new sea walls, roads, fuel pumps, cranes, and ramps for boat access, completely fills and dominates the protected cove below the terrace. The community water-polo facility and its bleachers are gone, and the quiet corner of the *slastičarnica* has been replaced by a three-story boating club of incongruous design and a new bus station of equally inappropriate character and scale. Finally, beyond the small Hotel Park the promontory that completed the cove, just where flat slabs of stone and twisted pines had provided contact with undisturbed nature, had been obliterated by a large hotel. Harshly out of place, its glass facades and numerous decks, pools, and tropical indoor plantings introduce an artificial identity, a product of another culture without roots in Korčula itself. This is one of eighteen marinas and clubs built on the coast from Istria south, financed by Italian boating interests; many of the projects have brought related development into choice, sensitive coves where no automobile access had reached.

The West Shoreline and Reflections from the *Kula*

Returning now to the crossroads, we begin reading the west shoulder of the Old Town's promontory. Lined with wall-to-wall stone dwellings, this steeper shoreline made possible the town's first expansion, in the seventeenth century. Korčula's second shipyard was located here. The tight facade of stone houses facing the sea evokes a sturdy, sober feeling, one that reflects the work-oriented lifestyle of the early shipbuilders in contrast with the amenities of the eastern shore. The substantial dwellings, with their doors on the ground floor wide enough to admit wine barrels or small craft, were surely those of the shipbuilders who supervised the work going on below. We walk down to a flat area a few feet below the road where boats are still hauled up for drydock work.

The stately row of spaced date palms, with benches at frequent, regular intervals, suggests aspirations to urban amenity and a grander urban scale than in the spartan earlier centuries. The impressive mass of the Old Town, rising above the sea, dominates the entire scene, more so than when viewed from the east. From the west we see the Korčula of the Korčulani, whereas the east shoulder, with its hotels, beach, and water-polo area, invaded by international yachting facilities, is clearly the Korčula of "outsiders." As we approach the end of this distinctly urban waterfront, small villas begin to predominate, one of them dated 1897, the year when the new sea wall improved boat access, as confirmed by a plaque on the wall of the Dominican nuns' large convent:

BOGU REDU I NARODU 1905
[To God, Order, and People, 1905]

Resting, and in thought, we feel alone here, struck by the absence of people, nodes of attraction, and a flow of pedestrians or cars. This shoreline holds its identity from us, but when we reach the end, where a dramatic promontory juts boldly out into the sea, we find strong new sources. There the huge Dominican Monastery and the church of Sveti Nikola dominate the landscape—as do the Dominican Monastery in Bol and the Franciscan in Hvar—but this one is semiabandoned and devoid of vitality. Strategically located, this site was clearly chosen for its view straight up the channel between Korčula and Pelješac. Like the Franciscan Monastery on the heights above Orebić on the opposite shore, it served both to guard the channel for Venice and to provide spiritual guidance for the Korčulani.

Sharing the promontory, a single large villa of rustic character built in recent years stands on the sea side of the road, magnificently placed and graced with a handsome landscape. Yet at the time of this reading it was devoid of activity and looked out vacantly to the vast panorama comprising Orebić, the profile of Pelješac, and the Old Town of Korčula. An inspiring site, the villa has taken on new life as an exhibit hall and meeting place for the scholarly Croatian Academy of Science and Art.

Urban development ends sharply just beyond the Monastery. Towering above all the urban and landscape scene we have traversed, the Kula Sveti Vlaho awaits us. The hilltop site of this installation, built in 1813 by the British, is the perfect place to end our "reading" of Korčula. We retrace our steps, walking back along the shore to a broad stairway tightly wedged between the houses to take us up the hill to the fort. After the horizontalness of the seaside walk, climbing step by step up the face of an almost vertical urban settlement is exhilarating. As we place one foot above the other, we can feel the steepness of the hillside, and we watch the urban panorama and the sea setting widen out and drop below us, lower and lower. A sense of power and sheer physical vigor pervades the experience as we gain an overview of much of the route we have traversed using only the energy of the human body.

The rhythm of the steps suddenly stops at a diagonal ramp that sweeps in a deliberate line across the flowing slope—a short cut—back to the crossroads and the Old Town. The magic energy generated by the challenge of these great steps to reach the *kula*'s heights sharply contrasts with the deadening pragmatism of the shorter, diagonal route, oblivious to contours. Was this the route of the British when they built the tower in 1813 or is it of more recent origin? Later we find it to be the work of the Partisans in 1943 to assure control of the peak. So we give in to its arrow-straight logic, which takes us cutting relentlessly through the pine woods made spicy by the hot sun of late afternoon. The landscape becomes increasingly wild as we reach the *kula*.

Apparently, not a soul is to be found on this vast hilltop! Again we find ourselves alone amidst images of momentous activity at critical points in history. And we realize that they were indeed critical when we consider the labor and will that had to go into placing stone upon stone to construct a tower to rise four stories high out of a broad footing laid deep in the earth. These kinds of reflections evoke a sense of awe coupled with some anxiety at the thought of possessing, alone for an hour, an entire hilltop that was of strategic importance for centuries.

We take the critical mental step of taking possession of the formidable structure and enter the dark, recessed opening, hoping that it will lead to the top. But here our illusion of isolation is quietly dampened by the discovery of a sole attendant seated by a sign saying "Bife." At once guard of the *kula* and purveyor of welcome cold drinks, he also provides personal insights on environmental history. Our sudden arrival stirs him out of a lethargy imposed by the rarity of visitors. He is quick to comment: "Great people—the British—doers! They and the French did more for Korčula in a decade and a half than the Venetians in four hundred years." And as a courtesy to the Yugoslavia of the 1980s, he added praise for the self-initiated works of Tito's Partisans.

With this reassurance and perspective—even, we might say, a sense of comaraderie with a friend found in mutual solitude—alone I ascend the spiral staircase to the top. There the vast panorama of the mountains of Korčula and the channel across to the Pelješac ridge again reaffirm nature's dominance over man's urban works. And there our reading of Korčula comes to an end. Alone on these strategic heights, we contemplate the irony that man has used an environment of quiet peace and inner spiritual light for purposes of war and revolution. And what uses might the strategic tower have served had the attacks of 1991 on the Dalmatian coast included the Korčula we now know?

CHAPTER SIX

Mountain Villages Linked to the Sea

From times remembered, below the steep, stony slopes of Mount Sveti Ilija and its naked peak, "The Viper," nothing grew, nor water flowed. The man of Pelješac knew only of oars and sails; he laid stone upon stone or left for far-off lands. —Josip Splivalo, 1964

PELJEŠAC: MOUNTAIN SETTLEMENTS VERSUS SEASIDE VILLAGES

Our exploration of identity with place now takes us to another type of urban place: those villages that form interdependent networks on mountain sites and yet are linked to the sea. The peninsula of Pelješac, with Kuna as its focus, and Konavle, adjacent to Dubrovnik, offer excellent examples. The self-contained and fertile valleys of Pelješac, ringed by ridges that drop steeply down to the sea, allow footholds for few shoreline settlements. This geographical restriction has induced connectedness within groups of villages of Pelješac and has been a source of collective identity among dwellers, who are both sturdy mountain people and adventurous seagoers.

Almost entirely surrounded by the sea, this precipitous peninsula of Pelješac comes close to being an island, or a "half-island," as *poluotok*, the Croatian word for *peninsula*, translates. Yet, this long narrow jut of land paralleling the Adriatic coast can also be identified as "half-sea" for the strong maritime identity of the dwellers. This configuration has denied settlers the advantages of the direct communication with the sea that the shoreline villages enjoy, but it has taught them a sense of interdependence, unlike the self-sufficiency of the seaside places we have come to know.

We will focus on about a dozen villages in the western third of Pelješac, beginning with Orebić, one of the villages' two main links to the sea—Trpanj is the other. We will then go to Potomje and Kuna, the two principal villages among those nestled in a high mountain valley, and return to the sea to conclude our journey in a reflective mood in the village of Podobuče. To intensify this experience, we stretch our imaginations and carry out the reading on foot for the most part, just as the residents did during the peninsula's formative centuries. It was only in

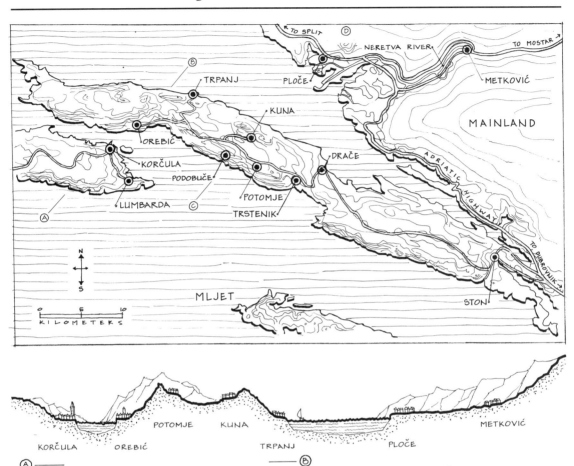

The peninsula of
Pelješac

the 1960s that cars and buses began to take over the traditional roles of the pedestrian and the *mazga*.

This final set of readings will be distinguished further by my own personal identity with Kuna, demonstrating how one experiences an environment more fully when the family ties with that environment are strong. This was less the case in Pučišća because family members on my grandmother's side had all but faded away since 1937. Each of us has a hometown somewhere, a place that holds special meaning. When the continuity across generations reaches as far back as does mine, human association enriches identity with place.

Turning Points in Evolving a Village Network

Pelješac stretches forty-seven miles up the coast, measures only five miles across at its widest point, and is connected to the mainland by a single mile of land at its stem, at Ston, thirty-five miles from Dubrovnik. Rugged terrain has made the upper half a particularly isolated living environment over the centuries. Indeed, Orebić, so distant from Ston, has

Pelješac's landscape by Celestin Medović, Kuna's illustrious painter, 1857–1920. *Split Art Gallery*

had far closer ties to Korčula across the narrow separating channel. Pelješac's severe shoreline, with few natural harbors, has limited free travel around the perimeter by sea as well as the direct access to the Adriatic that was possible for Korčula and Brač. Its almost vertical geography prevented the formation of a dominant urban center like those that developed on Hvar and Korčula. Its village system, however—essentially a function of the highland interior—remains remarkably intact and serves as a continuing source of identity for its residents and family members abroad.

Igor Fisković determined that the settlement pattern of Pelješac, with its fertile plateaus and sparsely wooded slopes, had its beginning in prehistoric times, as shown by remains such as caves on hilltops where visibility assured security against hostile tribes. Kuna and Potomje provide examples of such settlements. By the second century B.C. the Illyrians had unified early permanent settlers by established communication routes; remains of inhabited nodes in small elevated villages have been identified. The number and power of these places is indicated by remains in the Kuna area of some twenty *gomile* built at strategic sites as places to keep watch for invaders. Because of the number of elevated valleys in remote locations, these settlements were little affected by the Romans' few footholds on the steep coast. Thus, all through Dalmatian urban history this relative isolation of the Pelješčani in their elevated, inland valleys, kept them free of changing rulers.[1]

With the coming of the Slavs, Pelješac was sought after by the Serbian, the Croatian, and the Bosnian rulers for its strategic position. The name Kuna, which is the Croatian name for the small, foxlike marten, first appeared in historical records in A.D. 1248. Since marten skins were used in early times as coinage, Croatia in 1994 adopted the name *kuna* for the national monetary unit that replaced the Yugoslav dinar. Dubrovnik exploited the conflicts between the rulers of Serbia and Bosnia over ownership of Pelješac and in 1333 took control of the island for itself. Records show that by 1396 the area from Oskorušno to Janjina constituted a single administrative unit; churches were identified, as was vineyard ownership.[2] Zdravko Šundrica, long the keeper of the Dubrovnik archives, explains in detail how during the next sixty-seven years these rulings laid the groundwork for five centuries of stability, growth, and development.[3] Although Dubrovnik exploited the inhabitants through a feudal system, it helped to make Pelješac a productive territory in spite of its isolation and rugged terrain. This framework served well for Bosnian settlers, my forebears among them, fleeing from the Turks in the sixteenth century. Those settlers who possessed a sense of self-awareness and independence contributed further to Pelješac's developing its own strong degree of identity as a subregion.

A critical event in the shaping of the collective sense of identity of this end of the peninsula was the founding in the late seventeenth century of the Franciscan Monastery at Kuna, possibly through the intervention of Dubrovnik. The religious and educational needs the brothers met defined the geographical area as its parish, or *župa,* and continue to do so today. By the nineteenth century the area had officially become the *općina* and even to this day the Kuna area is popularly known as the Župa.[4]

Pelješac as a Region in Itself

Approximately seventy-two hundred persons live in thirty-two settlements on Pelješac, about one-tenth of them in Orebić, the largest. About three-fourths of the population lives in the middle third of the peninsula, where its width narrows down sharply toward to Orebić on the south shore and to Trpanj on the north. Having no strong urban center, Pelješac identifies with Dubrovnik.[5] During the Middle Ages, Dubrovnik's feudalistic policies antagonized the Pelješčani, and the peninsula developed its own, separate sense of identity and became prominent in maritime affairs. With the fall of Dubrovnik in the nineteenth century, Orebić became an important center because of its proven productivity in shipping and service to the inland communities.[6] Today, however, Pelješac's geography provides a new main route in the regional highway system from Dubrovnik to Orebić, a valuable link for Pelješac's social and cultural development.

Stjepan Vekarić attributes Pelješac's strong identity to its independent spirit in the face of long-term oppression by Dubrovnik and to the self-reliance that comes from living in a mountainous environment over

Model of a sailing ship
in the Orebić
Maritime Museum.
Photo by author

many generations. For example, Pelješčani men learned sailing through
working for the leaders of Dubrovnik, but later they started their own
enterprises. With few facilities at hand, they built smaller, faster, and
safer ships. And their major motivation for working on their own was
that once they left the territorial waters of Dubrovnik, their serf status
ceased. Recognized for their services and bravery at sea, they became
known for identifying with Pelješac rather than with Dubrovnik. They
developed knowledge of other cultures through their shipping activities in the eighteenth and nineteenth centuries—both as sailors and as
owners—which took them to many lands: Turkey and Bulgaria, Italian
and French coastal cities, India via the Cape of Good Hope, and the
United States, Australia, and New Zealand. They also learned how to
maintain high standards of health care so as not to transmit communi-
cable diseases to their mountain home villages.

One example is the Fisković family, of Orebić, whose members were
sea captains and ship owners for two hundred years, from the late seventeenth century to the end of the nineteenth. Mato Kristov Fisković left
a lively, detailed account of the sailing experiences at the end of the
seventeenth century. Letters, contracts, ship logs, even sonnets written
to commemorate launchings provide us with a real sense of the breadth
of personal development brought about by seafaring.[7] Bariša Krekić has
told me of his father's also being part of this seagoing life. Born in 1881,
he circumnavigated the globe between March 1898 and December 1899,
leaving from Kučišće. Krekić has two paintings of the ship, one in California and one in Dubrovnik.[8]

In 1800 of the 524 small vessels operated jointly by Dubrovnik and the

Orebić and the
Franciscan Monastery
across from Korčula.
Drawing by author

villages of Pelješac for coastal trade alone, 265 were owned by Pelješčani, one ship for each twenty-five inhabitants. Ownership was concentrated mainly in Orebić, Trpanj, Kuna, and Janjina. At one time in the nineteenth century, when the peninsula had 8,000 inhabitants, Pelješčani owned as many as ninety ships for overseas commerce. After the coming of steamships, the Pelješčani began to reduce the size of their fleet. By the end of the century it numbered only thirty-seven ships, and by 1907 the last ship had been sold.[9]

Their maritime identity contributed to the Pelješčani villages' economic well-being and to their obtaining full ownership of the land from the remaining Dubrovnik families. Sailors and sea captains became settlers in the New World. The peninsula's isolation conditioned its people to take advantage of shifts from prosperity to decline and to accumulate capital to get them through lean years. The more prosperous shipping families made loans to those employed in less successful activities, to be repaid with interest. Through their involvement in sea transportation, the Pelješčani broadened their understanding of political forces and economic markets and how to conduct business affairs.

OREBIĆ: VILLAGE OF SEA CAPTAINS' MANSIONS

Let us now approach the west end of Pelješac on the ferry from Korčula to begin a brief reading of Orebić, Kuna's link to the sea. This "village of sea captains' mansions" stretches out serenely, as if at rest, along the waters at the base of Mount Sveti Ilija, which rises protectively. The village's lineal structure becomes clear as we read the low skyline of large homes, mansions really, belonging to generations of sea captains at the water's edge. The campanile of the nineteenth-century parish church marks the village center and dominates the skyline as seen from the dock but diminishes gradually during the ten-minute crossing, while the town's landscape unfolds in greater detail. As we draw closer, the mansions take over the enlarged foreground. The maritime leaders who lived here were in largely responsible for the uniqueness of this village and its prime source of identity. Its identity with the sea captains and the simplicity of the single "main street" of Orebić contrast sharply with the highly structured compactness and authoritativeness of Korčula. Recognizing these differing images adds clarity and richness to the experiencing of each.

Tracing the shoreline to the left and raising our eyes to the Franciscan Monastery far above the sea, we terminate the west end of the otherwise low skyline. Lush green columns of Italian cypresses, exotics now, naturalized at random over centuries, stud the steep slopes to the monastery.

The "village of sea captains' mansions."
Photo by author

Standing stalwartly at attention, they symbolize the guardian role of the Brothers of St. Francis, who provided the spiritual context the people of Pelješac needed to face the daily challenges of their hazardous life over generations.

Orebić's lineal landform without coves resembles that of Bol, but enriched by the handsome facades of the mansions lined up in a continuous row. In the days when they flourished the shipowners had direct access from their own dock to larger sailing vessels in deeper water, which the sea captains could see from these homes. The island of Korčula, now only a few minutes away by ferry, offered protection from the open sea and the benefits of major shipyards. However, the new asphalted road connects Orebić to Trpanj—a twenty-minute trip—and its ferry to the mainland, as well as to Potomje, Ston, and Dubrovnik. This has generated new development on the gentle slopes above the traditional settlement, changing the lineal pattern to an elongated loop.

Sir J. Gardiner Wilkinson, our intrepid and perceptive observer of Dalmatia, conducted his own reading of these places in 1848. A "four-oared boat" took him from Korčula to Orebić in a half-hour. There he found "the houses well-built and the inhabitants wealthy," the gardens of the sea captains brimming with "oleanders, ilex oak, bay, rosemary, and juniper." On tilled land, the olives were abundant, along with figs, pomegranates, apples, pears, and almonds. "The beauty of the land was due to the dark foliage and columnar form of the cypresses," especially on the steep slopes rising up to the Franciscan Monastery. From Orebić to Trpanj he traveled three hours on foot, crossing mountains that formed ridges, where juniper, arbutus, heath, and oak grew. Then a four-oared boat took him in an hour and a half to the mouth of the Neretva River, on the mainland, the original home of many who fled Bosnia and the Turks to become settlers of Pelješac long ago.[10]

Orebić and Bol have similar linear systems; for that reason a true

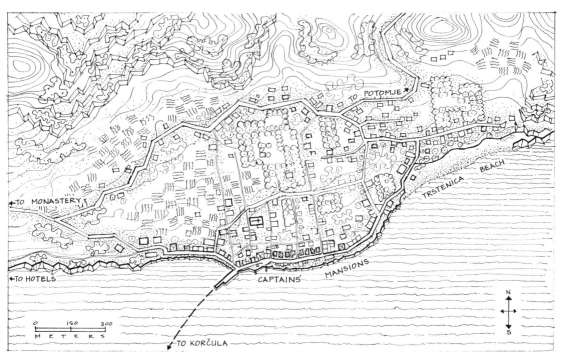

Orebić: A lineal
structure based on a
single route

reading of Orebić would be redundant. We have only to bear in mind
that Bol's two parallel routes and much larger scale introduce a far
greater variety of experiences and therefore a richer basis for identity.
Although the single route of Orebić makes for a certain monotony of
people-place relationships, nonetheless, the elegant row of sea captains'
mansions gives the village a unique source of identity worth exploring.
These seaside villas set in Victorian gateways and gardens visible to all
passers-by evoke images of a flourishing and cosmopolitan past. They
contrast sharply with the stark block of stone dwellings, a relic of more
spartan times that stands as a historical feature of Orebić's beginnings.

POTOMJE: A PLANNED WORKERS' VILLAGE

Hillside Site, Grid Streets

We now go by car directly to Potomje, a journey that for a millennium took several hours on foot and today takes only twenty minutes,
overcoming the precipitous terrain that for so long caused the isolation
of the fertile uplands of Kuna. In 1937 that route consisted of zigzag
ramped and stepped paths wide enough for *mazge.* Even in 1968 I saw
remnants of those paths, reportedly built under Napoleon's takeover to
replace earlier, more primitive paths. Now, high above Orebić, we see
Korčula jutting into the basin shaped by its convex shores, a spectacular
interplay of geographical forms. At one point a level road on the edge of

Mansion gardens
planted from overseas.
Photo by author

the vertical cliffs leads off to Postup, well known for its wines, and to Podobuče, at the end of the road, where we will end this reading of villages
linked to the sea.

The long roadway clinging to the mountainside leaves the sea below,
and we enter the plateau land of Kuna. Potomje, where we again become pedestrians, stretches parallel to the road on an inland-facing shelf
of the ridge that drops sharply to the sea. This level, livable site about
fifteen hundred feet above the sea stands slightly above the vineyards
that cover the rich red earth of the valley bottom lands. Kuna occupies a
similar though higher position on the opposite slope of the ridge, on the
mainland side of the peninsula. This configuration revives the illusion
of remoteness; we can sense how the villages identified with mountain
ranges as strongly as the shoreline sites we have experienced identified
with the sea.

The pattern of streets and crossways of Potomje, with some four hundred dwellings, resembles the geometry of Dubrovnik, only on a tiny
scale. Deliberately laid out all at one time by Dubrovnik's planners during the second half of the fourteenth century, it attracted workers from
the mainland to work the agricultural lands so necessary to that city-
state's livelihood. They laid out other settlements on Pelješac in rectangu-

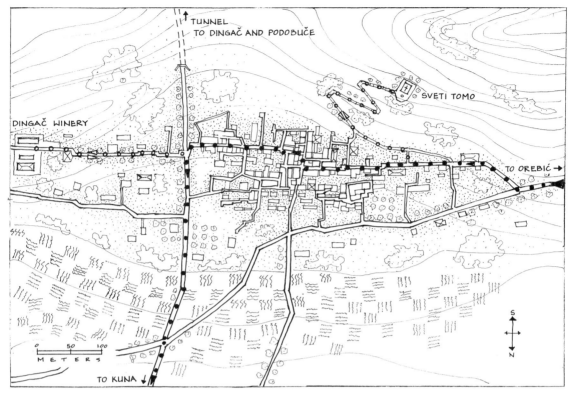

Potomje: A grid system
planned by Dubrovnik

lar blocks and parcels; areas were designated for cultivation, residences, and defense purposes according to the character of the land. Ston, at the link with the mainland, was developed for its proximity to age-old oyster beds and for its potential part in the defense of all the peninsula.[11]

Sveti Tomo, Protector of the Dingač Winery

The relatively flat site of Potomje was chosen for its location just above the extensive fertile lands. The tiny church of Sveti Tomo stands above and free of the efficient, utilitarian pattern of the village, seeming more a part of the timeless mountain range than of the transitory affairs of urban life. The church's elevated position is the perfect place to begin our reading; from here we can comprehend the planned regularity of Potomje's street system. Overlooking the town, the church embodies the spiritual origins of the name Potomje. Literally, the planned village lies "under Sveti Tomo," that is, under the protective aura of the church. In a mind-clearing silence, the sacred building stands with us on a platform of rock looking over the village below and the vineyards spread out across the valley to Kuna, backed by the slopes of Mount Rota. We sense the relentless accumulation of time in the processes that produced this place through the geologic works of nature and later by the inventiveness of men and women. We can feel the presence of the four centuries

The uphill site conserved rich vineyard land

in which the people of these houses, in their straight, logical rows, have been looked after by this modest place of worship, so firmly anchored to the slopes of the mountain ridge.

The only link to the village itself is the stone, ramped steps we climbed to reach this natural terrace. Surely this has been the path of countless processions marking the saints' days of family members, christenings of newborns, first communions, marriages, and burials. On one of my visits to Kuna, Stijepo, the great-grandson of my father's sister, proposed that I attend the ritual for Saint Steven, his own name saint. Recalling that three centuries ago Potomje had five families totaling forty-three persons bearing the name Violić, I am again aware of the depth of my own identity with Sveti Tomo and. In this tangent of contemplation and remembrances the powerful role that associations can play in giving life to the environmental sources of identity with place is once more evident.

Descending to the village, we follow elongated ramps cut into the mountain side and reach the west end of Potomje's main street to begin our reading. Rather than following a single straight line, this street actually jogs at several points just enough to block the view of it entire length. The result is a sense of containment and completion of a single segment. As we move around each of these disjointed intersections, the view of the next unit of the main street brings a fresh experience. The pattern of jogging intersections becomes an element in Potomje's individuality.

A chance incident at one of these intersections changed the nature of this reading. I took note of this jogging pattern while standing in front of a house with a carefully tended front garden that set it apart from the

Jogged streets in a grid system

My grandfather's boyhood home. *Photo by author*

neighboring drab facades. An elderly man who had been watching me from his front doorway came across the narrow street and asked who I was and what I was doing. When I explained, he quickly became quite animated and interrupted to ask my name. On discovering my personal connection with Potomje, he dashed into his little house crowded with the past and returned with a photograph of my father sent to his family from San Francisco about the time of my parents' marriage in 1902!

This chance encounter produced a day of collaborative reading that became rich in intimacy as a condensed life story of Potomje and the houses within it where Violić family members had lived. As we conversed—Ante using his broken English and I, my meager Croatian—he helped me mentally map the street pattern, explaining how the jogged street corners had served as refuge for Partisans defending themselves against the fascist forces during World War II. Through his reading, Potomje became a scene of violence. The joint experience introduced me to the idea of joint urban reading, a fruitful variation of my own individual method of intuitive reading.

As we neared the east end of the main street and the Dingač winery, Ante pointed out the house in which my great-grandfather Baldo Violić had lived. It was from there my grandfather moved in the early nineteenth century to Vidoš, the hamlet that became Zagujine, where he built the house my father was born in. Nearby, a tall young grandson of Teta Frane, my namesake, whom I had met in 1937, brought out a large, elegant portrait from her home, another example of the intergenerational continuity of identity in the context of The Bridge.

KUNA: THE MOTHER VILLAGE

Approaching from Potomje

Like Potomje, Kuna, on the opposite side of the valley, stretches out horizontally along the lower slopes. The slopes form a cradle wherein rest the green vineyards. Leaving Potomje on foot, we approach Kuna on a straight line to Pijavičino, cutting through the low-lying *polja* (fields), which are broken by jagged limestone outcroppings. Even from a distance we can tell by the stepped pattern of red tile rooftops that Kuna's parallel streets rise sharply one above the other, quite a contrast to the layout of Potomje, which is situated on a single shelf.

As the road turns westward, we see the hamlet of Zagujine resting with an air of self-assurance in its niche on the stony, pine-studded hillside; its vineyard, ripe with age and care, stretches out before us. It was in Zagujine that my grandfather, a builder of houses, raised his own family, and it was Zagujine that my father left to come to America. Only three dwellings and their farm buildings, in a single row, stand above a

Kuna: An urban village
set in vineyards

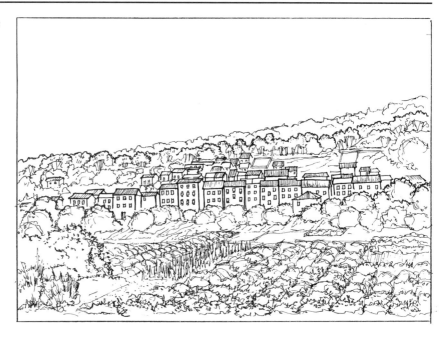

long retaining wall and broad terrace shaded by giant pine trees that my
father had helped plant. The irregularly placed buildings of upper and
lower Pijavičino rise to the east of Zagujine on higher ground. Nearby,
the silhouette of an imposing tower situated on high ground among the
fields appears ominous. Built by the nobility of Dubrovnik for overseeing the serfs in the fields, the tower remained a symbol of oppression
even after serfdom was banished and into the long period of Austrian
domination. Even in 1990, when I paced off the dimensions of this immense structure with my young cousins, it was a bold reminder of the
hard-won freedom from domination by external forces.

Moving westward, we see straight ahead, rising out of the low stony
ridges and green vineyards, Kuna's principal identifying feature, the
Campanile and the imposing Franciscan Monastery of Gospa Delorita.
About a quarter of a mile beyond, the parish church, called the Matica,
or mother church, marks the entry point to the village. These monuments stand out from almost any vantage point, whether one is arriving
at, living in, or leaving Kuna. Their setting among the life-giving vineyards, independent of the confinement of the village, reminds us of the
interdependence of towns and their countrysides throughout history,
which Spiro Kostof has called our attention to,[12] and speaks for the important role the rural environment plays in the life of Kuna's people.
Over the centuries the villagers have had to leave behind their houses
and worldly possessions and go into the inspiring open space of their
valley, the source of their livelihood and their stabilizing collective identity. Sensing this power, we are drawn to the Monastery. There we are

Zagujine: The family home my grandfather built

received in a courtyard almost enclosed on three sides. On the principal wall of the chapel an inscription in stone tells us that it was built in 1681, that the foundations of the Monastery itself were laid in 1706, and that the *coperto* (roof) was finished in 1738. For nearly sixty years the forces of religion and education inspired the remarkable human effort that went into the cutting, hauling, and laying up of stone in this remote, sparsely peopled setting. In 1971 the Monastery's contribution to Kuna's collective identity over three hundred years was celebrated by the people of Pelješac. An exhibition was held in the revered monument of the many paintings of Kuna-born Celestin Medović owned by Pelješac families, including a portrait of my grandfather. The paintings were catalogued in a history of the Monastery.[13]

From the broad terrace of the Monastery we take a path through the vineyards that leads us to the Groblje Sveti Spasa (Cemetery of Saint Saviour), where Kuna's families have placed their forebears within view of their homes. The cemetery's vineyard setting provides a third component outside of the village proper that brings together the realm of the spirit and the world of agriculture.

The road turns abruptly toward the village now, and the parish church, set in a stately grove of dark green pines and cypresses, rises above the road. The straight length of road leads our eyes directly to the side wall of the church, where a pleasant park with shade and benches humanizes the ecclesiastical architecture. Since we do not see the main facade of the church until the road jogs to the left past the park, the park dominates the landscape. An impressive, larger than life-size bust of Celestin Medović, who was not only a painter but also a priest, stands under the pines on the axis of the road.

How the sculpture of Medović, which is prized by the people of Kuna, came to be placed there demonstrates the way identity with place can be reinforced by selecting a location that all can see. The sculpture was created by Ivan Meštrović, Croatia's most renowned sculptor, as a gift to Kuna from several families who had emigrated to America, led by Rudy Palihnich of San Francisco. Meštrović completed the bust in Syracuse, New York, in 1954, during his tenure as a professor of art there. Shipped to Kuna the next year, the sculpture remained crated until 1971.

In linking these two great names the donor gave contemporary distinction to Kuna. In 1979, in an effort to call public attention to Kuna's outstanding native son, a committee was formed to choose a site for this monument. The committee invited Cvito Fisković, my "mentor" at Orebić and a leading authority on regional cultural resources, his son Igor, and me to Kuna to review possible locations. One faction with Marxist leanings advocated a site across the road from the church that was oriented to the "workers" in the vineyards. Those who identified with the church and Medović's role as a priest urged a location in front of the side wall of the church, where it would be framed by the pines and cypresses.

The group then turned to me for my opinion. Neither position appeared to me to relate fully to the landscape context as a whole. Aware of the politics underlying the decision, I spoke for Medović himself, suggested that he might prefer a place in the sunshine at the entry to the park, where it would be visible to all visitors on the axis of the approach road. I also proposed widening the narrow steps up to the park to provide a more inviting setting. By my next visit, in 1972, all of that vision had become reality. The year before, there had been an unveiling and a celebration featuring a program of speakers and music, together with the exhibit of Medović's paintings. In a letter of gratitude to the overseas donors, the residents stated: "In all its history Kuna has never had such an unforgettable event." This sequence of events confirms the inborn sense of identity of the people of Kuna, a heritage that has evolved entirely in response to its own local resources and way of life.

Reading Kuna and Zagujine

Moving on down the road, we pass the terraced facade of the parish church and walk uphill to the higher ground where Kuna stretches out before us. "Piazza" is name used in jest for the loosely formed T-shaped intersection where the entry road and the level main street join. Unlike a true piazza, it does give a sense of both entry and outlook to the valley that Potomje lacked. The few shops and benches seem to be relics whose vitality is long lost. This core of Kuna comprises a series of con-

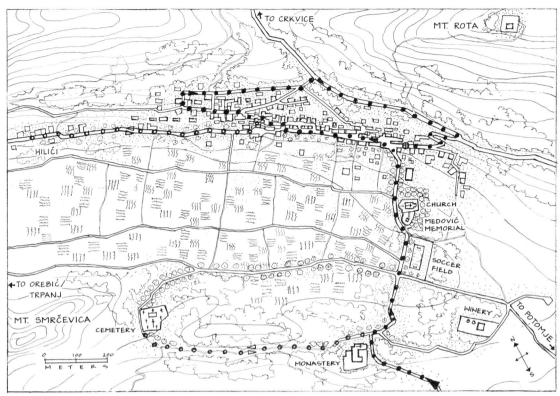

Kuna's structure integrates urban, rural, and structural elements

tinuous urban facades that seem to have been lifted out of a larger town, they do not fit the typical image of a rural village. Viewed from here, the rocky, pine-studded slopes below the Monastery suggest the rugged life of early settlers here. Kuna's identity is that of an urban culture set against nature's own countryside interwoven with vineyards.

Essentially, the village consists of one principal street plus a diagonal route that feeds into roughly parallel routes tiered up the hill. Kuna divides itself into two clusters of streets and buildings. The older, larger section, to our left, dates back to earliest time, whereas the newer part, to our right, is centered on the Piazza. Kuna's somewhat stark quality tells us of the destruction carried out by the Germans during World War II. Indeed, had we been standing in this same intersection in October 1943, we would have been caught in the crossfire between the Germans, with their heavy artillery, on Mount Rota, to the right, and the Partisans, on the smaller Mount Smrčevica, to the left, west of the town. Although Kuna has never fully recovered from that blow, people who identify with the place talk about making the Piazza a true gathering place through the moral and financial support of those overseas and with the coming of peace and a democratic order.

Both sides of the narrow street to the left of the Piazza are flanked by buildings several stories high, so that for a distance of about two blocks

Kuna's "Piazza"

we have the sense that we are in a truly urban place. Here are the post office, a store, buildings housing public services, and the county administrative headquarters, where in the 1880s my grandfather had his office when he served as mayor. A postcard photograph from that time (see the introduction) shows a gentleman reported to be him who epitomizes Kuna's "urban" look: with groomed beard and an umbrella in hand, he is wearing a felt hat, dark jacket, buttoned lapelled vest, and winged collar with tie.

As we move along, the street widens suddenly. The facades break where a walkway leads down to the verdant *vinogradi* (vinyards). Above, a ruin of a masonry residence rises, thrusting its three-story walls in the direction of the blue sky. This house was built by my father's older brother in the 1890s, and it is where I stayed in 1937. Kuna was almost leveled in the shelling of 1942, when first the Italians and then the heavily equipped Germans attacked. By sacking and burning the larger houses, they wiped out the community leadership.

Shortly, the facade on our left ends and the urban scene becomes rural, the street becoming a terrace lined with lindens and locusts for the dwellings on the uphill side. The afternoon sun washes the grape-arbored house fronts, warms the old people in the doorways, and dries the clothes hung in precise order on lines set on poles in the vineyards. Urbanity, amenity, utility, agriculture, and private and public life come together here. The rows of the vineyards stretch out across the valley to the Franciscan Monastery and the cemetery, which are outlined against Potomje's mountain slopes. This integration of the vineyards into the daily urban life pattern creates an open and inviting structural system

Main street as terrace over vineyards

that gives Kuna a stronger identity source than the internalized quality we experienced in Potomje provides for that place.

As we approach the second, older cluster the tight urban street that turned into a sun-drenched overlook becomes a quiet country lane leading to the outlying hamlet of Hilići. Looking back, we see how Kuna in cross section steps up the hill, in contrast to the horizontal character of Potomje. We return to the terrace above the vineyards, where the main street turns sharply to the right and goes up the open slopes that form Kuna's rising backdrop. At this intersection stands the village *dom*. Before the advent of cars, television, and ready access to Trpanj, Orebić, and Dubrovnik, the *dom* was a gathering place where people expressed their talents and creativity through song, dance, and poetry, binding successive generations into a collective identity.

Rising higher on this northern slope, we now look out over the dwellings fronting on the lower part of the road we have just walked and up to those even higher above. As the road grows steeper, the stepped pattern becomes a reality, and the increased variety and irregularity of structures and access ways create more of interest than we experienced below. We find signs of early settlement patterns in the primitive construction and haphazard relationships of buildings. This awareness allows us to imagine ourselves back in time as we move along the byways with their buildings crammed closely together, an experience that reveals the slow process of Kuna's formation over centuries.

On one of these byways we find the large home where Celestin Medović lived and worked until his death in 1920. The stone arches and the steps leading to the several levels of the castlelike building suggest his chosen contemplative way of living. I recognize this urban relic from the reproduction of a painting of the arched stone entry to the house in its prime that appears in the collection of Medović's works edited by Vera Kružić Uchytil.[14] Today, although the house is largely in ruins, its ample scale and stone construction reveal the high level of recognition Medović achieved in a world that combined art with the priesthood. The newer home of his descendants closer to the Piazza calls our attention to his celebrated status later in life.

We translate from Croatian a memorial plaque installed by the people of Kuna in 1971 when his statue by Meštrović was placed with my guidance. They had also attempted to convert the monumental home into a gallery where visitors could enjoy his many works owned by people living in the Pelješac region.

THE CROATIAN PAINTER

MATO CELESTIN MEDOVIĆ

ERECTED THIS HOME

TO BE ABLE TO LIVE AND WORK

IN THE HEART OF HIS BELOVED BIRTHPLACE

* * * * *

FROM THE PEOPLE OF KUNA

1971

Finally, we arrive at the uppermost road in Kuna. One large dwelling stands forth, distinguished from the others by its veranda and gabled roofline, which suggest a California influence. This is the home of the Palihnić family, a number of whose members have lived in America. A self-made footpath like those that sustained life in Kuna for centuries take us up to the pine-forested ridge, where we look across the gulf that separates Pelješac from the mainland. We see the mouth of the Neretva River and can envision the route inland to the lamentably war-torn Mostar and Sarajevo of the 1990s. Below us the narrow road from Kuna to Crkvice, barely wide enough for cars, clings to the steep mountainside. It was via that road that I left Kuna in 1937 mounted on a *mazga* under a bright spread of stars in the black predawn sky. At that time little could I have imagined my crossing The Bridge to Dalmatia to conduct this reading late in the century.

PODOBUČE: PRECIPITOUS MEETING OF LAND AND SEA

The Mazga *Route of Old*

We retreat now to the tranquility of Podobuče, where the land above Potomje plunges precipitously over the ridge into the sea. One of the few points that is accessible by boat and has buildable sites at the water's

Podobuče's precipitous niche on the sea. *Photo by author*

edge, the tiny village and its cove await us free of cars, just as they have for centuries. Precariously linked to Potomje only by cliffside footpaths, the site offered access to the sea and thus had a small role in the economic life of the Kuna area. Today, Podobuče serves as a permanent home for some, but it is also a place where people of the Župa, including my cousins, retreat to. For our dinners they had only to pull in the fishing net from a small boat. A few weekend homes await folks from the interior and occasional Europeans seeking a refuge from tourism. All are attracted by its rugged isolation and wild beauty.

Approached by car from Orebić, a distance of about seven miles, Podobuče marks the end of a single-lane roadway that passes through the terraced vineyards of Postup and then reaches intimidating heights on the cliffs above the sea. If we were to arrive that way, the car would stay high above Podobuče, looking down on the two clusters comprising some fifty dwellings. The upper cluster clings to the almost vertical slope; the lower one wraps around a cove of blue water and ivory-colored pebbles that has offered residents protection in the precarious transfer of people and goods from land to sea for centuries.

In order to establish in our minds the traditional linkage between mountain and sea, we return to Potomje and make the three-hour walk on an ancient footpath. In old times we would have departed from the tiny church of Sveti Tomo and taken the zigzag roadway of the *mazga* to the ridge above. But now there is a shortcut: in 1975, in response to pressure on the government by the old cooperative, which is run by all the villages in the Župa, a tunnel was bored through the ridge to facilitate the transport of grapes from the Dingač vineyards to the *vinarija* at Potomje. As we come through the coastal opening, the terrain breaks away abruptly and we see the entire Adriatic, all the way to Italy, stretched out before us. In the foreground the rugged island of Korčula points its many ridges westward, the town itself settled in the sea like a ship moored to its dock. To the southeast, in the distance lies the island of Mljet, and between the two, Lastovo.

From these heights the walk of treacherously loose stonework leads us gratefully downward through the terraces of well-tended, leafy vine-yards, then along steep ledges of bare limestone that hold Pelješac in place. In the pockets where earth and rainfall collect, stunted pines and *makija* provide sparse coverage. *Makija,* typical of the Dalmatian coast, represents what is left of ancient forests. As we walk through an endless variety of plant forms and species, from ilex oak to rockrose and rose-mary, our reading experience at close range contrasts with looking out to the distant horizon of the sea far below. Now and then these plant forms soften the wildness of the landscape by the shade they provide and their myriad textures of leaf and stem. Spicy fragrances enliven the air we breathe so deeply now, and our spirits are buoyed by the same ex-periences and challenges of direct contact with nature that shaped and made strong my own forebears. If we had made that long walk from Po-tomje to Kuna a century ago, Podobuče would have looked very much the same as it does today.

The dropping of our path's gradient announces that we are nearing Podobuče and the rocky shoreline. The landscape becomes wilder, the limestone curls into crag-studded walls. Below, a tiny cove with a lone cabin cuts back into the walls, a retreat from the mother village. Our path brings us to a point midway between upper and lower Podobuče. Perched on a series of huge twisted ledges, the upper cluster of solid stone dwellings is backed by enormous limestone crags of fantastic form and proportion. Surely some massive geologic process created this place for human settlement in such a way that no menacing intruder could threaten entry from above. Indeed, history tells us that settlement began with a cove cut in the rugged coast and a cave well above, set in steep, inaccessible cliffs. The name Obučen, for the forests that "dressed" this place of safe haven, was recorded officially in 1339. Only in the eigh-teenth century did this become Podobuče, meaning "under the protec-tion of cliffs, cave, and woods" [15]

Exploring upper Podobuče, we count some twenty houses; for all the irregularities of the natural setting, they are neatly arranged in quite a geometric system. Gigantic rust-stained crags overhang the veteran dwellings softened by faded tile rooftops and grape-arbored loggias. They step firmly down the hillside, like giant cubes of stone, maintain-ing an order that speaks for the purposeful and deliberate frame of mind of the original builders, holding their own against the forces of nature's

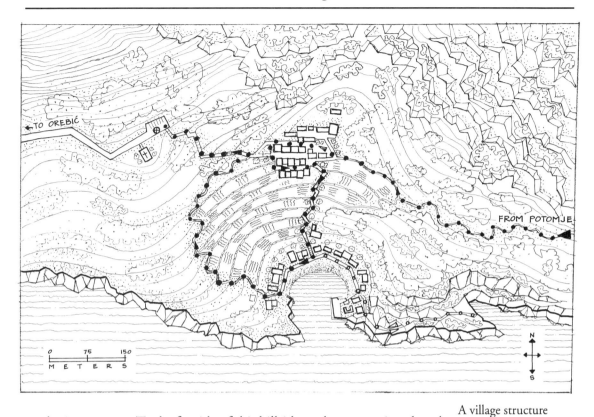

A village structure shaped by crags, vineyards, and sea

geologic processes. To the far side of this hillside settlement, a tiny chapel stands on a knoll jutting out from the precipitous site as if to distribute its benevolence and protective grace to both upper and lower Podobuče. The chapel's location transcends time in that the knoll is shared with the village's only parking space for cars, the sole link to our era.

Experiencing an Up-and-Down Village

After the exhilarating long walk from Potomje along the wild seacoast, we feel thoroughly detached from the more urban places we have read. Consequently, our entry into Podobuče, awaiting us at the far end of our *mazga* footpath, becomes a welcomed event. Here more than at any of the sites of our earlier readings we feel that we are fully independent of the motorized world, as if we were living in an earlier century.

Our aged path has brought us into the relatively open slopes that divide the village. Looking down onto the rooftops of lower Podobuče, we note that whereas the provocative crags high above delimit and dominate upper Podobuče, it is the sweeping curve of the pebbled cove, anchored in place by a great rocky point at each end, that defines lower Podobuče. The sense of foreboding and intimidation we get from the upper setting gives way to a light-hearted and joyous sense of entry into a settlement where people and the sea live most happily wed. Whereas the geometric placement of dwellings in upper Podobuče supersedes

Drawings by author
Upper Podobuče (top);
Lower Podobuče
(bottom)

nature, the dwellings in lower Podobuče faithfully follow the sweeping curve of the natural cove, tucked back in these fierce walls of limestone that typify Pelješac.

In Podobuče we experience a delightful contrast between the dancing qualities of the path we take downward and the stolid rows of upper Podobuče's dwelling. At times our path takes us literally between dwellings and under an archway. The difference between the sensual delight of these environmentally choreographed movements—descending around and down and under—and the somber mood of upper Podobuče is striking. Set in a grid pattern, the several rows of stalwart homes in upper Podobuče stand like soldiers above us, stiffened to maintain their position against the wild configurations of the massive, torturous cliffs. They stare blankly out to sea, reflecting the discipline of the rugged life of the people who built them, boldly challenging the tumultuous setting of nature, though on a minute scale—wondrous for us to experience.

The Grand Finale at the Sea

With upper Podobuče behind us, we come to the glorious terminus of this environmental sequence down to the sea, moving through a stretch of pure green vegetation. Following a contemporary planning precept, this natural greenbelt gives each of the village clusters its own identity. Then, almost as if we were in a tunnel, the ramped stone path buries itself between walls six feet high that cut off the anticipated view of our main goal, Podobuče's deep cove and inlet from the sea.

Shortly we arrive at lower Podobuče as the path pinches between two stone houses and turns into steps that take us to the row of dwellings fronting on the narrow beach. But this outlook is at first hidden by the tops of a row of Australian pines that follow the curve of the beach and the facade of houses. At the last flight of steps, the trees frame a close-up of this baroque crescent; its clear waters introduce us to the sound of the sea lapping timelessly and keeping in constant motion the millions of rounded stones. Jagged, upended slabs of rock mark the two ends of the crescent; like the plunging cliffs above, they depict the violence that geologic timespans can exert. The essence of Podobuče's identity lies in the way this meek little cove lies so quietly in the care of these wild formations of nature.

Even though the descent we have made seemed hazardous, we feel secure in a finite location where time seems to stop and the sea ends our long journey on Pelješac. We are face to face with nature at the critical meeting point of land and sea, sharing this solitude with the inhabitants of the two dozen houses that happily ring the crescent. Then, when our line of vision swings back up to the cliffs, the solemn cubes of upper Podobuče cloud our inspiration and hide the sky. Their eyes gaze down on eager outsiders such as ourselves, whether from Korčula, Vienna, or

California. The southern anchorage of the cove thrusts its rock and windblown pines well out into the sea, which in turn fights back with white water. Less rugged, the western point has been terraced into vineyards whose curving rows spring from the geometry of the upper village. But the joy of lower Podobuče returns in the placid, clear body of water that laps day and night in a quieting and contemplative repetitiveness onto the gently sloped beach of ivory pebbles. The dwellings facing this deep cove become one with the daily life of the sea as a source of food, access to nearby Orebić and Korčula, and, in recent years, reclamation of the spirit.

Drawings by author
The crags above (top);
the cove below
(bottom)

We now explore the harbor's eastern promontory, quite hidden from view so far. Behind a final cluster of dwellings a trail leads us through wind-blown pines scented with salt spray that frame the surging waters below. We experience nature in its primeval state. Rock walls rise as if to prohibit any other way of access to the riches of this untouched, sacrosanct spot. We scramble around a bend onto a gigantic, slanting platform of stone. On one side breaks the Adriatic Sea. On the other the stone face drops down to form a deep pool of moving water mingling tones from blue to green walled in by sheer cliffs of rock. A mass of limestone layered into strata tilts to the sky and gradually slopes into the pool. The open sea forces its way into this enormous, self-contained Adriatic swimming hole created without the hand of a human being. The line of the sun has moved toward the clear water below us, revealing the white bottom. To plunge into and swim in this place becomes a celebration of the perfection of nature left on its own and a source of reflections on the imperfection of the environmental works of man.

To complete this exhilarating reading, we turn back to experience the west arm of Podobuče's miniature harbor. Pausing at the spacious concrete dock, we observe the struggle between man and the sea as a boat has trouble unloading a few tourists from Korčula headed for the tiny cafe perched on the rock above. As we follow the curving shore past the dwellings, we note that the buildings weathered with age seem rooted to the base of this northern promontory, while others, newly built—surely by "outsiders"—are architecturally alien to Podobuče's identity. From the end of this gently sloping jut of land we can look back to the protected cove, the gleaming, pebbly beach, and the lower and upper villages and see them and the natural environment as parts of an integrated whole. Behind us the rough sea of high tide makes precious this stalwart foothold of dwellings.

Moving our eyes slowly upward, we reaffirm how the terraced vineyards so faithfully curve with the contours, in sharp contrast to the fixed, geometric look of the upper village. Through the vineyards we climb to the crest of the bowl-like slope, marked by the tiny village chapel, to rest and take in the panorama of Podobuče from yet another perspective.

Having passed through a sequence of experiences, from a fresh sense of arrival and discovery to feeling intimately a part of the internal makeup of Podobuče, we now grasp the wholeness of the place in a single, multi-faceted image.

After a few day's stay, we consolidate our accumulated sense of identity with Podobuče and come to the finality of experiencing departure. We climb the steep steps and stony paths to upper Podobuče, forgetting the ease with which we descended to lower Podobuče. Through this effort and our acquired identity, the village seems to be resisting our farewell, seeking to hold us and our baggage. The blue cove below shrinks as upper Podobuče receives us. We take its sharp turns, finding our way to the roadway entrance marked by the tiny chapel. The twentieth century awaits here with a car at the end of the narrow road along the cliffs that will take us back to Orebić. Possessed with an identity we believe to be indelible by reason of the depth of its penetration, we ask ourselves, How faithfully will Podobuče's magical powers of identity with place retain the freshness of this moment once we have returned to the cities we call home?

After bidding "dovidjenja" [until we meet again] to you "readers" of these dozen cities, towns, and villages, I spent several days in Podobuče as something of an insider, housed with cousins at the edge of the beach. The soft lapping of the waters and the gentle clinking of pebble on pebble in a rhythm set in motion eons ago formed an ever-present backdrop of sound. My cousin Vijeko's niece, children, and husband, a native of Podobuče, lived a dual life. They combined the traditions of farming and fishing from the limited terraced land and the abundant sea with hosting summer vacationers, people of an urban culture eager to experience the intimate connection to place offered in Podobuče. Sharing its wildness with others, they could experience the genuine quality of people, an experience less likely in more nonurban environments. One couple there from Vienna had returned year after year and built a loving identity with the place facilitated by their work in architecture. I witnessed firsthand the process of identity with place taking hold in them as we shared our similar urban backgrounds and trusted our intuitive responses to the uniqueness of the place. Breaking of ties each summer became a ritual. I recall their last evening meal with me, their last morning swim in the rock-bound pool, and their last showers with fresh water from a bucket out on the sun-drenched rooftop. These vignettes demonstrate the depth of our collective experience and the regret they felt as they ascended the steep winding path to the car, taking Podobuče's identity with them.

Our rich experience in Podobuče readily reveals the qualities offered the short-term visitor. Yet why people would live there permanently is not as clear. When I discussed this question with my cousins and neighbors, I found strong reasons for and against living there year round. Podobuče was the birthplace of my cousin's husband, who owns his family's land. His mother lives there, and his sense of security and belonging stems from generations of connectedness. Moving to another place has become unthinkable. On the other hand, the nearest school is many miles away in Kuna, there are no stores or doctors in Podobuče, and access to these amenities by their small Volkswagen is hazardous on the narrow road. They live isolated from people and lack intellectual and cultural stimulation.

In spite of this lack of modern services, Podobuče appears to hold a magnetic attraction that is rooted far deeper than our own incomplete reading experience allowed us to reach. This attraction clearly stems from the day-to-day focus on the richly varied environmental uniqueness of Podobuče, a vehicle for saturating the consciousness with intensely intimate experiences. Each day brings a sense of fulfillment that is not possible in a place with fewer features distributed in less direct juxtaposition. Its smallness allows one to experience Podobuče as an integrated whole. If we were to take away any of Podobuče's critical elements—the deeply indented cove, the pebbled beach, the crags above, the "cubes" of the upper cluster and the curves of the lower, the isolation and contemplation evoked by the tranquility, the directness of human contact and the self-containment of the place—it would lose its powerful hold on its people and on us.

Any of the places we have read might be evaluated by weighing their positive and negative qualities, but in Podobuče there seems to be a balance between the two: between the uniqueness of the natural and built environments and the fulfillment that comes from meeting the requirements of daily living, on the hand, and personal adaptation to reality, on the other. Podobuče's message could be that searching for this balance may be a way of carrying identity with place to a higher level: love of place and the distilling of its uniqueness into the spiritual response that is potentially contained in all of us.[16]

CHAPTER SEVEN

The Environment

Common Ground for Community Identity

Surely there are still peaceful countries and men of good sense who know of God's love? If God had abandoned this unlucky town on the Drina, He had surely not abandoned the whole world beneath the skies? . . . Who knows? . . . Anything might happen. But one thing could not happen; it could not be that great and wise men of exalted soul, who would raise lasting buildings . . . so that the world should be more beautiful and man live in it better, should everywhere . . . vanish from the earth. . . . That could not be. —Ivo Andrić, 1959

ANCHORS OF HUMAN IDENTITY, TARGETS OF WAR

A New Task: Rebuilding Meaning in Places

With ironic and foreboding significance, our search for human meaning in Dalmatia's hierarchy of urban places has come to a close with the evocative image of Podobuče's gentle, peace-inducing environment. Moving from the region to focus on cities, towns, and villages, we have gradually deepened our understanding of their varied nature. Through our own experienced responses, a uniqueness that sharply distinguishes each has been distilled.

Podobuče, the smallest and simplest of all, draws us nearest the human side of our inner consciousness, a realm of perceptive capabilities close to spirituality. There we have recognized the intimate bonding of people and place. A rugged landscape formed by grand strokes of earth and sea has been shaped into a living environment by a community fueled by love for this particular place. Wildness and human spirit have joined hands. From this we learn that places in themselves hold the power to awaken the spirit of unison with nature and secure peace with the environment that lies within us.

And now a painful impact on this spirit breaks our train of thought. We must now face the violent change that has been wrought on these urban places we have come to identify with. Attacks echoing those of the fourteenth century have shattered the human quality rooted in these places and threatened the independence and sustained peace they had achieved. In 1991 Dalmatia's revered symbols of human identity and love of place became targets as the walls of Dubrovnik itself—symbol of hard-earned freedom and self-sufficiency—were shelled. Zadar, Šibenik, Split, and dozens of villages suffered their share of medieval aggression carried out with twentieth-century military equipment.

In times of peace these symbols of culture, history, religion, ethnicity, and tolerance represent an accumulated heritage that binds the generations to a given place yet become taken for granted. But in times of war, especially when attacks are sudden and unprovoked, these targeted monuments, which give form and character to entire towns and villages, stand spotlighted by mind-piercing experiences that are destructive to the community's collective consciousness. Lasting wounds are imposed. Dark clouds of divisive aggression again settle over the connectedness of people and place that Dalmatia has demonstrated over the years.

For me the events of 1991 brought to mind images of the violence and bloodshed of 1914 and the 1940s. They forced me to look deeper into the role of identity with place than I had foreseen as I recalled the anguish of family members in both California and Dalmatia during the two world wars. The names Sarajevo, Bosnia and Herzegovina, and Montenegro came up constantly as synonymous with tragedy, human suffering, and irrational political and ethnic conflict. It was unthinkable for the adults then, and for me as a child, that these burning issues could be repeated a third time and on a greater scale now at the end of the century. Today, the capability of modern communication to instantaneously bring word to us from family, colleagues, and friends over there has brought a sense of reality to the abstractions of press reports on homes violated and people become refugees.

Relatives in Dubrovnik and Zadar had to seek refuge in Italy, just as others in World War II had to do in Egypt and South Africa. I also learned from these personal relationships that the strong Dalmatian sense of identity with place had not been extinguished and that ways would be found to overcome the physical damage done. From colleagues I received reports systematically prepared by monitoring and recording each and every hit, hole, and fire even as the attacks went on. A future role for identity with the environment became an illuminating vision. It took the current war to awaken in the collective minds of the stricken communities the significance of identity that lies deeply embedded in wars in general yet is not fully grasped in times of peace. The heartless bombing and eventual destruction of the old bridge of Mostar became an attack on my own cultural Bridge, a central theme of this work.

Inspired by these events, in October 1992 the corps of historic preservationists and environmentalists of Croatia announced an International Conference on the Effects of War Activities on the Environment. Held in April 1993 in Zagreb, this conference brought together people from numy countries, including some in Europe, representing numerous environmental disciplines and organizations. Considered an extension of the World Conference on the Environment held in Stockholm in 1972 and the conference in Rio de Janeiro in 1992, this was the first to examine the impact of war on the environment itself.[1] The main topics of research and policy making included the protection of forests, agriculture and water resources, the economy, and the cultural and natural heritage.

WHAT OUR SEARCH FOR THE MEANING OF PLACE HAS REVEALED

Goals Set and Paths Followed

Essentially, we have sought to determine the properties of identity with place in terms of daily living, the source of closeness to the environment that people once held. We aimed to establish clearly through firsthand experience the connection between traditional urban forms and the images of them held by people that generated their identity with place. We considered how this concept functioned in relation to the individual and the community as well as from generation to generation and from the local to the regional scale. Our focus has been on these often intangible qualities of the environment, yet we aimed to show how other, more tangible forms of identity, such as culture, religion, ethnicity, and family, relate to places in the physical sense. To give context and credibility to these aims, and drawing on Heidegger, we followed paths that took us from the personalized starting point and then examined the roles of geography, history, and social change in shaping the regional system and the form of individual places.

To interrelate these varied dimensions, we embarked on urban readings as a way of directly and personally experiencing selected sites; we carried on a dialogue with each place in order to learn what it could tell us about itself. By this phenomenological approach, introduced in chapter 1, we monitored our intuitive responses and our feelings and perceptions of a hermeneutic nature, gathering them into a holistic image of each place. In this way our urban reading revealed specific information on particular characteristics, from the overall structural system down to the details of buildings, streets, gathering places, and open areas, and how patterns of human use were determined by these.

By qualitatively experiencing the accumulation of history as manifested in a single, urban core used by people today, we gained a new awareness of the place as a whole. This perception of the physical environment became a vehicle through which the sequence of events over decades or centuries were integrated with the present. For planners and designers of environments, this awareness of the time dimension of space can help them to bring together the works of the past in a way that is sensitive to the continuity of identity for oncoming generations.

Using the Dalmatian urban places as case-study material for this method proved to be fruitful. Because the turning points in their history occurred over centuries, the differing ingredients in their evolution could be readily perceived. Today's highly accelerated process of growth, however, which is at work in Dalmatia as elsewhere, is so dominated by the self-propelled forces of economics and modern technology that time periods are compressed into decades and obliterated or blurred in the visual city. By comparison, the richness of Dalmatia's natural landscape and its wide range of urban patterns intimately related to people have generated clear images and metaphors. Examples such as Korčula's urban ship and Pučišća's urban amphitheater are not so readily revealed in most of today's cities. At the global level, Dalmatia's environmental attributes that are similar to those of California sparked the concept of The Bridge, upon which cultural identities could mingle with and enrich the diversity of the new homes of emigrants.

Identity as the Keystone of Meaning of Place

Through this experience all paths taken seemed to converge on the phenomenon of human identity as the keystone on which all aspects of meaning depend. By carefully sifting meaning from each of our experiential readings, I have defined ten basic properties or characteristics of identity with place. These properties may enable the reader to repeat the experience in places of his or her own identity, as I did in the hill towns of Italy, in Berkeley, and in San Francisco. In this concluding chapter, we shall use these properties to measure the impact of the war as well as the potential for renewal and reconstruction when peace and stability ultimately become established.

1. **Experiential Roots Deepest at the Local Level.** *Identity with place develops in its most profound sense out of the kinds of daily experiences that prevail at the local level.*

In our readings it became apparent that smaller places are generally shaped directly by the creativity of the local dwellers through their daily living experience and the intimacy of social and family connections that small scale fosters. As a result, people are more inclined to initiate and participate in activities related to resolving environmental issues. The sense of connectedness then becomes stronger and more fulfilling. An example is Sutivan's ability to maintain a popular consensus against building large hotels and in favor of converting larger homes into small inns and providing communal dining in the villa of a renowned nineteenth-century poet, thus allowing his cultural contributions to become known to.

Regional urban systems are formed in response to geography and the impacts of political and economic history. We found that larger places tended to be influenced by external forces, and their ultimate urban form came about through the collective experience of local government, as exemplified in our readings of Zadar, Split, and Dubrovnik. It became clear that turning points played a lesser role in the smaller places. We learned from our own intimate experience that the sources of distinction lay in both the structures of a place as a whole and in its individual elements.

2. **Hierarchy of Scale.** *Identity with place exists simultaneously in a hierarchy of environmental spaces, each with its own integrity of form and limits.*

In attempting to establish a sense of identity with any one place by experiencing its integrated form, we inevitably discovered that each place stood in the context of another, larger place with its own form and character. This awareness of a surrounding town, an island, or a region grew as we moved outward from the center of the place. Thus, a sense of connectedness to a given place matures fully only when our consciousness embraces the entire range of the hierarchy, from the smallest to the largest place. Our reading of Hvar's old town produced this concept. There, our intentionally prolonged stay in the highly confined complex that had been authoritatively planned in the Middle Ages for the urban elite produced a profound sense of imprisonment. This paved the way for the dramatic contrast we felt when we experienced the freedom and openness of the harbor and the quarter's strong connectedness to the town as a whole, which has its own integrity. Hvar's "arms open to the sea" spoke for the town's role on the island as a whole, in Dalmatia as a region, and upward in the hierarchy of identity as part of the Mediterranean and the world at large.

3. **Uniqueness of Urban Form and Quality.** *The particular physical pattern of an urban place plays a dominant role in generating identity through increased awareness of its uniqueness of form by individuals and the community as a whole.*

The physical pattern of an urban place and the way its people adapt to their natural environment result is a uniqueness and a powerful source of integrity of form and identity. In our reading of Sutivan, for example, we could not clearly determine this uniqueness until we compared it with Pučišća. The ability to recognize uniqueness through the capacity of the human eye itself is a potential we all share as users of a place. Through this common ability we can gain images of distinctive form and features by monitoring the pattern of our daily movements and responses to a given place. This visual awareness is more readily awakened in places with obvious elements of identity, such as the red tile roofs we have seen. However, when the unique qualities of a place are numerous, such as in Dubrovnik, large-scale tourism can threaten the genuineness of residents' identity with place.

4. **Common Ground for Other Identities.** *The spatial nature of urban places provides a common ground for diverse sociocultural identities.*

We have learned to distinguish the places where ancient Rome left its social and cultural marks, what the Croatian newcomers could and could not do, and how the Venetians introduced them to new urban refinements. In the street patterns of Zadar and of Split's core we sew evidence of the engineering and architectural capacity of ancient Rome, in the churches we read Rome's religious beliefs, and in the marketplaces we sensed its economic life. Dalmatia's uniquely varied geographical forms were a strong force in creating a remarkably homogeneous pattern of localities from diverse ethnic origins and cultural influences. Differences between nonphysical identities can be minimized when the collective goal is the well-being of a shared environment.

5. **A Dynamic Community-forming Force.** *Identity with place can become a dynamic collective force toward community formation.*

An individual's identity can spring from the physical form of any given place as his or her increased visual awareness broadens. Identity with place becomes a dynamic force, however, as more individuals share this identity through intensified experience and bring into being a collective identity, culminating in shared community action. The interaction between users of the place and the environment will grow as such ecological problems as depletion of resources or air and water pollution need to be dealt with and resolved. In this process the size and the pattern of a place play an important role. For example, a village the size of Sutivan is small enough to foster collective identity, and its radial pattern promotes frequent meeting of people at intersections of the harbor with five "stages." In the case of a city the size of Split, with its vast New Skyline, community formation is not likely to occur on its own.

6. **Insiders versus Outsiders.** *The depth of identity with place varies depending on the length and the intensity of the experience gained in the place.*

Even a person who visits a place for only a few days or a for months and does not have knowledge of its language, culture, or history can, with some preparation and open-mindedness, form a limited attachment to it or gain a limited understanding of it. Rarely, however, does a short-term visitor develop a sense of connectedness to match that of people who have, with a certain commitment to a place, spent the better part of their lives and even raised their families there. In my own case, at times—as in Kuna with family or in Split with colleagues—I felt myself to be something of an "insider" until my dual cultural identity as a Californian set me apart in spite of the facility of The Bridge. Clearly, there can be no precise boundary between the "outsider" and the "insider." Thus, the urban designer who is an outsider can rise to the level of an insider by deliberate involvement in the daily life of the place where his or her professional responsibilities are being carried out.

7. **Dual and Multiple Identities Fostered.** *Modern communication and transportation have made it increasingly common for persons to have an identity with two or more places.*

The automobile, the airplane, the telephone, television, FAX, e-mail, and, increasingly, the second home have made it possible, even on a global scale, to experience and relate to more than one locale. The coming of transatlantic steamship travel and the transcontinental railroad made possible The Bridge that has spanned two centuries. With the immigration taking place around the world, dual and multiple identities have become common and can serve to link families, cultures, and commerce for mutual benefit. As we have seen, dual identities have been firmly established between specific localities in Dalmatia and California. One example is between residents of Watsonville—Little Dalmatia—and Dubrovnik's Konavle area. A large number of the residents of Watsonville there came from Konavle, where some still have relatives that they bring to California or go to Konavle to visit; some even maintain homes in both places. This interchange between the two places became even more committed as the attacks began in 1991 and generated major community efforts to supply aid.

8. **Oneness of Place.** *A quality of oneness permeates our mental images of a given place as the accumulating experiences synthesize the sources of identity and uniqueness.*

Certain particularly rich themes became interwoven in such a way that they transcended the first superficial impressions of our readings. These individual places gained coherence and memorability drawing on a broad assembly of sources: mental patterns, images, associations, senses, and feelings, the discovery of contrasting and subtle differences, and the assembling of fragments of knowledge—all set in a context of geography and history. Beyond their visual images, they demonstrate an interdependence of parts, a wholeness that grows out of the not readily visible ecological nature of today's urban environment. This immersion into a place allows us to get outside the pressures of daily life, to enter a realm where we can fully experience human identity with a given place. Both my father and my grandmother maintained their deep sense of rootedness to their particular Dalmatian homes, but they worked to achieve a similar connectedness to California by fully embracing their new settings for family life.

Korčula and Dubrovnik both offer the opportunity to capture at first glance this level of oneness. In both, the visual form explicitly states the history of the place; both have patterns of streets that provide intimate experiences, but always in the context of the sea and their links to the world at large. Their appeal to travelers in recent decades has become worldwide. On the other hand, we found that places with less clearly defined form, like Split, Zadar, and Bol, required mustering greater depths of perception to fully experience the genuine, locally based forces that made them what they are.

9. Intergenerational Continuity. *Identity with a given place can become a heritage linking generations to one another.*

Our documenting of turning points in history and the several examples of family continuity, including my own, have shown that identity with place can be passed on from generation to generation. This value becomes a heritage to enrich and lend stability to people's lives, whether via family, village, town, or city, as well as to the urban places themselves. Obviously, this continuity of attachment to places as symbols of those who came before us is facilitated when the will exists to preserve those places of historical significance. In Dalmatia that will has been the motivation for scholarly studies and preservation movements that have made possible the richness and diversity of contact with generations that we enjoyed in our readings of Zadar, Split, and Dubrovnik. Surely this drive will reach full strength in the renewal to come after the war.

10. The Role of Spirituality. *Qualities of spirituality of places lie waiting to awaken the potential capacity within ourselves for creating a more human environmental identity.*

In our probing for the meaning of place, we have recognized two properties inherent in all places built by humankind: the physical makeup of the site per se and the human input that gave it shape. The first is governed by the nature of the land and the materials at hand, the second by the aims, motivation, beliefs, and skills that emanate from the human creativity that has fueled civilization. Places transcend their material being to the extent that we, as creators and users of environments, find meaning in their visual and nonvisual qualities. In his work on urban history Spiro Kostof has expressed this maxim in his central concern, "with form as a receptacle of meaning."[2]

We can recall how Zadar's jewel-like quality grew out of its variety of detail set in a clear, water-bound frame; Split's contemporary dialogue between cultures as distant in time as ancient Rome, Renaissance Venice, and the former Communist Yugoslavia; and Dubrovnik's ability to transport us to the intelligence and self-reliance of inspired Medieval urban life. In Hvar we found a free spirit expressed in its "open arms," and in Korčula, echoes of centuries of valiant shipbuilding and seagoing. The boldly contrasting patterns of Sutivan, Pučišća, and Bol spoke for the uniqueness of each, the result of human use and locally based collective identity.

These essences captured our minds and defined an intangible yet "vital core" of their being that we shall call "soul," an element that would evaporate were its symbols in the physical place to be obliterated.[3] We could no longer savor the way each experience of daily life within these places evokes its own admiration, beauty, wonder, or inspiration. The accumulative quality of our intimacy with the physical place has put us in touch with the intricacies that give coherence to the ecological nature of places.

This "soul," which is distinctive for each place, provides the key to awaken the spirit, or anima, within us and to anchor our sense of identity. While soul is the virtual, though inactive, personification of the place, spirit is the "animating force within human beings," acting on principle.[4] Thus, as Podobuče has revealed to us, what has been called the "spirit of place," or *genius loci,* resides not truly within places themselves but very much within ourselves, whether as creators of places or as analysts or users. This distinction between soul and spirit deepens, then, only through extensive and contemplative experience that sets forth the very soul of a place.

As established in chapter 1, identity with place is an achievement to be sought after and not to be found ready-made. This link between the physical and the spiritual generates our impulse to care for, protect, and renew the soul of a place. We can do this by activating our sense of ethical responsibility and social commitment, in response to the violence of war or other destructive forces, as an act of fulfillment of love of place.

THE 1991–1995 WAR: DALMATIA'S TRAGIC TURNING POINT

An Assault on Urban Landmarks from Land and Sea

My search for the meaning of place has produced as manifestations of identity with place the ten qualities named above, originally intended as guidelines for the cities, towns, and villages in the reshaped, democratic Dalmatia foreseen in 1990. Their usefulness will now have to be reexamined in light of the violation the people have suffered. But we are also impelled to explore how identity with place—now even more precious—can become a rallying theme for reconstituting the spiritual damage done to the minds and souls of the people so deeply connected to their places.

Even as I mentally retraced my steps through these cities, towns, and villages as I began to put these chapters into coherent form in the fall of 1991, these places were under fire from offshore vessels. Zadar, often besieged and demolished in part throughout history and bombed in World War II has always rebuilt. Split, never attacked before, endured bombing in that war. When Dubrovnik was shelled for the first time, by the Russians and Montenegrins in 1806, during the French occupation, damage was slight. Never in its long and peaceful history had the city received such devastation, except because of earthquakes, as it did in 1991. Ancient faded red tiles fell from caved-in rooftops that we had perceived as a dancing urban landscape. People fled from their homes, and the social and economic life came to a halt.

I asked a colleague from Sarajevo how this could happen and what this explosive act signifies in terms of identity with place. An insider to both Dalmatia and Bosnia by virtue of his mixed ethnic origin, he gave a bitter answer that came from the depths of his concern. These attacks, he said, have revealed that the culturally fertile places we have called symbols of love of place and a shared collective identity have also been symbols of materialistic envy and presumed injustice. In mountain valleys only twenty miles inland from the Dalmatian coast one finds groups of people of different ethnic and religious backgrounds living a rural life, out of touch with the stimulus of the sea, in a harsh and unproductive environment. As Dalmatia's economy and living conditions improved because of tourism and increased accessibility by sea, highways, and airports, many people further inland began to look to the coast and tourist dollars. They felt for centuries locked into a physical and cultural niche by circumstances they could not control.

Never before had Dalmatia been so sought after by foreigners and by residents of all social levels in the former Yugoslavia. High-speed urbanization—not without its own profit motives—was generated by a host of regional forces: the demand for a *vikendica*—that little weekend cottage by the sea; a high-rise, low-rent apartment in Split's New Skyline neighborhood; a job in a luxury hotel in Dubrovnik with access to the social life of the Stradun; or even the entrepreneurship of a tiny bar, cafe, or a *restauracija* in a hole in the wall of Diocletian's Palace. Furthermore, all of this took place without the regional planning that Dalmatia deserves to guide its growth and protect its unique environmental and cultural resources.

Misha Glenny, in *The Fall of Yugoslavia,* brings to life the "hidden symbols" underlying the attacks as he recounts his own arrival in Dubrovnik in November 1991. He describes the "hundreds of Montenegrin reservists . . . in space-backed lorries" returning from their day of plunder and burning in the area of Dubrovnik, Konavle, Čilipi, and Cavtat, more like a looting spree than the waging of traditional civil war. "These were part of the mob which had scorched the earth along the last tapering twenty kilometers of Croatia—Konavle." [5] Rather than love, Dalmatia as a place especially blessed by nature, history, and Mediterranean culture has generated envy and jealousy. In that sense, the assault and pillage are more like criminal acts than acts of war.

Record of Destruction and Damage

From the moment when artillery at sea and on land first began firing shells—1 October 1992—the people of Dalmatia steadfastly monitored the damage and destruction. Their local and regional offices of the Institute for the Protection of Cultural Monuments kept detailed records and itemized the cultural content of each building affected. Publicizing this information to the world at large was the first expression of outrage against an attack, not just on buildings, but on cultural identity itself. Made available in full through the Croatian Ministry of Environmental Protection and Spatial Planning, these "facts" shattered in specific ways the images assembled in my mind as a result of the readings.[6] We who have shared these images through our walks and respites can gain a more personal sense of the inhuman effects war activities of this kind anywhere can have on the sanctity and privacy of one's identity with place.

When I received the ministry's published listings, covering hundreds of cities, towns, and villages, early in 1992, I reviewed them in terms of the ten characteristics of identity with place . It became crystal clear that the military motivation and strategy had been to strike at those buildings that carried the cultural, historical, and religious meaning closest to the hearts of the dwellers of these places—a spiteful sort of "cultural cleansing," as it were. The ministry's records show that the attackers completely ignored the plaques signifying which buildings were to be exempted from war damage according to the Hague Convention of 1954. Indeed, rather than safeguarding the buildings, the designation highlighted them as priority targets for bombardment.

The records also point out that buildings besides those of cultural or similar merit were destroyed or damaged, such as homes, farms, and shops. These buildings are also important, since they fill out and explain the overall structure, continuity, and growth of these historic centers as integrated wholes, the concept of oneness we have stressed. This divisive motive works against the interests of all the countries of the former Yugoslavia since Dalmatia's is the only western coastline and all share in its history. In this sense, the impact of the attacks will reverberate in all the republics for years to come, exacerbating enormously the already difficult ethnic entanglements as reconstruction takes place. Now, conscious of the images of places attacked as of 1991, we revisit them to look at some of the urban features damaged in Zadar, Split, and Dubrovnik—among the far greater number in Dalmatia as a whole. The words of a Dubrovnik resident who experienced the bombing of that city on 6 December, St. Nicholas's Day, could describe what occurred in any one of the three cities: "The bombing began at 0540. About an hour later, the Old City became a priority target. What can be seen of it, through the clouds of dark gray smoke, is blazing. But the whole city, not just the old city, is being attacked by land and sea. . . . The army seems bent on completing the destruction of a world city. Immensely brave soldiers have protected this city against overwhelming odds. Who else can come to help it? It *belongs* to the world."[7]

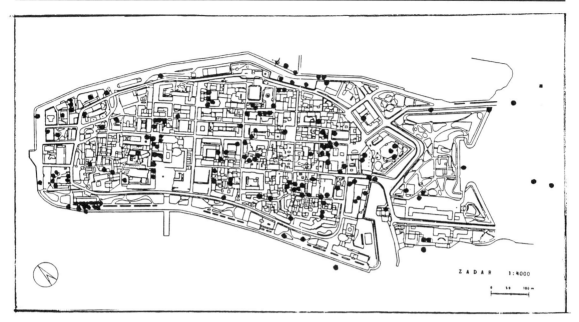

Reading Wounded Dalmatian Cities

Zadar

The records of Zadar's Historic Preservation Institute document a series of air raids on Zadar during eighteen days between 30 September 1991 and 7 April 1992 that accomplished a total of 150 damaging hits. During these attacks at least twenty buildings of the highest cultural significance and designated by the United Nations as having landmark status were hit. Many others of lesser importance, including eight churches, were also targeted. The detailed photographic record of damage to a wide range of buildings in the institute's 140-page book, ironically entitled *Zadar: Anno Domini (The Year of Our Lord) 1991,* tells far more about the townspeople's devotion than these statistics can possibly convey.

If we were to walk again our route beginning at the Trg Pet Bunara, we would see rising high above us the damaged Kapetanska Kula (Captain's Tower), one of the oldest remnants of structures by the people of Zadar from the thirteenth century. Directly hit by a shell in October 1991, roughly 150 square feet of its upper walls and a decorative cornice were destroyed. Nearby, the Land Gate from the Renaissance lost its cornice and balustrade; closer to the Fosa Harbor and the sea we would see deep scars in the Old Zadar's thirteenth-century walls.

As we head toward the Kalelarga, on our right is the Sveti Šime Church, Zadar's leading example of skilled historical restoration, which had reopened in 1989. After its roof was smashed in by a mortar shell, the building's structure gave way, opening to the sky some six hundred square feet of the roof and inner ceiling. In the residential blocks

Zadar suffered 150 direct hits by shells. *Institute for Protection of the Cultural Monuments and Natural Environment, Dubrovnik*

Damage in Old Zadar

on either side of the Kalelarga, damage to the Gradska Straža (Guard House) and the Gradska Loža (Town Loggia) speak for the intent to destroy Zadar's identity. We are drawn, as we were before, to the Forum, that silent yet powerful voice of ancient Rome, now hit by these vengeful shells of our era. A rocket destroyed part of the original Roman pavement. Turning to the right a few blocks ahead, we see the church of Sveti Krševan, which was hit three times, twice in October and again in November, at least once by direct shots intended for the neighboring Museum of Archaeology. That building, housing important archaeological collections, was badly damaged by repeated shelling.

Returning to the Kalelarga, as we near its opposite end we experience the climax in targeted destruction of cultural monuments: some eighteen hits within a few blocks, including the cathedral of Sveta Stošija, from the twelfth and thirteenth centuries. A direct hit badly damaged the roof structure in two places, demolishing a window and its walls, as well as a fourteenth-century painting in the nearby church of Sveti Donat, which in its earliest form dates back to the ninth century. Moving around to the west end of the *obala* and the sea, we come to the Croatian army headquarters, which had ten hits. At the east end, visible damage was done to the Scientific Library and the History Archive Building, where I researched documents on urban improvements of the Austrian period.

Other places in the vicinity of Zadar were struck, including Biograd, where forty buildings were damaged or destroyed, and Šibenik, which suffered significant damage. Šibenik, credited with having been founded by the Croatians well before the twelfth century, suffered a steady barrage of bombardment from guns on land and on warships anchored in the port that spared but few of the many buildings within the old city center. This occurred over five days in sequence in September, and there were further attacks in November. This deliberate attack on the city's most precious features hit the city walls, the municipal museum, a monastery, and the cemetery of the church and fortress of Sveta Ana, around which Šibenik developed in the fifteenth century.

The most damaging impact was suffered by the cathedral of Sveti Jakov, facing on the main square and together with the Town Hall framing the square. Completed by Juraj Dalmatinac, the Cathedral includes highly original sculptural works that make it a masterpiece of Gothic-Renaissance architecture. The damage to both the Cathedral and the Town Hall was a blow to the entire city. In all, sixteen landmark buildings, largely churches and monasteries, became targets. In 1992, shelling resumed in two areas from time to time, depending on conflicts at nearby Krajina, a focal point of Croatian-Serbian conflict, and shelling occurred as frequently as four to five times per day up to 1993. Only when the Croatians took back the region in 1995 did the shelling cease.

Split

The damage in Split was not as extensive as that in Zadar, but the shelling by the Yugoslav navy on 15 November 1991 following a blockage of the port was of major significance. Diocletian's Palace ranks among the highest on the world's heritage list, and this was the first time in its history that it was a target of direct bombardment. All in all, forty shells fell on houses, chapels, and other structures within Split's urban area at that time.

Standing on the *obala,* the starting point of our reading, with our backs to the Palace, we can imagine how our image of Split would be changed if we had seen warships in the port poised to fire on Dalmatia's most precious historical monument. Had the shells hit their targets, we would now see the Palace itself and the walls marked with shell holes. Emerging from the vacuous storage chambers below grade into the sunlit Peristyle, we can walk the *cardus* to the Zlatna Vrata and see the damage done to dwellings and shops along the *cardus* by shells probably intended for the Campanile. At that old north entrance we would see where shells fell close to the place where Diocletian entered from Salonae in the fourth century and refugees of that vandalized city entered to seek safety centuries later.

Outside the Palace, the firing reached back to the most ancient of times with the shelling of the Croatian Archaeological Museum on the *obala* and beyond the Yacht Harbor, a point close to the Meštrović Museum, which holds many of Croatia's finest sculptures. Returning to the *obala,* we see across from the nineteenth-century Italianate Trg Republike the popular church of Saint Francis with its Franciscan Monastery, damaged by shell fragments. Shelling hit the neighboring old residential area, Varoš, as well as Bačvice, on the east side of the Palace. There, a bomb fell without exploding within only a few feet of a colleague of mine who was running to a shelter. On the north side of the bay, the residential areas of Kaštela, as well as outlying places high on the ridges beyond Split, were bombed.

Dubrovnik

Dubrovnik suffered bombardment over twenty-five days between 1 October and 12 December. At times the firing was not targeted at particular monuments but randomly hit the tile roofs that give such a sense of uniqueness to the city. Hundreds of buildings were either damaged or destroyed, many of them of high quality, along with thousands of all types down the coast to Konavle and up the coast to Ston. The Institute for the Protection of Monuments estimated that 30 percent of Dubrovnik's historic buildings were heavily damaged and 10 percent were destroyed. With this in mind, we make our imagined return visit to the city with a heavy heart, experiencing the spirit Dubrovnik had evoked in us crushed by an immoral act against a great urban work of humankind.

Following the route of our urban reading of Dubrovnik, we first look at the Pile Gate environs, which on our first reading was bustling with the city's residents and their admiring visitors. Just across the street, the Inter-University Center, a large building from the Austrian period, is all but destroyed, and the hill and woods of Gradac have been fired on. From its height we looked back to the old city, nestled in its walls. Down the park's steep slopes toward the rock-bound neighborhood swimming cove, the peaceful seclusion of the Monastery of the Franciscan Sisters, together with the church and cemetery of Saint Mary's, was broken by the shelling of October and November, which damaged their structures. This act of insolence shattered the image of the quiet terraces where I had gathered my thoughts to prepare for my urban reading of Dubrovnik.

Further down the main street leading to the Pile Gate, the Hotel Imperial, a prime landmark of my 1937 and 1968 visits for its turn-of-the-century character, suffered shelling. Its roof was demolished and the upper floors burned. We see the giant tower of Fort Lovrjenac standing free of the city walls, and the attached fortified embattlement of Bokar, unharmed though both had been targeted. They were able to maintain their record of medieval self-sustainability since the shells fell short of their mark and ended up in the sea. Between them, down at the edge

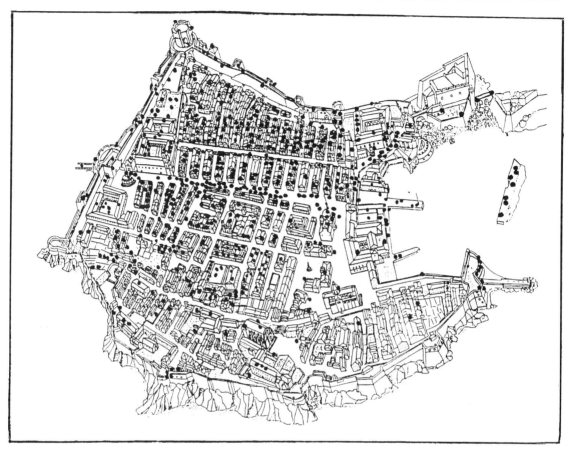

of the sea on Tabakarije Cove, my memorable image of earlier visits was marred when the top floor and roof of the seaside home of relatives of mine were smashed and burned, driving the family to refuge in Italy.

In our minds we now return to the Pile Gate. When we see the scars on Onofrio's Fountain, the city's waterworks constructed in 1438, and other damage in the area, that sense of entry onto a theater stage that we previously experienced vanishes. Struck during October and November 1991, the reservoir—so critical to Dubrovnik's life—was again heavily attacked on 6 December by one of the 150 shells fired from navy vessels that inflicted the most serious damage on historic buildings. The cupola of the handsome monument itself became a direct target, a projectile breaking through the dome and disfiguring the sixteen paneled walls with shell fragments. Indeed, it was on that day that the city suffered its most heartbreaking and traumatic destruction since the fatal earthquake of 6 April 1667.

Turning to our left, we see the effects of twelve shells that were fired deliberately on the Samostan Mala Brača (Monastery of the Little Brothers of Saint Francis). These hit the bell tower, the two wings of the

In Dubrovnik 30 percent of the buildings were heavily damaged and 10 percent were destroyed. *Institute for Protection of the Cultural Monuments and Natural Environment, Dubrovnik*

Dubrovnik under fire.
*Institute for Protection
of the Cultural Monu-
ments and Natural
Environment,
Dubrovnik*

building, and the colonnaded cloister, the monastery's most intimate element. Two buildings immediately next door were burned down. On our right, the former convent of the Order of Sveta Klara, the Little Sisters of Saint Francis, which was converted into a popular tourist restaurant, shows signs of its share of the shelling.

The Stradun stretches out ahead of us, stripped of tourists and vibrant urban life, its dignified medieval facades battered and shop windows smashed. Here is telling evidence that this, one of the great streets of the world, was the major target for demolition. Thirty-nine scars on the polished pavements show where grenades and rockets struck. Some upper-story windows are now gaunt holes in stone walls, their wooden frames and shutters burned. Color photographs show the flames that gutted the interiors rushing through them.

Just past the Franciscan Monastery we look up Celestin Medović's street to see the damage done to the Minčeta Tower. And up the sloping streets on the opposite side of the Stradun we see the massive Rupe Museum. Hit by shells on 24 October, it has a gaping hole in the roof around which a large area is damaged and numerous cracks in the walls.

As we move further along the Stradun it widens, and we can take in increasingly more of the view before us, which makes even more painful our witnessing of the damage wrought by artillery shells to the central feature, the city's Clock Tower. To the left of the Clock Tower stands the Sponza Palace, with its archives covering five hundred years of independence, containing not a single record of an onslaught of this scale. Even though the building was properly marked with the plaque of the Hague Convention, on 6 December fragments of cannon shell knocked off pieces of the facade structure and three shells hit the roofing and atrium. Across the square we see the church of Sveti Vlaho, who, ironically, has always been the protective and patron saint of Dubrovnik. Yet, to the dismay of the residents, the model of the city held in the hand of the statue of the saint, probably from 1714, was hit quite accurately and demolished.

Fortunately, to the right of the Clock Tower we find that the fragile beauty and stalwart institutional symbolism of the Rector's Palace suffered only minor damage. Blocking off the south end of this truly urban open space, however, the Cathedral of the Assumption was directly hit by two projectiles, which broke through its roofing and scarred the facades with shell fragments. Rising majestically above, the great mass of the eighteenth-century Jesuit church of Sveti Ignacijo and its adjacent Collegium Ragusim were bombarded in November and December, damaging roofing, a chapel, and a cloister.

To complete our image-fracturing tour, we move out to the Stara Luka, passing under the Clock Tower to survey the damage at this point, which was strategic to the success of a naval attack. To our right, Sveti Ivan's Tower, the angular sculpture of almost skyscraper scale that rises from the sea, received its share of bombardment. To our left, the Dominican Church and Monastery, home of the third religious order that for centuries lent stability and faith to the self-governed city-state, rises as an integral part of the soaring city walls. Numerous shells destroyed part of its roof and a sanctuary area and broke stained-glass windows.

Beyond the Dominican Church and Monastery and outside the walls, near the Ploče Gate, the massive Fort Revelin, focal point of Dubrovnik's military strategists in early times, served as the largest shelter during the attacks. Like all of Dubrovnik's fortifications, it could not retaliate against the attacks of three shells on its tower nor those fired elsewhere. The port structure was hit in October and over five days in November; resulting in heavy damage to the Kase Breakwater, the large and small arsenals, the main pier, and the fish market.

Offshore, in October and again in November, on the island of Lokrum, the place of relaxation for visitors over centuries, shells fell around Fort Royal and the Benedictine Monastery. And opposite this island, on the mainland shore, the cliffside Hotel Belvedere, built in the 1980s, which so harshly destroyed the rare and pristine shoreline site, was itself virtually demolished. Just outside the city walls summer villas in the vicinity of the Dubrovnik River were shelled, and other areas further up the coast were targeted; Ston, Osojnik, and Slano were especially devastated.

Since these acts of cultural vandalism took place, restoration has been carried forward consistently as resources became available; and there has been moral and material assistance from a broad spectrum of international sources. In human terms, the people of Dubrovnik readily awakened a hearty response from the world at large to the destruction of a heritage we all share. By early 1994 in monetary terms this amounted to some ten million dollars. Also by that time, of the two-thirds of the famed aged roofs that had been damaged, 70 percent had been replaced with faded red tiles brought from afar to match the old ones. Late that year, other reconstruction of critical architectural features and the laying of new limestone blocks to erase the shell holes in the Stradun had progressed far enough to assure meeting the 1996 target date for completion. This restoration program will reaffirm for centuries to come the spirited identity with place of the citizens of Zadar, Split, and Dubrovnik, and the environs of each, which together speak for Dalmatia as a whole.

The Impact of War on Identity with Place

In evaluating the impact of these attacks on Dalmatia's entire system, our concern is not with individual buildings but with human responses to each place seen as a whole. To measure these we can now use as criteria the properties of identity with place developed as the keystone of human meaning in the environment. Having come to understand these places through our broadly conceived readings as products of a centuries-long process, we can now see that a violent rupture has taken place in the phenomenon of environments' ongoing self-transformation, through human inventiveness, into places that give meaning to the community of their users. In that sense, the impact has been on the life-giving qualities of intergenerational continuity expressed in the epigraph and opening of chapter 1 and expanded on as one of our ten properties of identity with place.

Applying the property *hierarchy of scale,* when there has been armed aggression to any one settlement, the people of other places in the same region are also impacted with fear, disillusionment, and defensiveness. All of Konavle was damaged when many of the sixty villages were burned; the shelling of the town of Osojnik impacted the eleven villages in its immediate vicinity. On the larger scale, both were immensely affected by the relentless bombardment of Dubrovnik's streets and walls, unscathed for five hundred years. In turn, the destruction to that famed urban masterpiece degraded the image of the whole of Dalmatia. And, in all three examples, reverberating across The Bridge, outrage is felt among several generations of relatives in California who maintain a "dual identity" and by those in Chile, New Zealand, and Australia as well as the many visitors to Dalmatia worldwide.

Three of the properties of identity with place—*common ground for other identities, insiders versus outsiders,* and *dual and multiple identities fostered*—overlap and hold in common the quality of diversity. The human dimensions of identity with place demonstrate that a city, town, or village is much more than a physical place to be seen or lived in. The mind can only absorb and retain the richness of diversity a city contains when that city is free of prejudice, ethnic myths, and political exploitation, a freedom Sarajevo had so well achieved. Although Dalmatia has been characterized by openness and tolerance, the divisiveness caused by the war in the interior threatens to weaken the chances for a healthy, integrated community life. The presence of tens of thousands of refugees from "ethnic cleansing" in the interior who may never be able to return to their homes introduces the problem of a large number of "outsiders" who may have difficulty in assimilating. A clear division of communities and families could develop between the "insiders" and the "outsiders." Those who hold dual identities with Dalmatia and other parts of the former Yugoslavia will be restricted from maintaining ties to both places. In all three cases, the impact on identity with place is the rejection of the socially and culturally valuable role of the environment as a common denominator for mutual understanding and tolerance.

Applying the property *experiential roots deepest at the local level,* we find that the impact of armed assaults on religious and cultural monuments is most severe on villages and small towns. Shaped by a collective creativity and *intergenerational continuity,* their strong sense of connectedness to place gives them a more personal and heartfelt response to the shock of military mindlessness to human emotions. In larger cities, such as Split, the impact is shared by a larger population in which fewer residents have a deep identity.

Uniqueness of urban form and quality is the first property to spring to our minds upon hearing of attacks on Dubrovnik, the most precious of Dalmatian cities. Zadar and Split differ from Dubrovnik and from one another in terms of the varied record of history they offer and the quality of life their urban patterns foster. In each the most beloved architectural features are part of the whole that fixes images in the minds of "insiders" and "outsiders." Thus, for example, when we in California see photographs showing flames pouring out of the windows of buildings on Dubrovnik's Stradun, our anguish is far deeper than would be the case if the photographs were of any of the luxury hotels outside the walls. At the same time, when Diocletian's Palace is threatened, or when the architectural details of Zadar's compact street system, evoking two thousand years of history, are targeted by shells, the revered images we gained from our readings are shattered.

The early attacks on Dalmatian urban places have clearly reinforced the property *dynamic community-forming force*. The impact has revealed and increased the latent strength among the dwellers of cities, towns, and villages alike. Immediately aroused into a collective consciousness, each community became a vigorous force for restoration that, combined with the impetus toward local initiatives generated by the democratic reforms of 1990, offers promise for the future. On the other hand, in the inland scenes of turmoil, the war has badly damaged the collective identity that fostered community life. Families have been pulled apart, whole villages displaced to other regions or countries, and friends and neighbors whose individual identity with the place created strong ties, often regardless of ethnic origin, have become enemies. Yet reports from Sarajevo have shown that dwellers of apartment buildings and neighborhoods have formed working groups with designated responsibilities for meeting daily needs, evidence of a communal resource for the future.

A sense of *oneness of place* holds environments together in interdependent linkages that bind the separate urban elements and foster community integrity and fulfillment. The ruthless demolition of any single element, major or minor—Zadar's Campanile or just one of the five wells at the Land Gate, the sculpture of Juraj Dalmatinac on Šibenik's cathedral of Sveti Jakov, Split's colonnade of the Peristyle, or Dubrovnik's Pile Gate—would destroy the bonding of people to place and our responses of admiration, beauty, and wonder at the intergenerational creativity that has protected each place as a whole.

If we accept that *a place for spirituality* lies within us and that places can thus transcend their material being, then the confrontation of war becomes a challenge to our environmental awareness. We are called upon to broaden our perception of a place far more than the hurried visits of tourists allow. Something within us breaks when we hear of shells being fired at Zadar. It becomes a matter of ethics when the centuries-long dialogue on urban design principles is interrupted by shellings in the late twentieth century. Recalling Dubrovnik's productive period of medieval initiative, skill, and self-reliance, we rebel at the thought of demolition of the city those human qualities built. Our response to these disruptive acts on the "soul" of the old city confirms that our awareness of Dubrovnik's inner uniqueness has animated the spirit within us to rise above and beyond the pristine visual images we carry in our minds. Our innermost sense of loss confirms that rather than "spirit"'s residing in a place, the particular qualities of the place evoke the spirit within ourselves. And in this case that element of our being in itself has become a target of war.

To substantiate this fundamental meaning inherent in place, I offer a striking quotation from a newsletter sent from Mostar via "bosnet": "The identification of self with place, as exemplified by architectural symbols . . . such as the famous bridge over the Neretva built in 1566 is

Roofs on the Stradun: Demolished in December 1991, restored in June 1995. *Institute for Protection of the Cultural Monuments and Natural Environment, Dubrovnik*

an integral part of the human psyche. When such objects of self-identity are destroyed, much of the human spirit goes with it." One contributor's experience brought forth these eloquent words: "When I remember what is no longer there, I feel a spasm in my stomach, a knot in my throat. I feel death lurking in the absence. . . . Perhaps, it is because I see my own mortality in the collapse of the bridge . . . in all its beauty and grace, built to outlive us . . . an attempt to grasp eternity." [8]

AFTER THE WAR: REBUILDING BRIDGES AND ENVIRONMENTS

Looking Forward: The Year 2000 and the Next Millennium

Ultimately wars come to an end. And then the time to make peace with the ravaged environment can begin. Evidence is abundant that Dalmatia's strong sense of identity with place sparked the beginning of the renewal process just as the smoke of shells had cleared. The symbolic damage within ourselves served to stimulate a fresh start in urban design and planning even into the twenty-first century. The record of history has shown the need for fresh approaches based on local environmental identities to synchronize modernization with Dalmatia's rich regional heritage.

With this in mind, I direct my message principally to the younger generation of Dalmatia. At mid-century members of my own generation in California turned from the professional stalemate of World War II to make a fresh start toward planning for our unique shoreline. Our strong collective identity produced creative innovations inspired by the grandeur of the San Francisco Bay region, then in its untouched form, and by the need for harmony between the natural and built environments. Further, we saw this in the context of social betterment. Likewise, in Dalmatia the lessons of the 1990s should serve to vigorously launch the young people toward the opportunities for positive environmental change in the next millennium through the establishment of democratic processes at the local level of government.

The time is right for young professionals of Croatia to bring together environmentally oriented resources inherent in the fields of architecture, engineering, art, urban history, cultural anthropology, geography, ecology, and the social sciences. An inventive interdisciplinary approach to plotting out future directions of urban growth and environmental protection stands waiting to be implemented, especially with the vast new—and often misused—technology at hand. Collectively they could seek to reestablish the keystones of meaning in the hundreds of villages, towns, neighborhoods, and cities that make up the urban system of all Croatia.

Italo Calvino, in his posthumously published book *Six Memos for the Next Millennium,* has given us some clues about how we can go, in a single word, from a broad philosophical concept to specific ways of guiding our actions responsibly in urban planning. He develops lucid essays out of such one-word "memoranda" as "Lightness," "Quickness," "Exactitude," "Visibility," and "Multiplicity."[9] In the same way we could use as philosophical yet pragmatic points of departure for environmental purposes the properties of identity with place my search produced, highlighted as Experientiality, Hierarchy, Uniqueness, Commonality, Community, Intimacy, Duality, Oneness, Continuity, and Spirituality. Out of this could come a deeper meaning of the word *placeness,* as our reading of Podobuče revealed. A forward-looking, young-minded environmentalist in Dalmatia—or in California—could cull from the ten properties specific environmental goals for the twenty-first century. Indeed, this is what some of us did in California to advance the environmental fields in reaction to the setbacks of our own time—the Depression of the 1920s, World War II, denial of civil rights, Vietnam, and all the rest.

I call upon young people of Dalmatia to seek their own meaning in particular places. Even with only a limited sense of identity, deeply experiencing a street, neighborhood, or community in a big city or a hamlet, village, town, or ecological subregion can bring a wealth of personal rewards and public benefits. If the world is shrinking to a "global village," we need to celebrate, enhance, and revitalize small urban places in order to extend the understanding and intimacy they offer. Within existing structures, these experiences can serve as planning models for the world at large in which we can all live in ecological and human peace with the environment.

International Exchange and the Shrinking Globe

Cooperation on an international scale is critical to restoring environments that have been depraved by the rampant urbanization of peacetime and the ravages of war over ethnic origins. For, while global forces are pulling world regions together, local regions are becoming fragmented. Paradoxically, as the process of economic development becomes increasingly centralized, the local linkages and interdependencies needed for ecological sustainability are weakened.

Dalmatia, with its universal attractions, is such a region, yet it is subject to problems that are occurring worldwide—damage to the environment, social unrest, poverty, segregation, crime, ethnic conflict. However, Dalmatia's cities, towns, and villages hold some of the elements of the solution, thanks to their unique history of self-sufficiency and a geography friendly to strong local identity. California, which faces some of the same problems, is at the forefront of the global search for solutions to environmental imbalances. Having a unique bioregional integrity itself, the state of California has led others in working toward sustainable cities, self-directed communities, clean air and water supply, and environmental health. This leadership has come about thanks in large part to its superior universities, advances in technology, community awareness and citizen participation, and a balanced economy. Its expertise is particularly advanced in the fields of environmental design, urban and regional planning and management, human geography, and conservation of natural resources.

We in California could learn much about ecological and social sustainability from the Croatian coast's rich history of urban evolution. These two regions could establish cultural and educational exchanges—echoing the immigrational linkages of more than a century—to guide the environments at both ends of The Bridge. Participants would include university faculty, advanced students and scholars, community leaders and officials involved in improving the environment, as well as common citizens. Sponsorship could come from a variety of organizations in both regions, such as professional associations, foundations, museums, nongovernmental organizations, research institutes, and government and international agencies. Because there are a large number of Californians of Croatian descent, especially from Dalmatia, some linkages already exist and can be drawn on for existing programs and resources. Several universities offer some of the most advanced programs in urban planning and design in the country. The University of California at Berkeley has attracted numerous students from the former Yugoslavia. Indeed, through the informal collaboration on this work with my colleagues at the University of Zagreb an exchange program was established in 1996 between departments there and at Berkeley. In a sense, Dalmatia is the California of the Balkans, and California might be called the Dalmatia of the continental United States.[10]

Let us be guided by the foresight of the late Senator Fulbright of Arkansas, who said that "international educational exchange is the most significant current project designed to continue the process of humanizing mankind to the point that men can learn to live in peace—eventually even to cooperate in constructive activities, rather than compete in a mindless contest of mutual destruction."[11]

A Vision for the Coming Generation

In light of these open, collaborative thoughts images have formed in my mind of what I would do if I were once again a youth, though in Dalmatia and with my California experience to draw on. First, some key questions would need to be answered toward achieving Dalmatia's maximum environmental and human potential. What steps could be taken to provide a full range of social infrastructure—schools, clinics, and the like—accessible to all urban levels and subregions? How could an effective and diversified economy be developed based on local self-sustainability of both the built and the natural environment? How might the dependence on tourism and its exploitative policies of the past be altered in renewing war-damaged facilities and impacted environmental quality? How can the cultural heritage be preserved in the face of population increases and economic needs and yet be compatible with the everyday life of city dwellers and attractive for visitors? If we are to protect the natural environment and have a more decentralized distribution of population, how large should individual cities, towns, and villages be allowed to grow? What form of growth management can be established for each of the subregions to the mutual benefit of land and people?

A starting point would be to take an inventory of each subregion's needs in the areas of social, economic, physical-environmental, and administrative development with the full participation of the people in each city, town, and village. The self-knowledge gained by the residents of Brač, Hvar, Korčula, or Pelješac could launch a community-oriented style of local planning that is well recognized by all levels of government. Here hierarchy as a key property of identity with place comes into play. The higher administrative levels would be committed to respecting the distinctive environmental identities as determined by the local authorities through the collective self-expression of the citizenry.

In California a hierarchy of authority for planning at governmental levels from municipal to county to state has existed since the late 1920s and has been refined according to changing conditions. Such a framework requires that each city, small or large, have a continuous planning process that is related to that of the surrounding region and to the state. A gradual evolution of planning laws has taken place at all three governmental levels with constant participation and professional leadership from among the younger generation, many of whom are products of university graduate programs. Through an interchange program, this hierarchical model of California could be used to restructure Dalmatia's system for urban and regional decision making and environmental protection especially as it relates to Croatia's goal as a democratic state. From this fresh start, a native style of planning could evolve throughout Croatia guided by the concept of identity with place. Appropriately, underlying this goal is the need for continuity of municipal autonomy, a historical tradition that sets Dalmatia apart.

In the sociocultural area, Dalmatia lacks the full range of social infrastructure for education, health, and community stability. For example, children attending elementary school on Brač, Hvar, Korčula, or Pelješac should not have to leave home and live in Split or Dubrovnik in order to advance beyond six grades. The establishment of consolidated, up-to-date centers accessible by bus that included educational, recreational, health, community, and cultural facilities would reinforce young people's sense of identity with their own subregions and could contribute to their revitalization. Each such center, with a character of its own, could serve as a "growth pole" for advancing tourism based on local cultural and environmental characteristics, which in turn could finance diverse economic activities and services. In the fragmented geography of Dalmatia such centers could restrain current trends toward loss of subregional population to the mainland.

Both there and in California the unique variety of natural environments requires an ecological basis for relating to the built environment. In Dalmatia, identity with places we have come to know has been impacted by modern urbanization patterns and by the deterioration of the natural settings. The self-sustaining habitats of living organisms of sea and land have been severely damaged by water and air pollution and the destruction of wooded lands. The health of the biological life of a defined region is a criterion for determining where and how much land should become urban.

In California the concept "bio-regionalism" has provided a research and exploratory framework for balanced land use and population distribution. Accordingly, urban and regional planning processes take place within a predetermined ecological framework of "geographic areas having common characteristics of soil, watersheds, climate, and native plants and animals that exist within the whole planetary biosphere as unique and intrinsic contributive parts. . . . It includes all interdependent forms and processes of life, along with humans and human consciousness."[12] It is through this "human" component that a lasting identity with place takes root. Natural ecological regions such as San Francisco

Bay, Yellowstone, and the Sierra Nevada are losing their natural qualities through overdevelopment and encroachment. Studies toward maintaining a self-sustainable ecological balance between nature's ecosystems and those imposed by urbanization could also be done for Dalmatia's many geographical islands and peninsulas in relation to the pollution of the waters of the Adriatic. This knowledge may also be used to help coastal California achieve a more equitable and environmentally sound distribution of urban places and their economic bases.

In California leadership by an informed and organized citizenry has gained national recognition for its widespread advancement of environmental protection legislation and methods. Individual cities and counties are required by law to measure the impact of development on the environment and to inform the public of their findings. California's legislation for controlling coastal development over the past several decades could be studied for application to Dalmatia's varied shorelines, where identity plays such an important role.

In terms of the intergenerational property of identity with place, California has a variety of new developments designed to preserve the diverse landscape features and even some local history. Citizen groups have initiated legislation to give landmark status to buildings typical of the past and have supported proposals to separate pedestrians from cars, to install light-rail rapid-transit systems (or streetcars of the past, as in San Francisco), and to promote diverse shops. The old-fashioned main streets serve as precious sources of uniqueness around which to revitalize economic and social vitality and illustrate the work of earlier pioneering generations, my own grandparents' Sutter Creek, for example. Dubrovnik's Stradun and Zadar's Kalelarga could become models for California's main streets.

Dalmatia, because of its long experience with historic restoration and the variety of urban patterns of its cities, towns, and villages, could offer California stimulating examples for urban design on the coast and in its bays, such as those at San Diego, Monterrey, and San Francisco. The pedestrian scale of urban Dalmatia would lend support to the current movement to humanize existing new development in California through pedestrian-oriented "transit villages." The so-called neotraditional urban design models of this decade in the United States aim to reform the automobile-dependent suburbia. Dalmatia, in the wake of the war of the nineties, could well take heed of Thomas Sharp's vision for the preservation of the traditional villages of the English countryside after World War II as vehicles for local identity.[13] The islands and peninsulas of Dalmatia comprise a resource of villages that represent generations of pride and collective effort and that could, with revitalization, offset the cultural and environmental impact of large-scale tourism on local identity.

The dynamic nature of identity with place as common ground for community formation can be readily advanced in Dalmatia, where the inherent desire for self-management prevails in its towns and villages. Local leadership independent of higher authority has initiated movements to protect historical monuments. Such was the case in Zadar after World War II, in Sutivan to keep large hotels from being built, and in Kuna in locating in a public place of honor the statue of its native son Medović.

In light of ethnic differences in Dalmatia's urban places caused by the war of 1991–95 and its enormous displacement of people, the role of sociocultural diversity in identity with place will indeed be difficult to clarify. Inevitably woven into the fabric of larger places, increased demographic mobility is certain to change the homogeneous makeup of communities even without the impact of the war. This we have seen in the influx of new population from the interior into Split's New Skyline district in the absence of overall planning to encourage integration into traditional urban patterns. Community formation through participatory planning and sharing of the public open space and services would help diverse peoples to come together and even—with time and education— dissipate many of the differences revived by war. While it is difficult to envision under the clouds of war, collective caring for the local environment is the only path to follow in rebuilding human meaning and democratic urban life for the next generation.

The concept of insiders versus outsiders offers a particular opportunity for the younger professional urban designer and planner of Dalmatia. Likely to be an "outsider" in smaller towns and villages, the newcomer could become part of the place itself by intentionally working closely with the community as a whole, not just with its leaders. The newcomer should become engaged in the daily life and activities of the residents, debating issues and adopting local customs. In time, he or she could come to share the residents' feelings for the place, their problems and aspirations, and become an "insider, a voice of the place, able to extend that understanding to higher authorities in the hierarchy of identity.

As an "insider," the urban planner aware of the specific qualities of places revealed by my phenomenological reading method could become the vehicle for achieving oneness. In planning for postwar renewal, the ability to see the "whole picture" will strengthen the linkages between the interdependent elements: the urban core, neighborhoods, schools, shops, clinics, recreation areas, and the like. The young planner can thus breathe life into environmental designing and planning, which too frequently is thought to be a purely "technical" or "scientific" process.

Finally, the concept of spirituality has taken on new meaning with the realities of renewal after the war. It now falls to community leaders and those in the professional fields to reawaken people to their heritage of place identity. Faced with the breakdown of the economic footing of tourism for years to come and deterioration of the environment, they can turn to the underlying "soul" lodged in the uniqueness of these places. Again and again, in Zadar, Hvar, Korčula, Dubrovnik, and other places, the life-giving resource of community spirit has been rekindled by the beauty and grace of Dalmatia's built and natural environments. This belief in spirituality of place can become a motivation for leadership free of the narrowly economic-oriented directions taken in the "modernization" of the recent decades.

In 1786 Goethe expressed this thought in his own spiritual response to first experiencing in Italy the great urban works of "the ancients": "I walked up to Spoleto and stood on the aqueduct, which serves as a bridge to the other hill. The ten arches that span the valley have been quietly standing there through all the centuries, and the water still gushes in all quarters of Spoleto. . . . This is the third work of antiquity I have seen and it embodies the same noble spirit. A sense of the civic good, the basis of their architecture, was second nature to the ancients . . . anything that does not have a true *raison d'être* is lifeless and can never become great."[14]

Dalmatia, a passive intermediary between the Slavic and Western European cultures, has traditionally served as a bridge. Today, the new wave of urbanization, laden with materialistic values sweeping the world, puts both Dalmatia and California in a vulnerable yet potentially constructive position for the decades of peace ahead. Considering the threat of continued disorderly urbanization under a free-market system and fueled by large-scale tourism, Dalmatia needs to lead in securing for the environment the essence of its collective identity with place.

On the one hand, playing up the romantic images of its cities, towns, and villages, in their unique coastal settings, for economic advantage alone could make superficial urban Dalmatia's inspiring cultural values now spotlighted by the war. The chances for a genuine internal cultural evolution in the twenty-first century and within Croatia's new democracy would be lost. More ecological damage could come to the already overtaxed environment. With a return to the annual invasions of heedless worshipers of the sun and sea, Dalmatia's cultural heritage and identity with place could become fragmented and be relegated to its museums and monuments.

On the other hand, urban Dalmatia, situated between the Balkans and the Mediterranean, and with strong ties to Europe and the New World, can play an enriching role. The region has already shown itself to be a leader in the Mediterranean in the area of protecting the natural and built environments. As the unique and central part of the Croatian-Bosnian Confederation and potentially the European community, Dalmatia could become a model for environmental conservation and local identity with place and cultural heritage in balance with a diversity in decentralized economic enterprises. Dalmatia might lead other exceptional coastal regions in the world by setting viable environmental standards for its cities, towns, and villages befitting its own landscape, sea, and mountains. In this light, the metaphor of The Bridge becomes applicable to Dalmatia beyond my original intent based on family ties.

As exemplified in our experiential readings, the life-giving continuity that produced the human appeal of Dalmatia's urban places and its vitality has remained intact into this century. We have also witnessed a severe rupture with the productive sequence of the past. In numerous cases, new ways of building lack those qualities that could foster a continuity between traditional and new urban forms. The opportunity to give life to the renewal of Dalmatian urban places lies in a renewed concept of environmental planning and design, one committed to linking people to places. Coming generations can then work in a spirit of love of place, based on qualities of form, scale, diversity, and collective social purpose that identify the uniqueness of places, and again make peace with the environment.

The Arcadian peace, the mythological beauty of Dalmatia will not be lost; neither will its sons be lost in the fast rhythm of present-day life, nor will those be lost whose calling is to see that this cradle of south Slavic medieval and Renaissance culture, art, and letters achieves a harmony with progress that no nation can bypass (Cvito Fisković, 1962).[15]

Notes

INTRODUCTION   Crossing the Bridge to California

Epigraph: Louis Adamic, *My America* (New York: Harper & Bros., 1938), xi.

1. E. Shapiro, *The Croatian Americans,* ed. Daniel P. Moynihan (New York: Chelsea House, 1989).

2. Adamic, *My America;* idem, *The Native's Return: An American Immigrant Visits Yugoslavia and Discovers His Old Country* (New York: Harper & Bros., 1934).

3. See, for example, Louis Adamic, *From Many Lands* (New York: Harper Bros., 1939).

4. Lewis Mumford, *The Culture of Cities* (New York: Harcourt Brace, 1938).

5. Patrick Geddes, *Cities in Evolution* (London, 1915). For a comprehensive listing of Geddes' works, see Mumford, *Culture of Cities,* 522–23.

6. Stjepan Vekarić showed me a record from the Dubrovnik archive in which the name Violić appeared for the first time in the late 1500s. The original Bosnian name, Cvjetović, had been Italianized by changing the Croatian root, meaning "flower," to the Italian name of a specific flower, the *viola.*

7. See Jozo Tomasevich, *Peasants, Politics, and Economic Change in Yugoslavia* (Stanford: Stanford University Press, 1955); and idem, *War and Revolution in Yugoslavia, 1941–1945: The Chetnicks* (Stanford: Stanford University Press, 1955).

8. Jozo Tomasevich, "The Tomasevich Extended Family on the Peninsula of Pelješac," in *Communal Families in the Balkans: The Zadruga,* ed. Robert F. Byrnes (South Bend, Ind.: University of Notre Dame Press, 1976).

9. See Vera Kružić Uchytil, *Mato Celestin Medović* (Zagreb: Grafički zavod Hrvatske, 1978).

10. Francis Violich, "Historic Preservation and Cultural Exchange," "Three Houses: Three Grandmothers," and "The House in Sutter Creek," in *Zajedničar* (Pittsburgh: Croatian Fraternal Union of America, August–September 1981).

11. John V. Tadich, "The Yugoslav Colony of San Francisco on My Arrival in 1871," in *The Slavonic Pioneers of California,* ed. Vjekoslav Meler (San Francisco: privately printed, 1932), 40–42.

12. Robert Louis Stevenson, "R.L.S. on the Immigrant Train," excerpt from Robert Louis Stevenson, *Across the Plains* (1892), in *Railroads in America,* by Oliver Jensen (New York: American Heritage, 1975), 130–31.

13. Rev. H. B. Williams, "Sutter Creek," in *Amador County History* (Jackson, Calif.: Amador County Federation of Women's Clubs, 1927).

14. J. H. Cusanovich, "Memories of a Sutter Creek Boy," in *Amador Ledger* (Jackson), 1 February 1937.

15. Murray Morgan, *Skid Road: An Informal Portrait of Seattle* (Seattle: University of Washington Press, 1991), 107–15.

16. See Theodora Kroeber, *Ishi . . . in Two Worlds: A Biography of the Last Wild Indian in North America* (Berkeley: University of California Press, 1962).

17. The brotherhood tradition has its origin in the cooperative nature of Croatian village life, as a way of sharing responsibilities for well-being among men. The practice, transplanted to California, served well the needs of the newcomers by providing guidance in financing health care, life insurance, and the like. Few other ethnic groups had such an institution.

18. Curiously in the light of the current focus on specific ethnic identities in the context of the former Yugoslavia, the Dalmatians in California called themselves Slavs to distinguish them from the Anglo-Saxon, U.S.-born Americans. Even more inaccurate, used as an adjective the term became Slavonian, which meant a native of Slavonia, that region on the far eastern side of Croatia. For generic purposes, we were "Dalmatians," and we were further identified by our place of origin — Bračani, from Brač, or Konavljani, from Konavle.

19. See Francis Violich, "Cousin from America," in *Slavia* 14, nos. 6–8 (1939).

20. Andrew Trlin, *Now Respected, Once Despised: Yugoslavs in New Zealand* (Palmerston North, New Zealand: Dunsmore, 1979), 3.

21. Robert Lynd and Helen Merrell, *Middletown: A Study in Contemporary American Culture* (New York: Harcourt Brace, 1929).

22. See Francis Violich, *Cities of Latin America — Planning and Housing to the South* (New York: Reinhold, 1944); and idem, *Urban Planning for Latin America — The Challenge of Metropolitan Growth* (Cambridge, Mass.: Lincoln Institute of Land Planning, 1987).

23. See Francis Violich, "Evolution of the Spanish City: Issues Basic to Planning Today," *Journal of the American Institute of Planners,* August 1962.

24. Ivana Šverko, "Ranjene Školjke" (Wounded seashells), *Nedjeljna Dalmacija* (Split), 24 October 1991.

CHAPTER ONE   Identity: Key to the Meaning of Place

Epigraph: Christian Norberg-Schulz, *Genius Loci: Toward a Phenomenology of Architecture* (New York: Rizzoli, 1980).

1. W. I. Thompson, from his sermon at the Cathedral of St. John the Divine in New York City, 1 November 1981, on the occasion of the Missa Gaia (Earth Mass), by Paul Winter, dedicated to St. Francis on his eight-hundredth birthday.

2. Leonard Duhl developed this definition of Healthy Cities in the 1980s while establishing a comprehensive system of social improvement for cities under the World Health Organization.

3. The phrase "experiencing places" grew out of my observing the participatory and experiential nature of the evolution of urban form in the Dalmatian towns and villages between 1979 and 1983. I developed this theme in papers presented at symposia of the Environmental Design Research Association (EDRA)

in 1984 and 1985. The second, presented in New York, was entitled "Experiencing Places: The Aesthetics of the Participatory Environment"; I draw on both in this chapter. The reader is also referred to the refreshingly engaging example of Tony Hiss's *Experience of Place* (New York: Knopf, 1990).

4. Jack Lessinger, *Penturbia: Where Real Estate Will Boom after the Crash of Suburbia* (Seattle: SocioEconomics, 1991).

5. Francis Violich, *Urban Planning for Latin America—The Challenge of Metropolitan Growth* (Cambridge, Mass.: Lincoln Institute of Land Planning, 1987).

6. Martin Heidegger, "Building, Dwelling, Thinking," in *Poetry, Language, Thought* (New York: Harper & Row, 1971), 152–53.

7. Norberg-Shulz has shown how the natural environment has been a major source of the differences between Rome, Prague, and Khartoum, reading into this evidence sociocultural meaning that may no longer be supported in our times. His concern is mainly with cities in the historic sense and the form they gradually took before the advent of fast-moving modern technology and its ability to overcome limits of both nature and cultures (Norberg-Schulz, *Genius Loci*).

8. See Francis Violich, "The Search for Cultural Identity through Urban Design: The Case of Berkeley," *Berkeley Planning Journal* 2, nos. 1–2 (1985): 106–26.

9. See Judith Innes, "Knowledge and Action: Making the Link," *Journal of Planning Education and Research* 6, no. 2 (winter 1987); and idem, "Planning Theory's Emerging Paradigm: Communicative Action and Interactive Practice," ibid., 14, no. 3 (1995). These articles discuss in full this shift from planning theory based on models of how planners should make decisions to planning theory based on how planning actually functions in the real world.

10. Paola Coppola Pignatelli, *L'identita come processo: Cultura spaziale e progetto di architettura* (Identity as process: cultural space and architectural [urban] design) (Rome: Officina Edizioni, 1992), 172–77. In this work, the most lucid to date on sources of urban identity, Pignatelli uses this "triangolo" as the structure for the entire, thoughtful study. Each of the three elements—land, building, and people—is meticulously examined and exemplified in concise language and visual illustrations to demonstrate their roles and ecological interdependence in creating coherence and uniqueness of place.

11. William James, *The Essential Writings,* ed. Bruce W. Wilshire (Albany: State University of New York Press, 1984), 206.

12. Catherine Howett, "In Search of the Sexual Landscape: Reflections on Freudian Theory" (paper presented at EDRA 15, California Polytechnic State University, San Luis Obispo, June 1984). See also Gary Nabhan, *The Geography of Childhood* (Boston: Beacon, 1994).

13. Bernd Jager, "Body, House, and City: The Intertwinings of Embodiment, Inhabitation, and Civilization," in *Dwelling, Place, and Environment,* ed. David Seamon and Rugert Mugerauer (New York: Columbia University Press, 1989). A remarkable and intimate view of place relationships is Dennis Wood and Robert Beck, *Home Rules* (Baltimore: Johns Hopkins University Press, 1994).

14. See Martin Heidegger, "Origins of a Work of Art" and "The Thing," in Heidegger, *Poetry, Language, Thought,* 22–23 and 165–86.

15. Paola Coppola Pignatelli et al., *Roma: Esperienze di lettura urbana* (Rome: Experiences in urban reading) (Rome: New University Press, 1973).

16. Kevin Lynch, *The Image of the City,* 10th ed. (Cambridge: MIT Press, 1972).

17. See Anne Buttimer, *Geography and the Human Spirit* (Baltimore: Johns Hopkins University Press, 1993). This work speaks with fervor about the need today for society's spiritual response to the particular qualities of our everyday environment.

18. Appleyard's outline for *Identity, Power, and Place,* of some twenty chapters, opens with a theoretical and generic analysis of the elements involved in the phenomenon of identity with place. The position of people in relation to the many varieties of places is examined, and the kinds of dynamics and actions that affect identity are enumerated. Looking at the scale of the home, the neighborhood, and the community, Appleyard itemizes every conceivable attribute of identity that needs to be understood. At the city scale, he deals with symbolism expressed in the buildings, streets, and open spaces that make up our urban centers and relates these to each of the forces involved in establishing or changing urban characteristics. He concludes by discussing the complexities of decision making in today's large metropolitan areas in the light of the basic points made. The manuscript remains in the care of the Library of the College of Environmental Design at the University of California, Berkeley.

19. For a further valuable elaboration of this quotation, including extensive references, see David Seamon, "Phenomenology and Environment-Behavior Research," in *Advances in Environment-Behavior and Design,* ed. G. T. Moore and E. Zupe (New York: Plenum, 1987), 3–26, quotation on p. 4.

20. David Seamon, ed., *Dwelling, Seeing, and Designing: Toward a Phenomenological Ecology* (Albany: State University of New York Press, 1993), 16.

21. John Eyles and David M. Smith, *Qualitative Methods in Human Geography* (Cambridge: Polity, 1988), xii. The following chapters are especially relevant: Eyles, "Interpreting the Geographical World—Qualitative Approaches in Geographical Research"; Smith, "Constructing Local Knowledge—The Analysis of Self in Everyday Life"; Mel Evans, "Participant Observation—The Researcher as Research Tool"; and John Pickles, "From Fact-World to Life-World —The Phenomenological Method and Social Science." Equally important are the works of Nicholas Entrikin, especially his now classic book *The Betweenness of Place: Towards a Geography of Modernity* (Baltimore: Johns Hopkins University Press, 1991).

22. Professor Craik is an environmental psychologist at the University of California at Berkeley who worked closely with Donald Appleyard and reviewed early drafts of this book.

23. A brilliant analysis of the evolution of neighborhood identity is Alexander von Hoffman, *Local Attachments: The Making of an American Urban Neighborhood, 1850 to 1920* (Baltimore: Johns Hopkins University Press, 1994).

24. Appleyard, *Identity, Power, and Place,* chap. 5, p. 1.

25. Heidegger, *Poetry, Language, Thought,* 149–50.

26. Irwin Altman and Setha M. Low, eds., *Place Attachment—A Conceptual Inquiry* (New York: Plenum, 1992).

27. Setha M. Low, "Symbolic Ties That Bind—Place Attachment in the Plaza," in ibid., 165.

28. Pignatelli et al., *Roma: Esperienze di lettura urbana*, 9.

29. This sharpening of my ability to experience places deeply was further stimulated by Hillaire Belloc's *The Path to Rome: A Portrait of Western Europe before the World Wars* (Washington, D.C.: Regnery Bateway, 1987). This youthful work of "guided phenomenology" was written in 1901 as the author walked a diagonal "path" from eastern France to Rome. This steadfast commitment to a spiritually conceived route generated a very direct "reading" of landscapes and settlements experienced intimately in their pristine, prewar condition.

30. The material in this section is taken from Francis Violich, "Identity and Giove: Hill Towns Are Alive and Well in Umbria," *Traditional Dwellings and Settlements Review* 1 (1989): 65–82, the first issue by the International Association for the Study of Traditional Environments, University of California at Berkeley.

31. Since writing this, I have experienced the phenomenon of the "wildfire" that swept these hills in Berkeley and Oakland in 1991, destroying some 2,850 homes on 1,600 acres. In a matter of hours the identity of the numerous neighborhoods, each of individual character, was wiped clean and left with only the curving contours and the sweeping view to the Bay. For those dwellers who had spent decades building their lives into a place of distinctive identity, untold memories and sources of belonging were gone. Even after years of rebuilding, the starkness of devastation remains as a new identity takes over, replete with the marks of urgency, compulsion, and insensitivity to some overall coherence.

CHAPTER TWO   The Making of Dalmatia as a Regional Place

Epigraph: Cvito Fisković, "Dalmatia," *The Atlantic Monthly—Yugoslavia, A Special Supplement,* 210, no. 6 (December 1962).

1. For the sake of simplicity and according to customary practice, I have generally used the term *Croatian* to identify the people of Croatia, recognizing that the term *Croat* is more accurate. For a more precise usage, see Ivo Banac, *The National Question in Yugoslavia—Origins, History, Politics,* rev. ed. (Ithaca, N.Y.: Cornell University Press, 1984, 1993).

2. Henri LeFebre, *Il diritto alla città,* quoted in Paola Coppola Pignatelli et al., *Roma: Esperienze di lettura urbana* (Rome: Experiences in urban reading) (Rome: New University Press, 1973), 9.

3. See Francis Violich, *Cities of Latin America—Planning and Housing to the South* (New York: Reinhold, 1944); and idem, *Urban Planning for Latin America—The Challenge of Metropolitan Growth* (Cambridge, Mass.: Lincoln Institute of Land Planning, 1987).

4. See Kevin Lynch, *What Time Is This Place?* (Cambridge: MIT Press, 1972).

5. Bruno Milić's two volumes comprise a comprehensive interplay of text and maps, urban plans, old prints, diagrams, and drawings. This rich array offers as thorough a coverage of the shaping of urban forms in relation to cultures worldwide as can be found in any language (see Bruno Milić, *Razvoj Grada Kroz Stoljeća* [Urban development through the centuries], vols. 1, *Prapovijest–Antika* [Prehistoric–Ancient], and 2, *Srednji Vijek* [Middle Ages] [Zagreb: Školska Knjiga, 1988–95]).

6. See Lynch, *What Time Is This Place?*

7. Jože Kaštelić, "Illyrian Spring," *The Atlantic—Yugoslavia: A Special Supplement* 210, no. 6 (December 1962).

8. J. J. Wilkes, *Dalmatia* (Cambridge: Harvard University Press, 1969). This classic work, in a series on the history of the provinces of the Roman Empire, is a primary source for this book.

9. For the most authoritative clarification of the origin, history, and politics of the South Slavs from the earliest times to the 1920s, see Banac, *The National Question in Yugoslavia.*

10. For a concise history of the Croatian people in English, see E. Shapiro, "History of the Homeland," in *The Croatian Americans,* ed. Daniel P. Moynihan (New York: Chelsea House, 1989), 19–35.

11. Harriet Bjelovučić, *The Ragusan Republic, Victim of Napoleon and Its Own Conservatism* (Leiden: Brill, 1970), 119–21.

12. For fuller coverage of the material contained in this section and its sources, see Francis Violich, "An Urban Development Policy for Dalmatia," pt. 1, "The Urban Heritage to the Time of Napoleon," *Town Planning Review* 43, no. 2 (1972): 151–66.

13. See Andre Mohorovičić, *Architecture in Croatia: Architecture and Town Planning,* trans. Nedeljka Batinović (Zagreb: Školska Knjiga, 1994). This revered scholar of the Croatian Academy of Arts and Sciences in 1968 was one of the first to lend enthusiastic support to my search for my Dalmatian identity with place.

14. Exhaustive inventories of most of these urban complexes worthy of preservation in their entirety have been prepared by the Art History Institute of the University of Zagreb. Many of these are identified fully in the inventory, and their graphic documentation reveals a wealth of urban history; some of this documentation has been my source of illustrations for this work, thanks to Nenad Lipovac, of the university's Faculty of Architecture and Urban Planning.

15. See George Wheler, *Voyage de Dalmatia, de Grèce et du Levant* (Amsterdam: Jean Wolters, 1689); and Robert Adams, FRS, FSA, Untitled measured drawings of Diocletian's Palace in Split (privately printed, 1764).

16. For a full and authoritative understanding of Marmont in Dubrovnik and the French in Dalmatia in general, see, as suggested to me by Bariša Krekić, Lujo Vojnović, *Pad Dubrovnika* (The fall of Dubrovnik), vols. 1 (1797–1806) and 2 (1807–15) (Zagreb: Dionička tiskara, 1908) ("Old, but still the best"); Bjelovučić, *Ragusan Republic;* and *Maršal Marmont Memoari,* ed. Frano Baras (Split: Logos, 1984).

17. For a personalized view of Dalmatia in the 1930s and the special role of

Marshal Marmont, see Rebecca West, *Black Lamb and Grey Falcon: A Journey through Yugoslavia* (Harmondsworth: Penguin, 1982), 115–248.

18. The record of this program of modernization, long kept inaccessible even under Tito's administration, was opened to the public with Croatia's independence. A major exhibit was held in Split, and a handsome catalogue published (see Nataša Bajić, ed., *Hrvatska Obalne Utvrde u 19 i 20 Stoljeću* [Croatian seacoast fortification in the 19th and 20th centuries] [Split: History Archive, 1995]).

19. Carl E. Schorske, "The Ringstrasse and the Birth of Urban Modernism," in *Fin de Siècle Vienna* (New York: Alfred Knopf, 1980), 24–115.

20. Through colleagues, I was allowed to examine these remarkable maps in 1981, but with the fall of communism by 1990 they had become available for me to copy. In 1992 these maps were featured in an exhibit in Split and in its large, handsome catalog, published with English and Italian summaries, that describes their history. About sixty of the plans of better-known places were reproduced in the original colors (see Nataša Bajić et al., eds., *Blago Hrvatske iz arhiva za Istru i Dalmaciju* [The treasures of Croatia in archival maps for Istria and Dalmatia] [Split: Slobodna Dalmacija, 1992]).

21. Banac, *The National Question in Yugoslavia,* esp. the chapter titled "National Ideologies," 70–115.

22. Material in this section is based on the comprehensive work of Vinko Foretić, one of the rare studies of the critical nineteenth century, "Ekonomske i društvene prilike u Dalmaciji 1860–1914" (Economic and social conditions in Dalmatia, 1860–1914), *Zbornik o Hrvatskom Narodnom preporodu u Dalmaciji i Istri* (Zagreb: Economski Fakultet, 1969). See also Francis Violich, "An Urban Development Policy for Dalmatia," pt. 2, "Urban Dalmatia in the 19th Century and Prospects for the Future," *Town Planning Review* 43, no. 3 (1972): 243–53.

23. *Album Fotografico-Illustrativo del Viaggo de S.M. Imperatore/Giuseppe I in Dalmazzia* (Zadar, 1875); "The Visit of Emperor Franz Josef to Dalmatia in 1875," *Smotra Dalmatinska* (Dalmatian Review) (Zadar), 15 September 1906.

24. For the works of the writers referred to in this section, see the Bibliography of Works by Foreign Travelers in Dalmatia.

25. A highly relevant book on the effects of tourism on local cultures and economies is David Zurick's *Errant Journeys: Adventure Travel in a Modern Age* (Austin: University of Texas Press, 1995).

26. This type of specific criticism of political and economic policies dictated by Yugoslav political leaders was very much in the air among people with whom I spoke in my 1990 visit, just as Croatian parliamentary elections were being held. The housing question came from one of the candidates.

27. See Organization for Economic Cooperation and Development, *Environmental Policies in Yugoslavia* (Paris, 1986), chap. 2, in which the main chapter headings include "Environmental Policy: The Context," "The Environment of Towns and Cities," "Agriculture and Soils," "Forestry," "Air," "Energy," "Water," "Chemicals," "Waste," "Conservation of Nature and Monuments," "Tourism," and "International Technical Cooperation." The then director of the Environmental Committee, Gabriele Schimemi, had worked with other faculty mem-

bers of my department in connection with mutual interests between the University of Rome and California. I learned about this when I happened to be with him in Greece in 1981; at that time the Yugoslav study was being carried out along with that of Greece.

CHAPTER THREE   Cities of the Mainland

Epigraph: From Tomislav Marasović et al., "Zaštita, asanacija, i rekonstrukcija urbanističkog naslijedja u Dalmaciji" (Preservation and reconstruction of the urban heritage in Dalmatia), in *Urbs* (Split: Urbanistički Biro, 1958), 7.

1. Ivo Andrić, *The Bridge on the Drina* (New York: New American Library, 1967); Ismail Kadare, *Chronicle in Stone* (London: Serpent's Tail, 1987); Ann Bridge, *Illyrian Spring* (1935; reprint, London: Virago, 1990), the only novel in English that is set in Dalmatia; Ivo Vojnović, *Dubrovačka trilogija,* in *Drame Ive Vojnovića* (Dramas by Ivo Vojnović), ed. V. Pavletić (Zagreb: Školska knjiga, 1962); Italo Calvino, *Invisible Cities* (New York: Harcourt Brace Jovanovich, 1974).

2. Rebecca West, *Black Lamb and Grey Falcon: A Journey through Yugoslavia* (Harmondsworth: Penguin, 1982), 138–69 (Split) and 230–48 (Dubrovnik).

3. J. J. Wilkes, *Dalmatia* (Cambridge: Harvard University Press, 1969), 368. Extensive research by Mate Suić has produced a volume comparing in detail Roman Zadar with other works of that period, from Pula in the north to Bar in the south. His reconstruction of Zadar's grid system of streets and related land uses documents the enduring qualities of Roman urban planning up to our times (see Mate Suić, *Antički grad na istočnom Jadranu* (Ancient cities of the eastern Adriatic) [Zagreb: Sveučilišna naklada Liber, 1976]).

4. Nada Klaić and Ivo Petricioli, *Prošlost Zadra II: Zadar u Srednjem vijeku do 1409* (History of Zadar II: Zadar in the Middle Ages to 1409) (Zadar: Filozofski Fakultet, 1976), 8.

5. Thomas G. Jackson, *Dalmatia, the Quarnero, and Istria* (Oxford: Clarendon, 1887).

6. A. A. Paton, *Researches on the Danube and the Adriatic,* 2 vols. (Leipzig: Brockhaus, 1861).

7. Horatio F. Brown, *Dalmatia* (London: A. and C. Black, 1925).

8. The word *trg,* generally used for a public gathering place, may alternate with its Italian form *piazza,* depending on local usage.

9. Wilkes, *Dalmatia,* 387–91.

10. Grga Novak, *Povijest Splita* (History of Split), *Vol. II, 1420–1797* (Split: Matica Hrvatska, 1961).

11. Ibid., 158–59.

12. The incremental evolution of the palace into a strategic urban frontage of the natural harbor has been portrayed in old panoramic engravings and drawings made from the sixteenth to the early twentieth century. These include four from the sixteenth century, nine from the seventeenth, seven from the eighteenth, twelve from the nineteenth, and one bird's eye view of Split in 1926. See Vladimir Buljević, *Tragovima Staroga Splita* (The evolution of Old Split) (Split: Zavod za Zaštitu Spomenika Kulture, 1982).

13. Novak, *Povijest Splita,* 67.

14. Paton, *Researches on the Danube and the Adriatic.*

15. This perceptive point was made by Andrej Škarica, my colleague in the planning of Split's expansion.

16. See Vinko Foretić, *Povijest Dubrovnika do 1808* (History of Dubrovnik up to 1808), vols. 1 (to 1526) and 2 (to 1808) (Zagreb: Nakladni zavod Matice Hrvatske, 1980).

17. Bariša Krekić, *Dubrovnik in the Fourteenth and Fifteenth Centuries: A City between East and West* (Norman: University of Oklahoma Press, 1972), 5.

18. Ibid., 6–22.

19. Francis Carter, *Dubrovnik (Ragusa)* (London: Seminar, 1972).

20. Milan Prelog, "Dubrovački Statuti i izgradnja grada" (The Dubrovnik statutes and city building), *Peristil* (Zagreb), 1972, 14–15.

21. Henry Ellis, ed., *John Streater* (1511; reprint, London: Camden Society, 1851), 11–12. This significant reference came to me from Janja Ciglar-Žanić, who found it while carrying out research on Dubrovnik as a social and political model in seventeenth-century English thought on government. Another important reference is *Dubrovnik as a Model State,* by Jean Bodin (Paris: Republique, 1576), translated into modern English as *The Six Books of a Commonweale* (London, 1606).

22. Krekić, *Dubrovnik in the Fourteenth and Fifteenth Centuries,* 174.

23. Bruno Šišić, *Dubrovački Renesansni vrt i nastajanje oblikovna obilježja* (Dubrovnik Renaissance gardens, their evolution, design, and character) (Dubrovnik: Zavod za povijesne znanosti HAZU, 1991).

24. Paton, *Researches on the Danube and the Adriatic.*

25. This phrase is taken from Donald Appleyard's unpublished manuscript, described in chap. 1, n. 18.

26. Stijepo Mijović Kočan, *Konavle* (Dubrovnik: Society for Scientific and Cultural Activities, 1984).

27. Ellis, *John Streater,* 483.

28. Despite being shelled in 1991, the old palace still houses securely the complete archives of this remarkable city's municipal and commercial records.

CHAPTER FOUR   Three Seaside Villages of the Islands

Epigraph: Antun Cvitanić, *Srednjovjekovni statut Bračke komune iz godine 1305* (Statute of the Brač commune in the Middle Ages from 1305), Brački Zbornik, no. 7 (Supetar: Council for Education and Culture on Brač, 1968).

1. See Laurence Wylie, *Village in the Vaucluse* (Cambridge: Harvard University Press, 1974).

2. Thomas Sharp, *The Anatomy of the Village* (Harmondsworth: Penguin, 1946).

3. My direct contact with a number of these writers, in addition to my own identity with Brač, led me to probe further into the remarkable series Brački Zbornici (Brač Journals), which began to appear about 1950. Writers in other subregions—Hvar, Korčula, and Pelješac—followed suit later on, but with less coverage devoted to local history.

4. Dujam Hranković, "Opis otoka Brača" (A description of the island of Brač), in *Dujam Hranković i njegov "Opis otoka Brača" iz godine 1405* (Dujam Hranković and his portrayal of the island of Brač in the year 1405), ed. Andre Jutronić et al., Brački Zbornik, no. 2 (Split: Council for Education and Culture on Brač, 1954).

5. Andrija Cicarelli, *Osservazioni sull' Isola della Brazza* (Observations on the Island of Brač), 1405, in ibid.

6. Andrija Cicarelli, *Historični Prikaz Pučišća* (A historical review of Pučišća), trans. Hrvatski Skup (Pučišća, 1921).

7. Andre Jutronić, *Naselja i porijeklo stanovništva na otoku Braču* (Settlements and origin of population on the island of Brač), Zbornik za narodni žživot i običaje, 34 (Zagreb: Jugoslavenska Akademija, 1950).

8. Jutronić et al., *Dujam Hranković i njegov "Opis otoka Brača" iz godine 1405.*

9. Andre Jutronić, ed., with Dasen Vrsalović, Davor Domančić, and Kruno Prijatelji, *Kulturni Spomenici otoka Brača*, Brački Zbornik, no. 4 (Supetar: Council for Education and Culture on Brač, 1960). This work contains an extensive English summary.

10. Dasen Vrsalović, *Povijest otoka Brača* (History of the island of Brač), Brački Zbornik, no. 6 (Supetar: Council for Education and Culture on Brač, 1968).

11. Klement Derado and Ivan Čižmić, *Iseljenici otoka Brača* (Immigration from the island of Brač), Brački Zbornik, no. 13 (Zagreb: SIZ za kulturu Općine Brač, 1982).

12. Brian Bennett, "Sutivan: A Dalmatian Village in Social and Economic Transition" (Ph.D. diss., Southern Illinois University, 1971).

13. See Paola Coppola Pignatelli et al., *Roma: Esperienze di lettura urbana* (Rome: New University Press, 1973), 9.

14. For a more complete version of my Sutivan study, see Francis Violich, "Urban Reading and the Design of Small Urban Places: The Village of Sutivan," *Town Planning Review* 54, no. 1 (1983): 41–62.

15. Kevin Lynch first introduced the concept of nodes and revealed potentials for a human dimension in urban design. See his book *The Image of the City*, 10th ed. (Cambridge: MIT Press, 1972), 72–103.

16. Don Ivica Eterović, *Uspomena: Četiristo Godišnjica Osnutka Župe Pučišća 1566–1966* (Remembrance: The four hundredth anniversary of the founding of the parish of Pučišća) (Pučišća: Župski ured, 1966).

CHAPTER FIVE   Two Towns That Dominate Their Islands

Epigraph: Grga Gamulin, *Arhitektura u regiji* (Regional architecture) (Zagreb: Društvo historičara umjetnosti Hrvatske, 1967), 239.

1. Grga Novak, *Hvar kroz stoljeća* (Hvar through the centuries) (Zagreb: Izdavački zavod JAZU, 1962), 36, 43. The material in the following pages draws heavily on Novak's environmentally sensitive interpretation of Hvar's history; see esp. 159–65.

2. This social information was substantiated in field studies I participated in for a week in 1979. The field studies were carried out by students of historic preservation from various municipalities in the former Yugoslavia directed by Tomislav Marasović, of the University of Split, and faculty members of other universities. I became instinctively drawn to the power of intensely observing the wondrous environmental accumulation of centuries, which directed me to intuitive dialogue with urban form and from it to reading human meaning into Hvar.

3. Martin Heidegger, "Origins of a Work of Art," in *Poetry, Language, Thought* (New York: Harper & Row, 1971), 41–48.

4. Novak, *Hvar kroz stoljeća*, 73–93.

5. Vinko, Foretić, "Presjeci kroz prošlost Korčula—Povodom korčulanskih obljetnica" (Excerpts from the history of Korčula—on the occasion of Korčula's anniversary), in *Zbornik otoka Korčule*, ed. Marinko Gjivoje (Zagreb: Vlastita Naklada, 1972), 30–65.

6. Ivo Kaštropil, "Nekoliko Toponima iz zapadnog djela otoka Korčule" (Place names from the west side of the island of Korčula), in ibid., 255–57.

7. Sir J. Gardiner Wilkinson, *Dalmatia and Montenegro* (London: John Murray, 1848; reprint, New York: Arno Press and New York Times, 1971), 251.

8. Novak, *Hvar kroz stoljeća*, 162.

9. At this point the reader should understand that my reading of Korčula took place in two stages, first in the early 1980s then again in 1990. This and the following paragraph demonstrate how radical change can come about when preservation policies are allowed to lapse under the economic pressure of tourism promotion. A similar example, on a far greater scale, will interrupt our reading of the eastern shoreline.

10. Ivana Šverko, an architect and my liaison in Split, researched this bit of history to verify the origins and purposes of this monument.

11. Wilkinson, *Dalmatia and Montenegro*.

12. The trees planted early in this century below the road have completely obliterated the idyllic outlook. With careful removal of limbs to open windows and frame the unique view, the terrace might again serve its original social purposes as well as memorialize its joint Croatian-British origins in these trying years of new aggressions.

CHAPTER SIX   Mountain Villages Linked to the Sea

Epigraph: Josip Splivalo, *Kruh sa 7 kora* (Rijeka: Novi List, 1966), 207.

1. Igor Fisković, *Kulturno umjetnička prošlost Pelješkog kanala* (Cultural and art history in the Pelješac canal) (Split: Mogućnosti, 1972), 9; idem, "Pelješac u protopovijesti i antici" (Pelješac in prehistoric and ancient times), in *Pelješki zbornik* (Dubrovnik: Vinarija Dingač SIZ za kulturu, 1976), 9.

2. Vinko Foretić, "Kada je i kako Stonski Rat došao pod vlast Dubrovnika" (When and how the Ston Peninsula came under Dubrovnik), in *Pelješki zbornik* (Dubrovnik: Vinarija Dingač SIZ za kulturu, 1976).

3. Zdravko Šundrica, "Stonski Rat u XIV stoljeću (1333–1399)" (The Ston

Peninsula in the 14th century), in *Pelješki zbornik* (Dubrovnik: Vinarija Dingač SIZ za kulturu, 1980).

4. Nenad Vekarić, *Pelješka naselja u 14 stoljeću* (Pelješac settlements in the fourteenth century) (Dubrovnik: Zavod za povijesne znanosti JAZU, 1989), 86.

5. Ibid., 93.

6. Vekarić, *Pelješki jedrenjaci* (Split: Mornarički glasnik, 1960).

7. Cvito Fisković, "Putovanje pelješkog jedrenjaka iz kraja XVIII i početka XIX stoljeća" (Voyages of a Pelješac sailing vessel at the end of the eighteenth and beginning of the nineteenth century), *Pomorski zbornik* (Zagreb, 1962).

8. Bariša Krekić, communication to the author, 1994.

9. Stjepan Vekarić, communication to the author, 1970.

10. Sir J. Gardiner Wilkinson, *Dalmatia and Montenegro* (London: John Murray, 1848; reprint, New York: Arno Press and New York Times, 1971).

11. Marija Planić Lončarić, *Planirana izgradnja na području Dubrovačke republike* (Urban planning in the region of the Dubrovnik Republic) (Zagreb: Sveučilišna naklada Liber, 1980).

12. Spiro Kostof, "Junctions of Town and Country," *Buildings, Settlements, and Tradition: Cross-Cultural Perspectives* (journal of the International Association for the Study of Traditional Environments), 1989, 107–34.

13. Nedjeljko Subotić, *Crkva Gospe Delorite i Franjevački Samostan na Kuni 1681–1981* (The Franciscan Monastery in Kuna, 1681–1981) (Kuna: privately printed, 1981).

14. Vera Kružić Uchytil, *Mato Celestin Medović* (Zagreb: Grafički zavod Hrvatske, 1978), 93.

15. Vekarić, *Pelješka naselja u 14 stoljeću*, 87.

16. Early in my search for meaning in environments, the work that most stirred within me a consciousness of the nonphysical, spiritual qualities that Podobuče evoked was René Dubos, *A God Within: A Positive Philosophy for a More Complete Fulfillment of Human Potentials* (New York: Scribner's and Sons, 1972); see esp. the chaps. 4, "Individuality, Personality, and Collectivity"; 5, "Of Places, Persons, and Nations"; and 6, "The Persistence of Place."

CHAPTER SEVEN   The Environment:
Common Ground for Community Identity

Epigraph: Ivo Andrić, *The Bridge on the Drina* (New York: New American Library, 1967), 333.

1. Ministry of Environmental Protection and Spatial Planning, *Proceedings of the International Conference on the Effects of War Activities on the Environment* (Zagreb, 1993).

2. Spiro Kostof, *The City Shaped: Urban Patterns and Meaning through History* (Boston: Little, Brown, 1991), 4. No better source of documentation of his point has come forth than this expansive interpretation of the interplay between urban form and history.

3. In support of this concept of soul and environment in a contemporary sense see: Thomas Moore, *Care of the Soul: A Guide for Cultivating Depth and*

Sacredness in Everyday Life (New York: Harper, Collins, 1992), 168–284. "Vital core" is from the *American Heritage Dictionary,* New College Edition, in which *soul* is defined as "a central or integral part of something; a vital core."

4. For my distinguishing between the soul and the spirit (i.e., "object" and "human response"), see René Dubos, *A God Within: A Positive Philosophy for a More Complete Fulfillment of Human Potentials* (New York: Scribner's and Sons, 1972), in general and "Persistence of Place," 111–34, in particular. "Animating force within human beings" is from the *American Heritage Dictionary,* where *spirit* is defined as "that which is traditionally believed to be the vital principle or animating force with human beings."

5. Misha Glenny, *The Fall of Yugoslavia: The Third Balkan War* (Harmondsworth: Penguin, 1992), 130–37.

6. Institute for Protection of Monuments, *War Damages and Destruction Inflicted on the Cultural Monuments, Sites, and Historical Centers in Croatia* (Zagreb: Ministry of Education and Culture, March 1992).

7. From a typed statement by Catherine V. Wilkes, chair of the executive committee of the Inter-University Center of Dubrovnik, sent to me in 1992 by a colleague at the University of Zagreb.

8. This "bosnet" e-mail communication of 30 September 1994 was sent to Duško Bogunović (my collaborator, formerly of Sarajevo, now in New Zealand), who then sent it on to me by FAX. It described how a Bosnian architect and urban planner in 1994 assembled in Istanbul an international, month-long workshop called "Pilot Workshop—Mostar 2004." The purpose was to develop a methodology and timetable for reconstructing Mostar's urban core. In an inspiring surge of collective will, experts from twenty-five universities in fifteen countries and a number of international organizations collaborated and agreed to reassemble in Istanbul annually until 15 September 2006, the date for an international assembly in celebration of the rebuilt Mostar.

9. Italo Calvino, *Six Memos for the Next Millennium: The Charles Norton Eliot Lectures, 1985–86* (Cambridge: Harvard University Press, 1988). Unfortunately, Calvino died in Italy while preparing the sixth "memorandum."

10. Both Dalmatia and California could benefit from the agreement of the United Nations to empower local goverments—of cities, regions, and states— to actively participate in policy making on the international level. In 1994 this decision was formally adopted by recognizing local authorities and non-governmental organizations as well as partners with the national authorities. These groups collaborated in formulating issues discussed and recommendations made at the World Conference on Human Settlements—Habitat II— held in Istanbul in 1996. This significant decision came about in response to the remarkable increase in local elections following the end of centralized government in many regions of the world in the early 1990s. Newly found freedoms awakened through democracy in cities per se are giving a collective voice to urban places around the globe and may generate an inherent form of identity with place to stabilize and safeguard the human qualities of our environments in the coming century.

11. From a speech given by Senator J. William Fulbright on the occasion of the thirtieth anniversary of the founding of the Fulbright program in 1976.

12. See Carolyn Merchant, *Radical Ecology: The Search for a Livable World* (New York: Routledge, 1992), 218. This work is invaluable as an introduction to bioregional thinking and practice. See also Kirkpatrick Sale, *Dwellers in the Land: The Bioregional Vision* (Santa Cruz, Calif.: New Society, 1991), esp. its comprehensive bibliography, 197–203; and, particularly readable, Van Andruss et al., eds., *Home: A Bioregional Reader* (Santa Cruz, Calif.: New Society, 1990).

13. Sharp spoke of the village as "utilitarian, it often possesses a remarkable beauty. . . . A center of contemporary life, it is also a record of long history. The work of man, it is also a creation of time." Yet, he cautioned that "respect for tradition is excellent, provided the tradition respected is a genuine living tradition. . . . It is a flowing, eddying widening stream that is continually refreshed by new tributaries, a stream whose direction is subject to change by new currents created by new conditions" (see Thomas Sharp, *The Anatomy of the Village* [Harmondsworth: Penguin, 1946], 5).

14. Johann Wolfgang Goethe, *Italian Journey, 1786–1788,* trans. W. H. Auden and Elizabeth Myer (San Francisco: North Point Press, 1982), 111–12. En route to Dalmatia in 1990, I found this quotation in Spoleto boldly hand-lettered on a large panel mounted on the heights of LaRocca and overlooking the spectacular gorge of the Tessina River. Here Goethe had also stood and been inspired by the moving experience. This concept gave life to my own reading of Zadar, Split, and Dubrovnik and to my pursuit of this work. Goethe's comprehensive style of environmental inquiry into meaning beyond physical properties bordered on phenomenology. See also David Seamon, "Goethe's Approach to the Natural World: Implications for Environmental Theory and Education," in *Human Geography,* ed. D. Ley and M. Samuels (Chicago: Masroufa, 1978).

15. Cvito Fisković, "Dalmatia," *The Atlantic Monthly — Yugoslavia, A Special Supplement,* 210, no. 6 (December 1962), 86.

Glossary of Croatian Words and Place Names

The reader's comprehension of the text can be enriched by frequent reference to this guide to the pronunciation and meaning of words and place names in Croatia. He or she can thus incrementally gain a sense of cultural proximity toward the more subtle messages contained in the text and illustrations.

Words

ban (bahn): governor

blitva (blee-tvah): Swiss chard

Bračanin (brah-chah-nin), pl. *Bračani* (brah-chah-nee): inhabitant of Brač

bratinstvo (brah-tin-stvoh): brotherhood (obs.)

bura (boo-rah): chill winter north wind

čitaonica (chee-ta-oh-nee-tsah), pl. *čitaonici* (chee-ta-oh-nee-tsee): public reading room

ćevapčići (he-vahp-chee-chee): barbecued ground meat

dobar dan (doh-bahr dahn): "good day" (late morning to early evening greeting)

dom (dohm): community social club

dovidjenja (doh-vee-jehn-yah): "good bye," "see you," "so long"

gimnazija (ghee-sheem-nah-zee-ya): public high school

gomila (goh-mee-lah), pl. *gomile* (goh-mee-leh): dome of field stones

grad (grahd): town, city

Kako je u Kaliforniji? (kah-koh yeh oo kah-lee-fohr-nee-yee): "How is it in California?"

kapetanske kuće (kah-peh-tahn-skeh koo-cheh): captains' houses

karst (kahrst): limestone (dial.)

kavana (kah-vah-nah): coffee house

kolonat (koh-loh-naht): old form of sharecropping

komuna (koh-moo-nah): county, municipality

konoba (koh-noh-bah): wine cellar

korta (kor-tah): courtyard (dial.)

kotar (koh-tahr), pl. *kotari* (koh-tahr-ees): county

kumstvo (koom-stvoh): godparenthood

kupus (koo-poos): cabbage

lipi mali (lee-pee mah-lee): nice little boy (dial.)

luka (loo-kah): harbor, port

makija (mah-kee-yah): macchia, dense evergreen underbrush

mazga (mah-zgah), pl. *mazge* (mah-zgheh): mule

mjesna zajednica (myee-snah zah-yee-dnee-tza): local improvement committee

narodni (nah-rohd-nee): national
načelnik (nah-chel-neek): administrative district head, mayor (obs.)
nema problema (neh-mah prob-bleh-mah): "no problem"
obala (oh-bah-lah), pl. *obalae* (oh-bah-leh): waterfront, shoreline
odar (oh-dahr): pedestal for a casket
opanci (oh-pahn-tsee): strapped soft-soled footwear, type of moccasins
opat (oh-paht): abbot
općina (ohp-chee-nah): county, district, municipality
palača (pah-lah-cha): palace
polja (pohl-ya): rural field
poljana (pohl-yah-nah): open space in urban areas (Dubrovnik dial.)
rakija (rah-kee-yah): brandy
restauracija (reh-stah-oo-rah-tzee-yah): restaurant
riva (ree-vah): waterfront (It.)
sala (sah-lah): living room (dial.)
selo (seh-lo): village
slastičarnica (slah-stee-char-nee-tzah): pastry shop
stari kraj (stah-ree krahy, kreye): old country
šuma borova (shoo-mah boh-roh-vah): pine wood
tajnik (teye-neek): secretary
teta (the-tah): aunt
tramvaj (trahm-veye): streetcar
turist biro (too-rist bee-roh): tourist office
turističko društvo (too-ree-stee chkoh droo-sh-tvoh): tourist society
tvrdjava (tvrd-jah-vah): fortress
ulica (oo-lee-tzah): street
vinarija (vee-nah-ree-yah): winery
vinograd (vee-noh-grahd), pl. *vinogradi* (vee-noh-grah-dee): vineyard
vikendica (vee-ken-dee-tsah): weekend house
voda (voh-dah): water
zadruga (zah-droo-gah): community, cooperative; family enclave
zajednica (zah-yehd-nee-tsa): togetherness, community
zaselak (zah-seh-lak): hamlet
zbogom Kuna grade (zboh-gohm koo-nah grah-deh): "Farewell, town of Kuna"
zbornik (zbohr-neek): journal; collection of papers
župnik (zhoop-neek), pl. *župnika* (zhoop-nee-kah): parish priest, pastor

Place Names

Bačvice (bah-ch-vee-tzeh)
Brač (brahch)
Cavtat (tzahv-taht)
Cetinje (tzeh-tee-yeh)
Čilipi (chee-lee-pee)
Dingač (din-ga-hch)
Drače (drah-cheh)
Dubrovnik (doo-brohv-neek)

Gornji Humac (gohr-nyee-hoo-neek)
Gruž (groozh)
Hilići (hee-lee-chee)
Hum (hoom)
Hvar (h-vahr)
Janjina (yahn-yee-nah)
Jelsa (yehl-sah)
Konavle (koh-nah-vleh)
Korčula (kohr-choo-lah)
Kučišće (koo-cheesh-cheh)
Kuna (koo-nah)
Lovišće (loh-veesh-cheh)
Ljubljana (lyub-lyahn-ah)
Marjan (mahr-yahn)
Mirce (mir-tzeh)
Mljet (m-lyeht)
Orebić (oh-reh-beech)
Oskorušno (oh-skoh-roosh-noh)
Pelješac (pel-ye-shatz)
Pile (pee-leh)
Pijavično (pee-ya-veech-noh)
Podobuče (poh-doh-boo-cheh)
Potomje (poh-tohm-yeh)
Pražnica (prazh-nee-tza)
Pučišća (poo-cheesh-hah)
Sućuraj (soo-choo-ray)
Šibenik (shee-behn-eek)
Šolta (shohl-tah)
Trebinje (treh-been-yeh)
Trogir (troh-geer)
Trpanj (tr-pan-yeh)
Vidoš (vee-dohsh)
Viganj (vee-gahn-yeh)
Višegrad (vee-sheh-grahd)
Vrboska (vr-boh-skah)
Zagujine (zah-goo-yee-neh)

Bibliography
Writings by Foreign Travelers
to Dalmatia

The works listed below were published by foreigners who visited Dalmatia from 1689, when they first appeared, until 1963, with none to be found since I have been motivated to add mine. These books provide examples of what travelers at long intervals have read into the places they visited. In general these writers demonstrate their lack of a broad cultural understanding of the people of Dalmatia and their positive role in the accumulative formation of the built environment in spite of the powerful external political influences over the centuries.

The images conveyed by these writings have generated abroad an attraction to the region's natural landscapes and urban places, one so unlike those of nearby Italy and Western Europe. This response has in our times contributed to the form of tourism that has been damaging to the environmental and cultural identity in the second half of the twentieth century. This end result in a sense underlies the inability of the West to respond more promptly in 1991 to the wartime attacks on Dalmatia's urban treasures.

Horatio Brown's *Dalmatia* (1925), describing three visits spread over forty years, shows that he was himself influenced by the romanticism of earlier-nineteenth-century visitors. Given the broadened communication facilities of our times, future writings on Dalmatia may be able to build a more realistic cultural bridge between the West and this Slavic edge of the East, which is one of the aims of this book.

Wheler, George. 1689. *Voyage de Dalmatia, de Grèce et du Levant.* Amsterdam: Jean Wolters.

Adams, Robert, FRS, FSA (Architect to the King and Queen). 1764. Untitled measured drawings of Diocletian's Palace in Split. Privately printed.

Fortis, Alberto. *Viaggo in Dalmazia.* 1774. Translated into English as *Travels into Dalmatia.* 1978. London: J. Robson. Reprint. New York: Arno Press and New York Times, 1974.

Cassas, L. F., and J. Lavallee. 1802. *Voyage pittoresque de L'Istria et de la Damatie.*

Marmont, August Frederic Louis Viesse, duc de Raguse. 1830. *Vie et memoires du Marechal Marmont.* Paris: Libraire Ladvocat.

Wilkinson, Sir J. Gardiner. 1848. *Dalmatia and Montenegro, (With a Journey to Mostar and Hercegovina and Remarks of the Slavonic Nations, the History of Dalmatia and Ragusa).* London: John Murray. Reprint. New York: Arno Press and New York Times, 1971.

Paton, A. A. 1861. *Researches on the Danube and the Adriatic.* 2 vols. Leipzig: Brockhaus.

Carrara, Francesco N. 1864. *La Dalmazzia.* Zadar: Fratelli Batlara Tipografi Editori.

Freemen, Edward A. 1881. *Sketches from the Neighbor Lands of Venice.* London: Macmillan.

Evans, Arthur J. 1883. *Antiquarian Researches in Illyricum.* London: Nichols & Sons.

Jackson, Thomas G. 1887. *Dalmatia, the Quarnero and Istria.* Oxford: Clarendon.

Munro, Robert. 1894. *Rambles and Studies in Bosnia-Herzegovina and Dalmatia, with an Account of the Proceedings of the Congress of Archeologists and Anthropologists in Sarajevo, August 1894.* Edinburgh: W. Blackwood.

Jackson, Hamilton. 1908. *The Shores of the Adriatic—Austrian Side.* London: John Murray.

Frottingham, A. L. 1910. *Roman Cities in Northern Italy and Dalmatia.* London: John Murray.

Holbach, Maude M. 1910. *Dalmatia, the Land Where East Meets West.* London: John Lane.

Herbrard, E., and J. Zeiller. 1912. *Le palais de Diocletien.* Paris, Libraire generale de l'architecture et des arts decoratifs.

Moque, Alice. 1914. *Delightful Dalmatia.* New York: Funk & Wagnalls.

Brown, Horatio F. 1925. *Dalmatia.* London: A. and C. Black. This book covers visits in 1883, 1910, and 1924 and provides a history and evaluation of writers from Adams in 1764 to Jackson in 1887 This suggested my own sequences in the text and in this bibliography.

West, Rebecca. 1941. "Dalmatia." In *Black Lamb and Grey Falcon: A Journey through Yugoslavia,* 2 vols., 1:115–248. London: Viking.

Goldring, Douglas. 1951. *Three Romantic Countries: Reminiscences of Travel in Dalmatia, Ireland and Portugal.* London: MacDonald.

Whelpton, Eric. 1954. *Dalmatia.* London: Hale.

Rhodes, Anthony. 1955. *The Dalmatian Coast.* London: Evans.

Dennis-Jones, Harold. 1963. *Your Guide to the Dalmatian Coast.* London: Redman.

Index

Page numbers in italics refer to illustrations.

Library of Congress Cataloging-in-Publication Data

Violich, Francis.
The bridge to Dalmatia : a search for the meaning of place / Francis Violich ;
cartography and drawings in collaboration with Nicholas Ancel.
p.   cm.
"Published in cooperation with the Center for American Places, Harrisonburg,
Virginia" — Copr. p.
Includes bibliographical references and index.
ISBN 0-8018-5554-3 (alk. paper)
1. Human geography — Croatia — Dalmatia.   2. Cities and towns — Croatia —
Dalmatia.   3. Urban anthropology — Croatia — Dalmatia.   4. Dalmatia (Croatia) —
Geography.   5. Dalmatia (Croatia) — Description and travel.   I. Title.
GF642.C872D358   1997
304.2′3′094972 — dc21
97-8754
CIP